Protein Physics

Protein Physics
A Course of Lectures

Alexei V. Finkelstein
Oleg B. Ptitsyn

Institute of Protein Research, Russian Academy of Sciences,
Institute of Protein Research, 142290 Pushchino,
Moscow Region, Russia

ACADEMIC PRESS

An imprint of Elsevier Science

Amsterdam   Boston   London   New York
Oxford   Paris   San Diego   San Francisco
Singapore   Sydney   Tokyo

Academic Press
An Imprint of Elsevier Science
84 Theobold's Road, London WCIX 8RR, UK
http://www.academicpress.com

Academic Press
An Imprint of Elsevier Science
525 B Street, Suite 1900, San Diego, California 92101-4495, USA
http://www.academicpress.com

ISBN 0-12-256781-1

Library of Congress Catalog Number: 2002101251

A catalogue record for this book is available from the British Library

Typeset by Newgen Imaging Systems (P) Ltd., Chennai, India

Printed and bound in the United Kingdom
Transfered to Digital Printing, 2011

CONTENTS

Main functions of proteins. Amino acid sequence determines the three-dimensional structure, and the structure determines the function. The reverse is not true. Fibrous, membrane and globular proteins. Primary, secondary, tertiary and quaternary structures of proteins. Protein biosynthesis, protein folding *in vivo* and *in vitro*. Post-translational modifications.

Stereochemistry of natural L-amino acids. Covalent bonds and angles between them. Their vibrations. Rotation around the covalent bonds. Peptide group. *Trans*- and *cis*-prolines.

Van der Waals interaction: attraction at long distances, repulsion at short distances. Allowed conformations of amino acid residues (Ramachandran plots for glycine, alanine, valine, proline).

Necessity of protein structure prediction from amino acid sequences. Recognition of protein structures and their functions using homology of sequences. Key regions and functional sites in protein structures. Multiple alignments. Detection of stable elements of protein structures. "Templates" of elements of protein structures. Predicted structures are unavoidably judged from only a part of interactions occurring in the chain. As a result, we have probabilistic predictions. Interactions that stabilize and destroy secondary structures of polypeptide chains. Calculation of secondary structures of non-globular polypeptides. Prediction of protein secondary structures.

Overview of approaches to prediction and recognition of tertiary structures of proteins from their amino acid sequences. Protein fold libraries. Recognition of protein folds by threading. Prediction of common fold for a set of remote homologs reduces uncertainty in fold recognition. Structural genomics and proteomics. Bioinformatics. Protein engineering and design. First advances in design of protein folds.

Protein function and protein structure. Elementary functions. Binding proteins: DNA-binding proteins, immunoglobulins. Enzymes. The active site as a "defect" of globular protein structure. Protein rigidity is crucial for elementary enzyme function. Catalytic and substrate-binding sites. Inhibitors. Co-factors. Multi-charged ions. Mechanism of enzymatic catalysis. Example: serine proteases. Transition state theory and its confirmation by protein engineering. Abzymes. Specificity of catalysis. "Key–lock" recognition.

Combination of elementary functions. Transition of substrate from one active site to another. "Double sieve" increases specificity of function. Relative independence of protein fold from its elementary catalytic function. Visible connection between protein fold and protein environment. Combination of elementary protein functions and flexibility of its structure. Induced fit. Mobility of protein domains. Shuffling of domains and protein evolution. Domain structure: kinases, dehydrogenases. Allostery: interaction of active sites. Allosteric regulation of protein function. Allostery and protein quaternary structure. Hemoglobin and myoglobin. Mechanism of muscle contraction.

PREFACE

This course of lectures is devoted to protein *physics*, i.e., to the overall topics of structure, self-organization and function of protein molecules.

The course is based on lectures given by us (earlier by O.B.P., and later by A.V.F.) first at Moscow PhysTech Institute and then at Pushchino State University and the Pushchino Branch of Moscow State University. Initially, our students were physicists, then mainly biologists with some chemists. That is why, by now, the lectures have not only been considerably up-dated (as research never stops) but also thoroughly revised to meet the requirements of the new audience.

Since what you are reading is not a monograph but a course of lectures, repetition is hardly avoidable (specifically, we have repeated some figures). Indeed, when delivering a lecture one cannot refer to "Figure 2 and Equation 3 of the previous lecture"; however, we have done our best to minimize repetition.

The following comments will help you to understand our approach and the way in which the material is presented.

On the "lecturer". All lectures are presented here in the way they are delivered, i.e., from the first person – the lecturer.

On the "inner voice". The personality of the "lecturer" comprises both authors, O.B.P. and A.V.F., although this does not mean that no disagreement has ever occurred between us concerning the material presented. Moreover, sometimes each of us felt that the problem in question is disputable and subject to further studies. We made no attempt to smooth these disputes and contradictions over, so the "lecturer's" narration is sometimes interrupted or questioned by the "inner voice". Or the "inner voice" may simply articulate the frequently asked questions or elaborate the discussion.

Why "protein physics"? Because we are amazed to see how strongly biological evolution enhances, secures and makes evident the consequences of the physical principles underlying molecular interactions in the protein. It is also striking how much our understanding of biological systems, proteins in particular, has developed through the application of physical methods. We see this from mass spectrometry to electron microscopy, X-ray crystallography and NMR studies of proteins. There is hardly any other area of contemporary science in which the traditional boundaries between disciplines and philosophies have been so clearly breached to such great profit.

On the physics and biology presented in these lectures. In the course of lecturing, we shall take the opportunity to present physical ideas, such as some elements of statistical physics and quantum mechanics. These ideas, to our minds, are absolutely necessary not only for an understanding of protein structure and function but also for general scientific culture, but usually a "normal" biology graduate has either completely forgotten or never known them. On the other hand, among a myriad of protein functions we will discuss only those absolutely necessary to demonstrate the role of spatial structures of proteins in their biological, or rather biochemical, activities.

On "in vivo" and "in vitro" experiments. The terms "*in vivo*" and "*in vitro*", as applied to experiments, are often understood differently by physicists and biologists. Strictly speaking, between the pure "*in vivo*" and the pure "*in vitro*" there are a number of ambiguous intermediates. For example, protein folding in a cell-free system (with all its ribosomes, initiation factors, chaperones, etc.) is unequivocally an "*in vivo*" experiment in the physicist's view (for a physicist, "*in vitro*" would be a separate protein in solution; even the cell-free system contains too many biological details). But for a biologist, this is undoubtedly an "*in vitro*" experiment (since "*in vivo*" is referred to a living and preferably intact organism). However, structural studies of a separate protein in an organism are hardly possible. Therefore, reasonable people compromise by making biologically significant "*in vivo*" events accessible for experimental "*in vitro*" studies.

On experiment, physical theory and calculations. Experiment provides the basic facts underlying all our ideas of phenomena, as well as a lot of refining details. Theory allows us to understand the essence and interrelation of the phenomena and is helpful in planning informative experiments. Calculations connect theory with experiment and verify key points of the theory. However, not everything that can be calculated must be calculated: for example, it is easier just to measure water (or protein) density than to calculate it from first principles. And not only is this easier but it yields a more accurate result, since a detailed calculation requires many parameters that can hardly be accurately estimated.

Some basic physical theories – in a simplified form, naturally – are included in our lectures, not only because they permit us to put in order and comprehend from a common standpoint the vast experimental material, but also because they are elegant. Besides, we believe that knowledge of basic physical theories and models is essential to human culture.

On physical models, rough estimates and computer (in silico) experiments. In these lectures, we will often discuss simple models, i.e., those drastically simplified compared to reality, and use rough estimates. And we want you, after reading these lectures, to be able to make such estimates and use simple models of the events in question. The use of simple models and rough estimates may seem to be quite old fashioned. Indeed, it is often believed that with the powerful computers available now one can enter "all as it is in reality": water molecules, salt, coordinates of protein atoms, DNA, etc., fix the temperature, and obtain "the precise result". As a matter of fact, this is a utopian picture. The calculation – we mean a detailed one (often made using so-called "molecular dynamics") – will take days and cover only some nanosecond of the protein's life since you will have to follow the thermal motions of many thousands of interacting atoms. In any case, this calculation will not be absolutely accurate either, since all elementary interactions can be estimated only approximately. And the more detailed the description of a system is, the more elementary interactions are to be taken into account, and the more minor errors will find their way into the calculation (not to mention the increased computer time required). Eventually, you will obtain only a more-or-less precise estimate of the event instead of the desired absolutely accurate description of it – with days of a supercomputer's time spent. Meanwhile, what really interests you may be a simple quick estimate such as whether it is possible to introduce a charge into the protein at a particular site without a risk of protein explosion. That's why one of our goals is to teach you how to make such estimates. However, this does not mean that we will simply ignore computer experiments. Such experiments yield a lot of useful information. But the computer experiment is a real experiment (although *in silicio*, not *in vitro* or *in vivo*). It involves highly complex systems and yields facts requiring further interpretation, which, in turn, demands simplified but clear models and theories.

On equations. We are aware that mathematical equations are a difficulty for biologists, so we have done our best to refrain from using them, and only those really unavoidable have survived. Our advice is: when reading these lectures, "test words by equations". It is certainly easier to read words only, but they are often ambiguous, so word-verifying by equations and *vice versa* will help your understanding. To aviod going into insignificant (and unhelpful) detail we shall often use approximate calculations; therefore you will often see the symbols "$\approx$" ("approximately equal to") and "$\sim$" ("of the same order of magnitude as").

On references and equations. We avoided references in the text to make it easier to read, but the appropriate references to original works, reviews and textbooks are given in legends to figures and tables. Protein structures are drawn using the program MOLSCRIPT (Kraulis P.J., *J. Appl. Cryst.* **24**: 946–950, 1991) and WHAT IF (Vriend G., *J. Mol. Graphics* **8**: 52–56, 1990), and coordinates taken from the Protein Data Bank (described in: Bernstein F.C., Koetzle T.F., Meyer E.F., Jr., Brice M.D., Kennard O., Shimanouchi T., Tasumi T., *J. Mol. Biol.* **112**: 535–542, 1977). Most of other figures are purposely schematic.

On tastes. These lectures, beyond doubt, reflect our personal tastes and predilections and are focused on the essence of things and events rather than on a thorough description of their details. In the main they contain physical problems and theories, while only the necessary minimum of experimental facts are given and experimental techniques are barely mentioned. (Specifically, almost nothing will be said about the techniques of X-ray crystallography and NMR spectroscopy that have provided the bulk of our knowledge of protein structure.)

Therefore, these lectures are by no means a substitute for regular fact-rich biophysical and biochemical courses on proteins, and they must not be used as a reference book. When referring to specific proteins we merely give the most important (it goes without saying – from our viewpoint) instances; only absolutely necessary data are tabulated; all values are approximate, etc.

On small print. This is used for helpful but not essential excursus, additions and explanations.

On the personal note in these lectures. We shall take the opportunity of noting our own contributions to protein science and those by our co-workers and colleagues from the Institute of Protein Research, Russian Academy of Sciences. This will certainly introduce a "personal note" into the lectures and perhaps make them a bit more lively...

The majority of these lectures are intended to last for one hour. However, lectures 5, 6, 8, 16, 17 and 24 require two hours.

FOREWORD

In June 1967 Oleg Ptitsyn became the Head of the Laboratory of Protein Physics at the new Institute of Protein Research at Puschino. Three months later he was joined by Alexei Finkelstein; first as research student and then as a colleague. Their approach to the study of proteins was different from that common in the West, being strongly influenced by the Russian school of polymer physics. One of its most distinguished members, Michael Volkenstein, had been Ptitsyn's PhD supervisor. Together Ptitsyn and Finkelstein created at Puschino one of the world's outstanding centres for the study of the physics and chemistry of proteins.

Certain areas of their work, particularly that on protein folding, have become well known in Europe, India and America either directly through their papers or indirectly through the elaboration of their work by two of their former students: Eugene Shakhnovich, in America, and Alexei Murzin, in England. However, it became obvious when Finkelstein or Ptitsyn talked with colleagues that the range of their work went far beyond what was commonly known in the west. We would find that fundamental questions that were our current concerns had already been considered by them and they had some elegant calculation that provided an answer. Usually, they had not got round to publishing this work. Now, to make their overall achievements available to more than just their friends and the students of Moscow University, Finkelstein has written this book on the basis of lectures he and Ptitsyn gave to these students.

In the breadth of its range, the rigor of its analysis and its intellectual coherence, this book is a *tour de force*. Of those concerned with the physics and chemistry of proteins, I doubt if there can be any, be they students or senior research workers, who will not find here ideas, explanations and information that are new, useful and important.

Cyrus Chothia, FRC
MRC Laboratory of Molecular Biology
Cambridge, UK

ACKNOWLEDGMENTS

We are grateful to all co-authors of our works that we have used to illustrate this book. We are grateful to our colleagues from the Institute of Protein Research, Russian Academy of Sciences, and especially to former and current co-workers from our Laboratory of Protein Physics: this book would never have been written without scientific and friendly human contact with them.

We thank our university students and participants of various summer schools for their questions and comments that helped us to clarify (primarily, to ourselves) some points of this course and were helpful for better presentation of the lectures.

We are most grateful to A.B. Chetverin, D.I. Kharakoz, V.E. Bychkova, Yu.V. Mitin, A.S. Spirin, V.A. Kolb, M.A. Roytberg, V.V. Velkov, A.V. Efimov, I.G. Ptitsyna, G.I. Gitelzon, D.S. Rykunov, G.V. Semisotnov, A.V. Skugaryev and D.N. Ivankov for reading the manuscript, discussions and most useful comments.

Special thanks to G.A. Morozov, D.S. Rykunov, A.V. Skugaryev, D.N. Ivankov, M.Yu. Lobanov, N.Yu. Marchenko, I.V. Sokolovsky, A.A. Shumilin, M.G. Dashkevich, T.Yu. Salnikova, A.E. Zhumaev and A.A. Finkelstein for their valuable assistance in preparation of the manuscript and figures, and to E.V. Serebrova for preparation of the English translation of this book.

And, last but not least, we are grateful to the Russian Foundation for Basic Research and to the Howard Hughes Medical Institute for the financial support that enabled us to complete this book.

Part I
INTRODUCTION

LECTURE I

Proteins are molecular machines, building blocks, and arms of a living cell. Their major and almost sole function is enzymatic catalysis of chemical conversions in and around the cell. In addition, regulatory proteins control gene expression, and receptor proteins (which sit in the lipid membrane) accept intercellular signals that are often transmitted by hormones, which are proteins as well. Immuno proteins and the similar histocompatibility proteins recognize and bind "foe" molecules as well as "friend" cells thereby helping the latter to be properly accommodated in the organism. Structural proteins form microfilaments and microtubules, as well as fibrils, hair, silk and other protective coverings; they reinforce membranes and maintain the structure of cells and tissues. Transfer proteins transfer (and storage ones store) other molecules. Proteins responsible for proton and electron transmembrane transfer provide for the entire bioenergetics, that is, light absorption, respiration, ATP production, etc. By ATP "firing" other proteins provide for mechano-chemical activities – they work in muscles or move cell elements.

The enormous variety of protein functions is based on their high specificity for the molecules with which they interact, a relationship that resembles a key and lock (or rather, a somewhat flexible key and a somewhat flexible lock). Anyhow, this specific relationship demands a fairly rigid spatial structure of the protein. That's why the biological functions of proteins (and other macromolecules of the utmost importance for life – DNA and RNA) are closely connected with the rigidity of their three-dimensional (3D) structures. Even a little damage to these structures, let alone their destruction, is often the reason for loss of or dramatic changes in protein activities.

A knowledge of the 3D structure of a protein is necessary to understand how it functions. Therefore, in these lectures, the physics of protein function will be discussed after protein structure, the nature of its stability and its ability to self-organize, that is, at the very end.

Proteins are polymers: they are built up by amino acids that are linked into a peptide chain; this was discovered by E. Fischer as early as the beginning of the 20th century. In the early 1950s Sanger showed that the sequence of amino acid residues (a "residue" is the portion of a free amino acid that remains after polymerization) is

3

unique for each protein. The chain consists of a chemically regular backbone ("main chain") from which various side chains (R_1, R_2, . . . , R_M) project:

$$-NH-CH-CO-NH-CH-CO-\ldots-NH-CH-CO-$$
$$\quad\;\; | \qquad\qquad\quad\; | \qquad\qquad\quad\;\; |$$
$$\quad\;\; R_1 \qquad\qquad\quad R_2 \qquad\qquad\quad R_M$$

The number M of residues in protein chains ranges from a few dozens to many thousands. This number is gene-encoded.

There are twenty main species of amino acid residues. Their position in the protein chain is gene-encoded, too. However, subsequent protein modifications may contribute to the variety of amino acids. Also, some proteins bind various small molecules, serving as co-factors.

In an "operating" protein the chain is folded in a strictly specified structure. In the late 1950s, Perutz and Kendrew solved the first protein spatial structures and demonstrated their highly intricate and unique nature. However, it is noteworthy that the strict specificity of the 3D structure of protein molecules was first shown (as it became clear later) back in the 1860s by Hoppe-Zeiler who obtained hemoglobin crystals: in a crystal each atom occupies a unique place.

The question whether the structure of a protein is the same in a crystal and in solution had been discussed for many years (when only indirect data were available) until the virtual identity of these (apart from small fluctuations) was demonstrated by NMR spectroscopy.

Proteins "live" under various environmental conditions, which leave an obvious mark on their structures. The less water there is around, the more valuable the hydrogen bonds (which reinforce the regular, periodic 3D structures of the protein backbone) are, and the more regular the stable protein structure ought to be.

According to their "environmental conditions" and general structure, proteins can be roughly divided into three classes:

1. Fibrous proteins form vast, usually water-deficient aggregates; their structure is usually highly hydrogen-bonded, highly regular and maintained mainly by interactions between various chains.
2. Membrane proteins reside in a water-deficient membrane environment (although they partly project into water). Their intramembrane portions are highly regular (like fibrous proteins) and highly hydrogen-bonded, but restricted in size by the membrane thickness.
3. Water-soluble (residing in water) globular proteins are less regular (especially small ones). Their structure is maintained by interactions of the chain with itself (where an important role is played by interactions between hydrocarbon ("hydrophobic") groups that are far apart in the sequence but adjacent in space) and sometimes by chain interactions with co-factors.

Finally, there are some small or hydrocarbon group-poor polypeptides, which can have no inherent fixed structure by themselves but obtain it by interacting with other molecules.

The above classification is certainly extremely rough. Some proteins may comprise a fibrous "tail" and a globular "head" (like myosin, for example), and so on.

To date (2001), we know hundreds of thousands of protein sequences (they are deposited at special computer data banks, e.g., SwissProt) and about ten thousand protein spatial structures (they are compiled at the Protein Data Bank, or simply PDB). What we know about 3D protein structures mostly concerns water-soluble globular proteins. The solved spatial structures of membrane and fibrous proteins are few. The reason is simple: water-soluble proteins are easily isolated as separate molecules, and their structure is relatively easily established by X-ray crystallography and by NMR (nuclear magnetic resonance) studies in solution. That's why, when speaking about "protein structure" and "protein structure formation" one often actually means regularities shown for water-soluble globular proteins only. This must be kept in mind when reading books and papers on proteins, including these lectures. Moreover, it must be kept in mind that, for the same experimental reason, contemporary protein physics is mainly physics of small proteins, while the physics of large proteins is only starting to develop.

Non-covalent interactions maintaining 3D protein architecture are much weaker than chemical bonds fixing a sequence of monomers (amino acids) in the protein chain. This sequence – it is called the primary structure of a protein (Fig. 1.1) – results from biochemical matrix synthesis according to a gene-coded "instruction".

Protein architectures, especially those of water-soluble globular proteins, are complex and of great diversity, unlike the universal double helix of DNA (the single-stranded RNAs appear to have an intermediate level of complexity). Nevertheless,

Primary ...-Gly-Val-Tyr-Gln-Ser-Ala-Ile-Asn-...

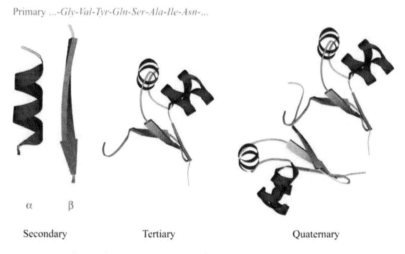

α   β

Secondary   Tertiary   Quaternary

Figure 1.1. Levels of protein structure organization: primary structure (amino acid sequence), regular secondary structure (α-helix and one strand of β-structure are shown), tertiary structure of a globule folded by one chain, and quaternary structure of an oligomeric protein formed by several chains (here, dimeric *cro* repressor).

in proteins, certain "standard" motifs are detected as well, which will be discussed in detail in these lectures.

In the first place, proteins have regular secondary structures, namely, the α-helix and β-structure; α-helices are often represented by helical ribbons, and extended β-structural regions (which by sticking together form sheets) by arrows (see Fig. 1.1). Secondary structures are characterized by a regular periodic shape ("conformation") of the main chain with side chains of a variety of conformations.

The packing of the secondary structures of one polypeptide chain into a globule is called the tertiary structure, while several protein chains integrated into a "super-globule" form the quaternary structure of a protein. For instance (another example in addition to the dimeric *cro* repressor shown in Fig. 1.1), hemoglobin consists of two β- and two α-chains (this has *nothing to do* with α- and β-structures). A quaternary structure formed by identical chains usually appears to be symmetrical (*cro* repressor and hemoglobin are no exception). Sometimes a quaternary structure comprises tens of protein chains. Specifically, virus coats can be regarded as such "superquaternary" structures.

Among tertiary structures, some can be distinguished as the most typical, and we will consider these later. They often envelop not the entire globular protein but only compact subglobules (so-called domains) within it (Fig. 1.2). A domain (like a small protein) usually consists of 100–200 amino acid residues, i.e., of ~2000 atoms. Its diameter is 30–40Å.

The *in vivo* formation of the native (i.e., biologically active) tertiary structure occurs during biosynthesis or immediately after. However, it is noteworthy that a 3D structure can result other than from biosynthesis: around 1960, Anfinsen showed that it could also be yielded by "renaturation", i.e., by *in vitro* refolding of a somehow unfolded protein chain, and that the process goes spontaneously, unaided by the cellular machinery. This means that the spatial structure of a protein is determined

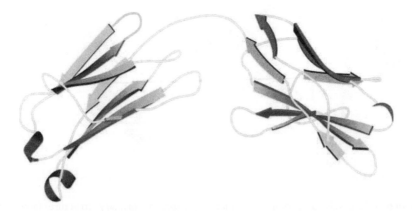

Figure 1.2. The domain structure of a large protein is similar to the quaternary structure built up by small proteins. The only difference is that in large proteins the compact subglobules (domains) pertain to the same chain, while the quaternary structure comprises several chains.

by its amino acid sequence alone (provided water temperature, pH and ionic strength are favorable), i.e., the protein is capable of *self*-organizing.

Inner voice: Strictly speaking, this has been shown mainly for relatively small (up to 200–300 amino acid residues) water-soluble *globular* proteins. As concerns larger proteins, especially those of higher organisms, it is not that simple: far from all of them refold spontaneously...

Lecturer: Thanks for the refinement. Yes, it is not that simple with such proteins. Part of the reason is aggregation, part is post-translational modification, especially when it comes to higher organisms, eukaryotes. Still less is known about spontaneous self-organization of membrane and fibrous proteins. In some proteins it happens, but mostly they do not refold. Therefore, let us agree right away that when speaking about protein physics, protein structure and its formation I will actually discuss (if not specified otherwise) relatively small, single-domain globular proteins. This convention is quite common for biophysical literature, but the fact that it is insufficiently articulated causes frequent misunderstandings.

Anfinsen's experiments provided fundamental detachment of the physical process, that is, the spatial structure organization, from biochemical synthesis of the protein chain. They made clear that the structure of protein is determined by its own amino acid sequence alone and is not imposed by cellular machinery. It seems that the main task of this machinery is to protect the folding protein from unwanted contacts (among these are also contacts between remote regions of a very large protein chain), since *in vivo* folding occurs in a cellular soup with a vast variety of molecules to stick to. But in diluted *in vitro* solution a protein, at least a small one, folds spontaneously by itself.

Strictly speaking, proteins are capable of spontaneous refolding provided they have undergone no strong post-translational modification, i.e., if their chemical structure has not been strongly affected after biosynthesis and initial folding. For example, insulin (which loses half of its chain after its *in vivo* folding has been accomplished) is unable to refold.

Post-translational modifications are of a great variety. As a rule, chemical modifications are provided by special enzymes rather than "self-organized" within protein. First of all, I should mention cleavage of the protein chain (proteolysis: it often assists conversion of zymogen, an inactive pro-enzyme, into the active enzyme; besides, it often divides a huge "polyprotein" chain into many separate globules). The cleavage is sometimes accompanied by excision of protein chain fragments (e.g., when deriving insulin from proinsulin); by the way, the excised fragments are sometimes used as separate hormones. Also, one can observe modification of chain termini, acetylation, glycosylation, lipid binding to certain points of the chain, phosphorylation of certain side groups, and so on, and so forth. Even "splicing" of protein chains (spontaneous excising of a chain fragment and sticking the loose ends together) has been reported recently. Spontaneous cyclization of protein chains or their several portions is also occasionally observed.

Particular attention has to be given to formation of S—S bonds between sulfur-containing Cys residues: "proper" S—S bonds are capable (under favorable *in vitro* conditions) of spontaneous self-formation, although *in vivo* their formation is catalyzed by a special enzyme, disulfide isomerase. S—S-bonding is typical mostly of secreted proteins (there is no oxygen in the cell, and consequently, no favorable oxidizing potential for S—S-bonding). The S—S bonds, if properly paused, are not at all harmful but rather useful for protein renaturation.

In contrast, "improper" S—S bonds prevent protein renaturation. That is why, by the way, a boiled egg does not "unboil back" as the temperature decreases. The reason is that high temperature not only denatures the egg's proteins but also initiates formation of additional S—S bonds between them (like in a gum). As the temperature is decreased, these new chemical bonds persist, thereby not allowing the egg's proteins to regain their initial ("native") state.

Thus, the amino acid sequence of a protein determines its spatial structure, and this structure, in turn, determines its function, i.e., with whom this protein interacts and what it does.

Here, I have to make some comments.

First, Fig. 1.1 apparently shows that there is ample empty space in the interior of the protein, and can create an impression that the protein is "soft". In fact, this is not true. Protein is hard: its chain is packed tightly, atoms against atoms (Fig. 1.3a). However, the space-filling representation is inconvenient for studying protein anatomy, its skeleton, its interior; these can be seen using the wire model with "transparent" atoms and a clear pathway of the protein chain (see Fig. 1.3b and especially Figs. 1.1 and 1.2 where the side chain atoms are stripped off and secondary structure elements stand out). The space-filling model (Fig. 1.3a) gives no idea even of the polymeric nature of protein; it only shows the surface of a globule looking potato-like. However, this model is useful for studying protein function, since it is the physico-chemical and geometrical properties of the globule surface, the "potato skin", that determine the specificity of the protein activity, whereas the protein skeleton is responsible for the creation and maintenance of this surface.

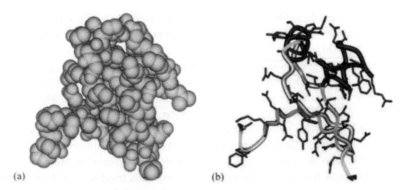

(a)   (b)

Figure 1.3. Space-filling representation of a protein globule (a) and its wire model (b). In the wire model, side chains are blue; the backbone is red in α-helices, yellow in β-strands, gray in irregular loops.

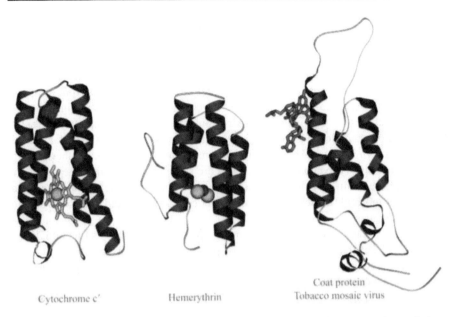

Cytochrome c′   Hemerythrin   Coat protein
                                                Tobacco mosaic virus

Figure 1.4. Three α-helical proteins similar in overall architecture (comprising four α-helices each), but different in function: cytochrome c′, hemerythrin, and coat protein of tobacco mosaic virus. Protein chain is shown as a ribbon; co-factors are shown as follows: wire models, heme (in cytochrome) and RNA fragment (in viral coat protein); orange balls, iron ions (in cytochrome heme and hemerythrin); a red ball, iron-bound oxygen (in hemerythrin).

Second, apart from the polypeptide chain, proteins often contain *co-factors* (Fig. 1.4), such as small molecules, ions, sugars, nucleotides, fragments of nucleic acids, etc. These non-peptide molecules are involved in protein functioning and some-times in the formation of protein structure as well. The co-factors can be linked by chemical bonds or just packed in cavities in a protein globule. Also, many water molecules are usually tightly bound to the protein surface.

Third, a solid protein ("aperiodic crystal" in Schrödinger's wording) behaves exactly like a crystal under varying conditions (e.g., at increasing temperature), that is, for some time it "stands firm" and then melts abruptly, unlike glass, which loses its shape and hardness gradually. This fundamental feature of proteins is closely allied to their functional reliability: like a light bulb, proteins become inoperative in the "all-or-none" manner, not gradually (otherwise their action would be unreliable, e.g., cause low specificity, etc. We will discuss this later on).

And finally, as concerns hardness, we have to distinguish between relatively small single-domain proteins that are really hard (they consist of one compact globule) and larger proteins that have either a multi-domain (Fig. 1.2) or quaternary (Fig. 1.1) structure. The component subglobules of larger proteins can move about one another.

In addition, like a solid body, all globules can become deformed (but not completely reorganized) in the course of protein functioning.

Proteins with similar interior organization ("anatomy") usually have the same related function. For example, many (though not all) cytochromes look like the one shown in Fig. 1.4, and many (though again not all) serine proteases from different species (from bacteria to vertebrates) look like chymotrypsin shown in Fig. 1.5.

But sometimes very similar spatial structures may provide completely different functions. For example, cytochrome, one of the three proteins with similar spatial structures shown in Fig. 1.4, binds an electron, while hemerythrin, another protein of this shape, binds oxygen (these functions are somewhat alike since they both are involved in the chain of oxidation reactions) and viral coat protein associates with much larger molecules, such as RNA and other coat proteins, and has nothing to do with oxidation.

We have already said that the structure of a protein determines its function. Is the reverse true? That is, does the function of a protein determine its structure?

Although some particular correlations of this kind have been reported, in general the influence of function on the structure has been detected mainly at a "rough" structural level, that is, the level connected with the "environmental conditions" of

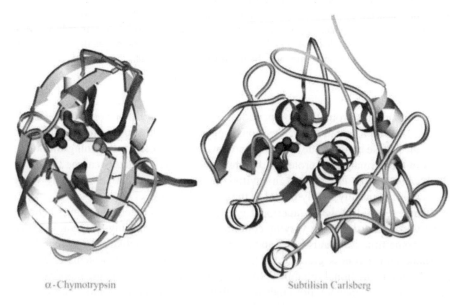

α-Chymotrypsin   Subtilisin Carlsberg

Figure 1.5. Two proteins structurally different but almost identical in function (serine proteases): chymotrypsin, formed by β-structure, and subtilisin, formed by α-helices (some of which pertain to the active site) along with β-structure. In spite of drastically different chain folds, their catalytic sites comprise the same residues similarly positioned in space (but not in the amino acid sequence): Ser195 (orange), His57 (blue) and Asp102 (crimson) in chymotrypsin and Ser221 (orange), His64 (blue) and Asp32 (crimson) in subtilisin.

protein functions (for example, proteins controlling the structural function, like those building up hair or fibrils, are mostly fibrous proteins, receptors are membrane proteins, etc.). But most frequently, we see no influence of function on protein anatomy and architecture. For example, two Ser-proteases, chymotrypsin and subtilisin, have the same catalytic function and even similar specificity, whereas their interior organizations have nothing in common (Fig. 1.5). (Their similarity is no greater than that between a seal and a diving beetle: only the proteins' "flippers", i.e., their active sites including half a dozen of amino acid residues (from a couple of hundreds forming the globules), are structurally alike, while in every other respect they are completely different.) Moreover, there exist structurally different active sites performing the same work (e.g., those of Ser-proteases and metal proteases). These, and many other examples, show that the function of a protein does not determine its 3D structure.

But, while saying this, account must be taken of size.

If the treated molecule (the molecule with which the protein interacts) is large, then a large portion of the protein may be involved in the interaction, and hence nearly the entire architecture of the protein is important for its functioning.

If the protein-treated molecule is small (which is more common), then it is minor details of a small fraction of the protein surface that determine its function, while the rest of its "body" is responsible for fixing these crucial details. Hence, the main task of the bulk of the protein chain is to be hard and provide a solid foundation for the active site.

Inner voice: The non-trivial and piquant facts that one and the same function is performed by proteins of utterly different architectures, and different functions by architecturally similar proteins, should not overshadow the absolutely correct commonplace that architecturally close proteins are often homologous (genetically related) and have identical or similar functions . . .

Lecturer: This is true – but trivial. And what I wanted to emphasize is the idea, important for protein physics (and protein engineering), that active sites may depend only slightly on the arrangement of the remaining protein body.

And the common feature of "the native protein body" is its hardness, since there is no other way to provide active-site specificity.

In due time we will consider the structures of proteins, their ability to self-organize and the reason for their hardness; we will discuss their functions and other aspects of interest for a biologist; but first we have to study amino acid residues and their elementary interactions with one another and the environment. These will be the subject of the next several lectures.

Part II

ELEMENTARY INTERACTIONS IN AND AROUND PROTEINS

LECTURE 2

Amino acids that build up polypeptide chains (Fig. 2.1) can have either the L or the D steric form.

The forms L and D are mirror-symmetric: the massive side residue side chain (R) and the H-atom, positioned at the α-carbon (C^α) of the amino acid, exchange places in these forms (arrows show atoms projected into the plane):

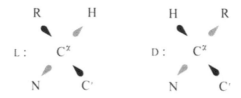

Glycine (Gly), with only a hydrogen atom as a side chain, has no difference between the L- or D-form.

Protein chains are built up only from L-residues. Only these are gene-coded. D-residues (sometimes observed in peptides) are not encoded during matrix protein synthesis but synthesized by special enzymes. Spontaneous racemization (L ↔ D transition) is not observed in proteins. It never occurs during biosynthesis but often accompanies chemical synthesis of peptides, and its elimination is highly laborious.

In the protein chain, amino acids are linked by peptide bonds between C'- and N-atoms (Fig. 2.2).

In protein structure an important role is played by the planar rigid structure of the whole *peptide unit*:

$$
\begin{array}{ccc}
O & & C^\alpha \\
\diagdown & & \diagup \\
& C' = N & \\
\diagup & & \diagdown \\
C^\alpha & & H
\end{array}
$$

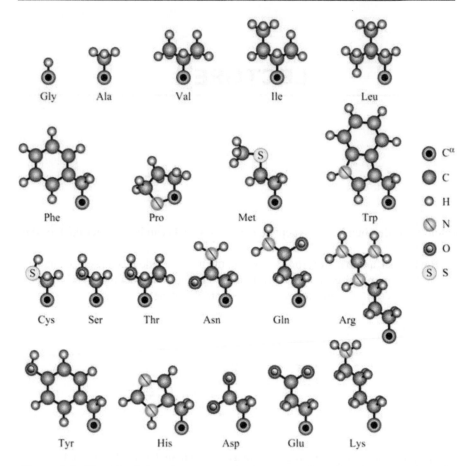

Gly   Ala   Val   Ile   Leu

Phe   Pro   Met   Trp

C^α
C
H
N
O
S

Cys   Ser   Thr   Asn   Gln   Arg

Tyr   His   Asp   Glu   Lys

Figure 2.1. The side chains of twenty standard amino acid residues (projecting from the main-chain C^α atoms). Atoms forming the amino acids are shown on the right.

Its planar character is provided by the so-called sp²-hybridization of electrons of the N- and C′-atoms. "Hybridization" of electron orbitals is a purely quantum effect. sp²-hybridization turns one spherical s-orbital and two "8-shaped" p-orbitals into three extended (from the nucleus) sp²-orbitals. These three orbitals involve the atom in three covalent bonds pertaining to one plane ($-\bullet<$). A covalent bond is created by a "delocalized" electron cloud covering the bound atoms.

The peptide group is rigid owing to an additional bond formed by p-electrons from the N- and C′-atoms uninvolved in sp²-hybridization. These electrons of N-, C′-, and O-atoms also bond and "delocalize", thereby creating an electron cloud that envelops all these atoms (that is why the bonds C′==N and C==O are drawn as equal "partial" double bonds, ==). And since the "8-shaped" p-orbitals are perpendicular to the

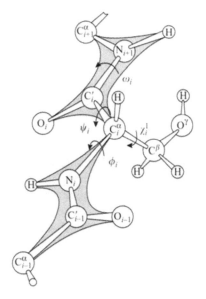

Figure 2.2. Diagram (adapted from [12]) showing a polypeptide chain with a side group (Ser; "i" is its number in the chain). The peptide units are outlined. The main-chain angles of rotation (ϕ, ψ, ω) and that of the side chain (χ^1) are presented. Arrows show the direction of rotation of the part of the chain closest to the viewer about its remote part that increases the rotation angle.

plane of the sp^2-orbitals ($-\bullet<$), the additional covalent bond of these perpendicular p-orbitals

blocks the rotation around the C'—N bond.

I would like to remind you that chemical bonds are caused mainly by delocalization of electrons, their transition from one atom to another. This follows from Heisenberg's *principle of uncertainty*:

$$\Delta p \Delta x \sim \hbar \tag{2.1}$$

Here Δp is the uncertainty in impulse of the particle, Δx is the uncertainty in its coordinate, and $\hbar \equiv h/2\pi$, where h is Planck's constant. Since the direction of electron movement within the atom is unpredictable, $\Delta p \approx |p| = mv$, where v is the velocity and m is the mass of the particle. Hence,

$$v \sim \frac{\hbar}{m\Delta x} \tag{2.2}$$

At the same time, the kinetic energy of the particle $E = mv^2/2$, i.e.,

$$E \sim \frac{\hbar^2}{m\Delta x^2} \tag{2.3}$$

Hence, owing to the delocalization, the energy of the particle decreases with increasing Δx, and thereby the particle adopts a more stable state. As seen, light particles (electrons) are those mostly affected. This is how electron delocalization causes chemical bonding.

The length of a chemical bond is close to the van der Waals radius of atoms, i.e., it amounts to 1–2 Å (to be more exact, it is 1 Å for C—H, N—H and O—H bonds, about 1.2–1.3 Å for C=O, C==O, C==N and C=C, 1.5 Å for C—C and about 1.8 Å for S—S).

Typical values of covalent angles are approximately 120° and 109°. The former are at sp^2-hybridized atoms like —C'<, —N<, where three covalent bonds are directed from the center to the apexes of a planar triangle, and the latter are at sp^3-hybridized atoms, like >C$^\alpha$<, where four bonds are directed from the center to the apexes of a tetrahedron, as well as at O< or S< atoms having two bonds each.

Now let us consider typical values of vibrations, i.e., thermal vibrations of covalent bonds and angles. These can contribute to the flexibility of the protein chain.

Vibrational frequencies manifest themselves in the infrared (IR) spectra of proteins. Typical frequencies are as follows: $\nu \sim 7 \times 10^{13}\,\text{s}^{-1}$ for vibrations of the H atom (for example, in the bond C—H; the corresponding IR light wavelength $\lambda = c/\nu \sim 5\,\mu\text{m}$, where c is the speed of light, 300 000 km s^{-1}). For vibrations of "heavy" atoms and groups (for example, in the bond CH$_3$—CH$_3$), $\nu \sim 2 \times 10^{13}\,\text{s}^{-1}$ (then $\lambda = c/\nu \sim 15\,\mu\text{m}$).

Are these vibrations excited at room temperature?

To answer this question, we have to compare heat energy per degree of freedom ("heat quantum", kT) with vibration energy. Let us estimate kT at "normal" temperature. Here, T is absolute temperature in Kelvin ($T = 300$ K at 27 °C, i.e., at about "room" temperature; K denotes Kelvin), and k (sometimes written as k_B) is the Boltzmann constant (equal to ≈ 2 cal mol^{-1} K, or 0.33×10^{-23} cal K^{-1} per particle, since one mole contains 6×10^{23} particles). Hence, at room temperature the "heat quantum" $kT = 600$ cal mol^{-1}, or (600 cal)/(6×10^{23} particles), i.e., 10^{-21} calories per particle.

The frequency ν_T corresponding to this heat quantum can be derived from the well-known equation $kT = h\nu_T$ (where Planck's constant $h \equiv 2\pi\hbar = 6.6 \times 10^{-34}$ J s $= 1.6 \times 10^{-34}$ cal s; let me remind you that 1 calorie is equal to 4.2 joules (J)]. So, the characteristic frequency of thermal motions, ν_T, is equal to 7×10^{12} s^{-1} at $T = 300$ K, i.e., at 27 °C.

The "heat quantum" cannot induce vibrations with a higher frequency than its own.

Thus, at room temperature, covalent bonds are "hard" and do not vibrate: their vibrational frequency $\nu \sim 2 \times 10^{13}$–$7 \times 10^{13}$ s^{-1}, i.e., an order of magnitude higher than $\nu_T = 7 \times 10^{12}$ s^{-1}.

However, vibrations of covalent bonds can be induced by infrared light; this underlies the IR spectroscopy of proteins. It is IR spectroscopy that provides information about the vibrations of atoms, covalent bonds and covalent angles. The properties of these vibrations are first derived from experiments on small molecules and then used for protein investigations.

Covalent angles are less rigid than bond lengths, and therefore, they vibrate at room temperature; their vibrational frequency ranges from 10^{12} to 10^{13} s^{-1}. However, their typical amplitude amounts only to $5°$.

Thus, covalent bond vibrations do not contribute to the flexibility of protein chain, and the contribution of vibrations of covalent angles is minor.

In fact, this flexibility (which implies the ability to fold into α-helices and globules) is provided by rotation (although not a completely free rotation – see below) *around* the covalent bonds. That is why the chain structure is often described simply in terms of angles of rotation around covalent bonds – then it is called the "conformation". It should be noted, that the terms "structure" and "conformation" are often used as synonyms.

The relative position of atomic groups linked by a covalent bond is described by a dihedral angle (that is formed by two planes: in Fig. 2.3, one includes points **a**, **O′**, **O″** and the other **O′**, **O″**, **c**).

Figures 2.2 and 2.3 illustrate the measurement of this angle. The measurement is made as described in school trigonometry, if we assume that the covalent bond closest to the viewer is the "rotating arrow", the far bond is the "coordinate axis", and the central one is the "rotation axis". As in trigonometry, the arrow's turn in a counterclockwise direction increases the angle of rotation, while its clockwise movement decreases this angle.

The major information on rotation of atomic groups around covalent bonds is also provided by IR spectroscopy; again, I will discuss only the basic results obtained.

Figure 2.4 illustrates the typical variation of energy in the case of rotation around the bond between two sp^3-hybridized atoms ($H_3C—CH_3$ and $CH_2C—CH_2C$ serve as examples). These bonds are typical of aliphatic side chains. Side-chain angles of rotation are called χ ("chi") angles (Fig. 2.2). The maximums of such triple (in accordance with symmetry of rotation about the $sp^3—sp^3$ bond) potentials (i.e., potentials that have three maximums and three minimums within $360°$) correspond to three "eclipsed" conformations ($0°$, $120°$ and $240°$) that result in approach (and repulsion) of the electron clouds. The repulsion occurs because these electrons have been already involved in covalent bonding. The resultant potential barriers of rotation

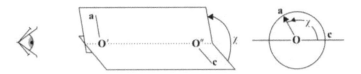

Figure 2.3. Measurement of the dihedral angle (angle of rotation) shown in both axial (right) and transverse (left) views. The central bond **O′—O″** serves as the "rotation axis", the covalent bond **O″—c** serves as the "coordinate circle axis", and the closest (to the viewer) bond **O′—a** serves as the "arrow on the coordinate circle" in trigonometry. When the dihedral angle is measured in the chain, the atoms **a** and **c** belong to the heaviest atomic groups attached to the **O′—O″** bond.

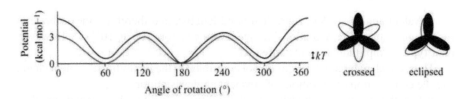

Figure 2.4. Typical potential of rotation around a single bond between two sp^3-hybridized atoms: around $H_3C—CH_3$ (red curve) and $CH_2C—CH_2C$ (blue curve). The major energetic effect results from repulsion of electron clouds that is at a maximum in the "shaded" conformations (at 0°, 120° and 240°) and at a minimum in the "crossed" ones (at 60°, 180° and 300°). Repulsion of small H-atoms is negligible. However, repulsion of heavy C-atoms surrounded by large electron clouds occurring around 0° (in the chain C—C—C—C) yields an additional energetic effect that distinguishes rotation around the $(C—CH_2)—(CH_2—C)$ bond from that around the $H_3C—CH_3$ bond. For comparison, $\updownarrow$ shows the magnitude of kT.

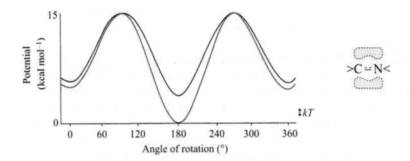

Figure 2.5. Typical potential of rotation around a peptide bond between two sp^2-hybridized atoms (C′ and N); p-orbital-bonding at 0° (or 180°) is shown in yellow on the right. For all (except pre-proline) peptide bonds, their energy (red curve) is higher at 0° than at 180° owing to the repulsion between massive C^α-atoms at 0°. This difference in energy is small for the peptide bond preceding Pro (blue curve): Pro has not one but two C-atoms bonded to the N-atom. See below for the text and the figure illustrating the structural formula of proline.

around $H_3C—CH_3$ amount to about 3 kcal mol^{-1}, and the typical range of thermal fluctuations about these minimums (i.e., deviations accompanied by energy increasing by kT) is 15–20°.

When more massive atoms are sp^3-bonded instead of some H atoms, repulsion of these contributes to the barrier in the region where they become too close to one another. This is exemplified (see Fig. 2.4) by the rotation around the central bond in $(C—CH_2)—(CH_2—C)$.

Figure 2.5 illustrates the typical variation of energy in the case of rotation around a peptide bond between two sp^2-hybridized atoms (C′ and N). The angle of rotation around this bond is denoted as ω (Fig. 2.2). The potential is double (i.e., it has two maximums and two minimums within 360°) in accordance with the symmetry of rotation about the sp^2—sp^2 bond. The potential barriers are high owing to the involvement of additional p-electrons in the peptide bond (as discussed at the beginning of this lecture). The potential minimums are at 0° and 180° (where the p-orbitals pulling together C' and N atoms are at their closest), and its maximums are at 90° and 270° (where these p-orbitals are farthest apart and, hence, least connected with one another). High barriers mean that the typical range of thermal fluctuations of the angle of rotation around such bonds is small (5–10°).

It is noteworthy that repulsion of massive C^α-atoms makes the *cis* conformation ($\omega = 0°$) rather unfavorable energetically; therefore, in proteins, almost all peptide groups are in the *trans* conformation ($\omega = 180°$).

$$
\begin{array}{llcccc}
 & & O \quad\quad C^\alpha & & O \quad\quad H & \\
 & & {\backslash\!\backslash} \quad\ / & & {\backslash\!\backslash} \quad\ / & \\
\text{Typical:} & trans: & C' =\!= N & cis: & C' =\!= N & \\
 & & /\quad\quad {\backslash} & & /\quad\quad {\backslash} & \\
 & & C^\alpha \quad\quad H & & C^\alpha \quad\quad C^\alpha &
\end{array}
$$

An exception is the proline-preceding peptide bond. Pro is an *imino* but not an *amino* acid: its N atom has not two but three similar massive radicals (C'—, —$C^\alpha HC_2$ and —CH_2C, see Fig. 2.1), and therefore its *trans* conformation has only minor advantage as compared with the *cis* one.

$$
\begin{array}{llcccl}
 & & \quad\quad C' & & & \\
 & & \quad\quad / & & & \\
 & & O \quad\ C^\alpha\!-\!C & & O \quad\ C\!-\!C & \\
 & & {\backslash\!\backslash} \quad / \quad\quad | & & {\backslash\!\backslash} \quad / \quad\quad | & \\
trans\ \text{Pro:} & & C' =\!= N & cis\ \text{Pro:} & C' =\!= N & \\
 & & / \quad\ {\backslash} \quad\quad\quad | & & / \quad\ {\backslash} \quad\quad\quad | & \\
 & & -C^\alpha \quad C\!-\!C & & -C^\alpha \quad C^\alpha\!-\!C & \\
 & & & & \quad\quad\quad {\backslash} & \\
 & & & & \quad\quad\quad C' &
\end{array}
$$

In both globular and unfolded (i.e., fluctuating, lacking a fixed structure) peptides there are about 90% of *trans* and 10% of *cis* prolines. I would like to draw your attention to this regularity – the more favorable some detail is in itself (individually), the more frequently this detail occurs in proteins. We will see it many times to come.

Lastly, let us consider the potential of rotation around the bond between sp^3-hybridized and sp^2-hybridized atoms. Angles of rotation around such bonds are denoted as ϕ (rotation around N—C^α) and ψ (rotation around C^α—C') (Fig. 2.2).

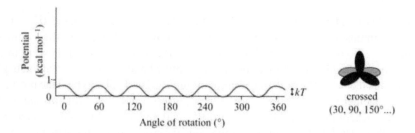

Figure 2.6. Typical potential of rotation around a single bond between sp³- and sp²-hybridized atoms (exemplified by rotation around H_3C—C_6H_5). The sp²-hybridized (light-blue) and the sp³-hybridized (black) electron clouds are shown in the "crossed" conformation.

Either rotation yields a six-fold (six minimums and six maximums within 360°) potential with rather low barriers ($\approx$1 kcal mol^{-1}) (Fig. 2.6) that are of the same order as the energy of thermal fluctuations (which, as we remember, amounts to 0.6 kcal mol^{-1} at room temperature). It is these nearly free rotations around such bonds (N—C^α and C^α—C') in the polypeptide main chain that ensure the flexibility of the protein chain.

LECTURE 3

In this lecture we shall focus on elementary non-covalent interactions between atoms.

First of all, let us recall in what circumstances no covalent bond is formed between approaching atoms. Let me remind you that chemical bonding basically results from delocalization of electrons, i.e., their "transition" from one atom to another and back: these electrons are often said to share the same ("common") orbital enveloping two or more atoms. However, according to the *Pauli exclusion principle*, not more than two electrons can share the same orbital, and they can do this only when their spins (moments of rotation) are oppositely directed (then they are "paired electrons"). Pairing in a common orbital of electrons coming from two different atoms results in a tight covalent bond.

If orbitals of two mutually approaching atoms already bear a pair of electrons each, no covalent bond can emerge. Otherwise there would be too many electrons in the common orbital – four. But, as stated by the *Pauli exclusion principle*, one orbital can bear no more than two electrons. Consequently, a "saturated" orbital with an electron pair on cannot accept extras. Therefore, atoms with saturated electron orbitals repel as they come near enough for their electron clouds to begin to overlap. Such atoms are impenetrable to one another at ordinary (though not at stellar) temperatures.

The same is observed when molecules with no vacant valency approach one another: they repel at a distance between their atoms as short as 2–3 Å.

However, at a greater distance (when electron clouds do not overlap) all atoms and molecules attract each other, unless they are charged (we will discuss this later). This attraction is purely quantum in nature. It is connected with coordinated vibrations of electrons in both atoms. The thing is, the coordinated ("in the same direction") shift of electrons results in attraction of the atoms (whereas atoms with non-shifted electrons do not attract or repel each other). This becomes clear from the following diagram showing two atoms, #1 and #2, with an electron "−" and nucleus "+" in each.

1	2		1	2		1	2
−	−	worse, than	−		or	−	
				−			−
+	+		+	+		+	+
1	2		1	2		1	2

When considering this diagram one must bear in mind that electric energy increases with decreasing distance r as $1/r$. The electron shift causes no change in electron–electron and nucleus–nucleus interactions, but the increase in attraction between electron 1 and nucleus 2, which become close (see the extreme right of the picture) is greater than the decrease in attraction between electron 2 and nucleus 1. Now recall that electrons are in constant vibration within an atom (they "orbit the nucleus" and cannot fall onto it because of Heisenberg quantum uncertainty). The above effect provokes a coordination of electron vibrations as the atoms come closer and hence causes attraction of these atoms.

At greater distances, the atomic interactions weaken. As a result, interaction energy decreases as a function of distance r between the centers of the two atoms; it can be shown that the energy decrease is proportional to $(1/r)^6$.

The total potential of atomic interaction (also called the energy of van der Waals interaction) is illustrated by Fig. 3.1 and approximately described by the Lennard–Jones potential of the form:

$$U_{LJ}(r) = E_0 \left[\left(\frac{r_0}{r}\right)^{12} - 2 \left(\frac{r_0}{r}\right)^6 \right]. \tag{3.1}$$

Here r_0 (as can be easily proved by taking the U_{LJ} derivative over distance r) is the distance at which the energy U_{LJ} is at a minimum, and E_0 is the depth of the minimum. The last term that decreases as $(const/\text{distance})^6$ gives the attraction ("minus" shows that the corresponding energy decreases with decreasing distance); the term of the 12th power gives the repulsion (it is positive, i.e., the corresponding energy increases with decreasing distance).

Equation (3.1) gives a precise description of the attraction at large distances (when $r \gg r_0$). The repulsion at small distances is described only qualitatively as "very strong and exceeding any attraction when r tends to zero". The approximate character of eq. (3.1) is demonstrated by the fact that atoms are implied to be spherical (since the described interaction is direction-independent), whereas actually the atomic electron cloud is not spherical because of projecting p-electrons. In general, only quantum mechanics can provide the correct description of interactions between atoms, but it can strictly calculate only very simple systems like the He atom, H_2^+ ion or H_2

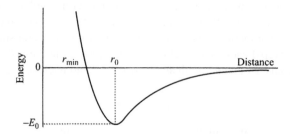

Figure 3.1. Typical profile of the van der Waals interaction potential.

molecule. All other systems have to be described using approximate "semi-empirical" equations like eq. (3.1), the form of which is based on only semi-quantitative physical considerations, and parameters (in this case, E_0 and r_0) on experiment. Some basic data are given in Table 3.1.

You should notice that Table 3.1 presents not only values of the optimal distance r_0 but also those of the minimum distances r_{min} that exist in the crystalline state. In Fig. 3.1, r_{min} approximately corresponds to the point where energy passes through 0 at a short distance between atoms. Values of r_{min} are helpful in estimating the possibility of a particular chain conformation.

Inner voice: There are a lot of works on deriving potentials of atom–atom interactions. Isn't this indicative of the questionable precision of all these potentials?

Lecturer: As to the *form* of the potential (Fig. 3.1), there's no particular disagreement. The estimates of r_{min} are also alike as they are directly measured in crystals. The difference in views concerns values of r_0 and especially E_0 ... These are mainly derived from crystals as well, that is, from their structure (distance) and sublimation heat (energy). However, crystals usually consist of not atoms but molecules, for example, CH_4, C_2H_6 ... So, when calculating the energy, a question arises as to the contribution of interactions $C \cdots C$, $H \cdots H$ and $C \cdots H$. Different authors answer it in different ways, so sometimes their potentials differ significantly. However, this difference is smoothed out as soon as they come back from atoms to molecules. But it is important to remember that when calculating molecular structures, the source of one parameter (say, the energy of the $C \cdots C$ interaction) should not differ from the source of the others (e.g., the $H \cdots H$ interaction energy): here the principle "all-or-none" must be followed to avoid mistakes.

The tabulated values are helpful for understanding why the *trans* conformation (180°) of the $C'{=}{=}N$ bond is allowed and its *cis* conformation (0°) is disallowed (for all amino acid residues, except Pro, as mentioned earlier): for the $C'{=}{=}N$ *trans*

Table 3.1. Typical parameters of van der Waals interaction potentials.

Interaction	E_0 (kcal mol^{-1})	r_0 (Å)	r_{min} (Å)	Minimum van der Waals radius of atom (Å)	
$H \cdots H$	0.12	2.4	2.0	H:	1.0
$H \cdots C$	0.11	2.9	2.4		
$C \cdots C$	0.12	3.4	3.0	C:	1.5
$O \cdots O$	0.23	3.0	2.7	O:	1.35
$N \cdots N$	0.20	3.1	2.7	N:	1.35
$CH_2 \cdots CH_2$	≈ 0.5	≈ 4.0	≈ 3.0	CH_2:	≈ 1.5

E_0 and r_0 values for interactions between atoms are from Scott R.A., Scheraga H.A., *J. Chem. Phys.* (1965) **45**: 2091; r_{min} values from Ramachandran G.N., Sasisekharan V., *Adv. Protein Chem.* (1968) **23**: 283. These values provided the basis for estimating $CH_2 \cdots CH_2$ interaction parameters. The interaction $CH_2 \cdots CH_2$ depends on the relative orientation of these groups; therefore, the tabulated results are approximate. Nevertheless, they are often used to calculate interactions in proteins when H-atoms are "invisible" to X-rays.

conformation, the distance between C^α atoms is 3.8 Å, while for its *cis* conformations (when these atoms are at their closest), it is only 2.8 Å, i.e., less than the minimum distance $r_{min} = 3.0$ Å allowed for the $C \cdots C$ pair.

$$
\begin{array}{ccc}
 & \overset{\displaystyle C^\alpha}{\diagup} & \\
trans: \quad C' == N & \qquad cis: \quad C' == N & \\
\diagup & \diagup \qquad \diagdown & \\
C^\alpha & C^\alpha \qquad C^\alpha &
\end{array}
$$

$$\leftarrow 2.8\,\text{Å} \rightarrow$$

C^α atoms of the sequence-neighboring amino acids are rather far apart in space owing to the rigid *trans* form of the $C' == N$ bond. This provides an opportunity for these neighboring residues to change their conformations almost independently of each other. But inside a residue, rotations over ϕ and ψ angles are interconnected. The "allowed" and "disallowed" conformations of a residue plotted in the (ϕ, ψ) coordinates are called *Ramachandran plots* (or, to be more exact, Sasisekharan–Ramakrishnan–Ramachandran plots).

Prior to drawing these maps, let us see what conformations are allowed (and what are not) in the case of ϕ (about $N — C^\alpha$) and ψ (about $C^\alpha — C'$) rotations separately.

As we already know, rotation around these bonds (between the sp^3-hybridized C^α atom and sp^2-hybridized N or C') is nearly free. However, in *cis* conformations (at $\phi = 0°$ or $\psi = 0''$), atoms rotating around these bonds (C'_{i-1} and C'_i for ϕ, and N_i and N_{i-1} for ψ; see Fig. 3.2; $i-1, i, i+1$ are numbers of consecutive residues in the chain) come too close to each other (especially considering the approach of C'_{i-1} and C'_i atoms), and, because of their repulsion, this conformation may be disallowed, or in other words, sterically prohibited.

What makes rotation over ϕ more difficult than rotation over ψ?

The next scheme shows that minimum distances between the atoms C'_{i-1} and C'_i (ϕ angle) and between N_i and N_{i-1} (ψ angle) are the same, 2.9 Å, that is a bit *less* than $r_{min} = 3.0$ Å for the $C \cdots C$ interaction (so the *cis* conformation over ϕ is disallowed) and a bit *more* than $r_{min} = 2.7$ Å (but less than the optimal $r_0 = 3.1$ Å) for the $N \cdots N$ interaction (so the *cis* conformation over ψ is not prohibited, though strained; however, as shown by calculations, even the minor flexibility of the covalent angle $N — C^\alpha — C'$ is sufficient to relieve the strain considerably).

$$2.9 \text{ Å} < r_{min}(C \cdots C) = 3.0 \text{ Å} \qquad 2.9 \text{ Å} > r_{min}(N \cdots N) = 2.7 \text{ Å}$$

$$
\begin{array}{cc}
\leftarrow 2.9\,\text{Å} \rightarrow & \qquad\qquad \leftarrow 2.9\,\text{Å} \rightarrow \\
C' \qquad\quad C' & \qquad\quad N \qquad\quad N \\
\diagdown\!\!\!\! \qquad \diagup & \qquad\quad \diagdown \qquad \diagup \\
N — C^\alpha & \qquad\quad C^\alpha — C \\
\phi & \qquad\qquad \psi
\end{array}
$$

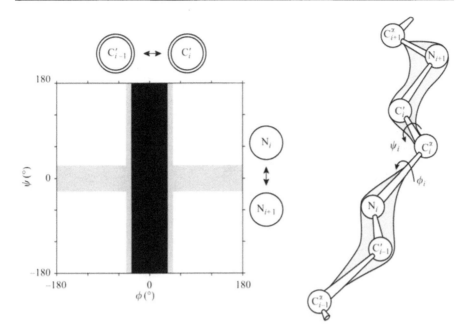

Figure 3.2. This is how Ramachandran plots of the disallowed (■), strained (▨), and fully allowed (□) (ϕ, ψ) conformations of the fragment $C^\alpha C'N—C^\alpha—C'NC^\alpha$ would look, provided all these atoms had no other atoms attached (right) and atoms of residues $i - 1$ and $i + 1$ had no interactions.

If we only had to consider these C' and N atoms, Ramachandran plots of the disallowed (■), strained (▨), and allowed (□) conformations would look as shown in Fig. 3.2, and the ϕ, ψ rotations would be independent of each other.

However, the C' atoms have, in addition, O and C^α atoms attached, and the N atoms are, in addition, bonded to H and C' atoms (and in water this H atom is quite rigidly linked by a hydrogen bond to a water molecule, as we will soon see). As a result, the Ramachandran map, i.e., plots of disallowed and allowed conformations of the smallest amino acid residue, glycine (with H as its side chain), looks as presented in Fig. 3.3.

Glycine has no massive side chain. All other amino acid residues do have such a chain, and its collision (or, rather, the collision of its C^β atom closest to the main chain) with the C'_{i-1}-atom accounts for the disallowed ϕ region, while its collision with the N_{i+1}- atom accounts for the disallowed ψ region (Fig. 3.4).

The map shown in Fig. 3.4 is for alanine that has a small side chain comprising the $C^\beta H_3$ group only. The side chains of all other amino acid residues are larger; they include one or two heavy γ atoms attached to the C^β atom. Since these "new" γ atoms (and the still more remote δ, ε, etc.) are far from the main chain, their effect on the Ramachandran map is only minor. More precisely, in a small region (left white in Fig. 3.5), γ atoms have no collisions at all with the main chain, whereas in other conformations allowed for alanine such collisions are possible for some side-chain

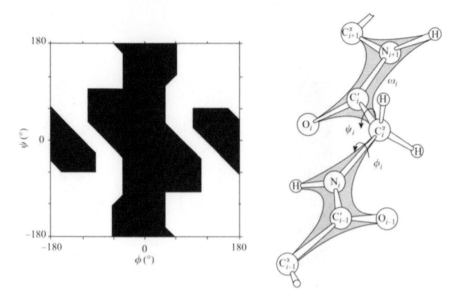

Figure 3.3. The map of disallowed (■) and allowed (□) ϕ, ψ conformations of glycine (Gly) in the protein chain.

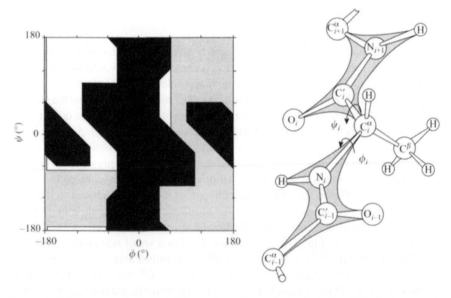

Figure 3.4. The map of allowed (□) ϕ, ψ conformations of alanine (Ala) in the protein chain; (▨), regions allowed for Gly only; (■), regions disallowed by main-chain interactions for all residues.

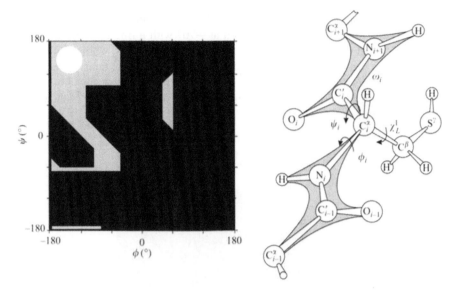

Figure 3.5. The map of disallowed (■) and allowed (□, ▨) ϕ, ψ conformations of larger residues in the protein chain. (□), the region where all side-chain χ^1 conformations are allowed; (▨), the region where some χ^1 conformations are disallowed.

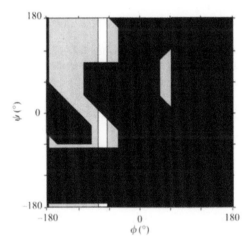

Figure 3.6. The map of allowed (□) Pro conformations plotted against allowed Ala conformations (▨); (■), the region of disallowed conformations for both residues.

conformations, and therefore, not all of them are allowed. These regions are slightly shaded in Fig. 3.5.

Collisions between γ atoms and the main chain are most significant for valine, isoleucine and threonine, which have two large γ atoms each.

As a final example, let us consider the Ramachandran plot for the imino acid proline. There, the ϕ angle is nearly fixed at $-70°$ with a ring built up by the Pro side group linked to the N-atom of its main chain (see Fig. 3.1), while rotation over ψ is similar to that of alanine. As a result, the allowed conformations of proline are accommodated by the white band in the Ala's Ramachandran plot, Fig. 3.6.

The Pro ring also diminishes the region of allowed conformations of the residue preceding proline in the protein chain (Fig. 3.7).

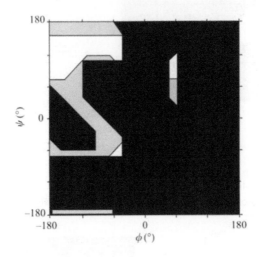

Figure 3.7. The region of allowed (□) conformations of an Ala residue that precedes Pro in the protein chain. If not the following Pro, ▦ would be included in the region of allowed Ala's conformations.

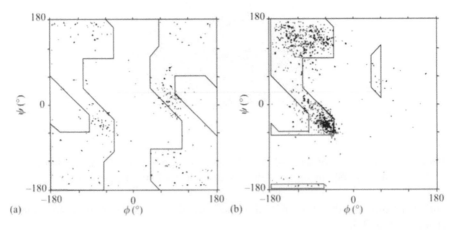

Figure 3.8. Observed conformations (dots) of glycine (a) and of other amino acid residues (b) in proteins. The sterically allowed regions are contoured. Adapted from [1a].

Lastly, let us see whether all these theoretical considerations agree with the observed (by X-ray crystallography and NMR spectroscopy) conformations of amino acid residues in proteins. In Fig. 3.8 these observed conformations are plotted against the contours of regions theoretically allowed for glycine, on the one hand, and for alanine and other residues, on the other hand. As seen, the agreement is quite good.

We see that the "sterically allowed" regions accommodate the majority of the experimental points. However, some points are in the "sterically prohibited" regions. This is not surprising, since the regions that we call "sterically prohibited" are those of high energy – not infinitely high, for sure, as that would cause complete prohibition – but higher than the minimum conformational energy by, say, a couple of kilocalories per mole. In other words, a protein has to spend some energy to drive its residue into such a region. We see that it is able to do so, although it rarely does.

In the course of these lectures we will learn the general rule: strained, high-energy elements are rare, though sometimes observed. This is not surprising since a stable protein must contain – mostly, but not exclusively – stable components.

LECTURE 4

So far we have not taken the aqueous environment of proteins into account. It's high time we bridged the gap.

Water is a peculiar solvent. First, it boils and freezes at abnormally high temperatures compared to those typical for substances of similarly low molecular weight. Indeed, water, H_2O, boils at 373 K and freezes at 273 K, while O_2 boils at 90 K and freezes at 54 K; H_2 boils at as low a temperature as 20 K and freezes at 4 K; CH_4 boils at 114 K, etc. The fact that ice and water structures are heat-resistant suggests some strong bonding among water molecules.

The bond responsible for this effect is specifically that between O- and H-atoms of H_2O (as individually, O_2 and H_2 boil and melt easily). This bond is called a *hydrogen bond.*

Hydrogen bonds are not only observed in water. They invariably occur when a hydrogen atom approaches some electronegative (i.e., electron attracting) atom while being chemically bonded to another electronegative atom, as exemplified by the O—H:::O, N—H:::N bonds. But, for instance, a C—H group is not involved in hydrogen bonding since the electronegativity of the C atom is insufficient.

The solvent properties of water are dominated by strong hydrogen bonds.

The hydrogen bonding between water molecules is electronic in nature. It is connected with electrons and charges, but not the nuclei of the hydrogen atoms, as is shown by the close similarity of boiling and melting parameters of light (H_2O) and heavy (D_2O) waters in spite of a two-fold difference in mass between D and H nuclei.

The water molecule is polar. This implies small ("partial") charges on its atoms: negative on O and positive on H. The distribution of charges and electron clouds at these atoms looks as follows:

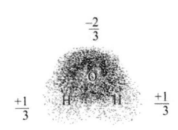

Here, the density of dots reflects the density of the electron cloud, and numerals indicate the partial charges on the atoms. These are expressed in fractions of the proton charge which, naturally, in these units amounts to $+1$ while the electron charge is -1.

An electronegative oxygen draws the electron clouds off the neighboring hydrogens, thereby causing polarization of atoms. As a result, H atoms acquire partial positive charges, while there is a negative charge on O in the water molecule.

As you may remember, in a vacuum, charges q_1 and q_2 at distance r have the energy of interaction

$$U = q_1 q_2 / r. \tag{4.1}$$

In a vacuum, interactions between charges are very strong. The energy of interaction of two single charges (i.e., proton or electron charges) at a distance of $1\,\text{Å}$ is nearly $330\,\text{kcal mol}^{-1}$ (keep this figure in mind: we will use it for different estimates later), while at a more realistic distance of $3\,\text{Å}$ (with van der Waals repulsion of atoms taken into account) this energy is about $110\,\text{kcal mol}^{-1}$. The energy of single-charge interaction is the typical energy of a chemical bond; it is hundreds of times higher than the typical thermal energy kT or the typical energy of van der Waals interactions between atoms.

The partial charges of water molecules are still lower, and therefore, their interaction is weaker: at a distance of $3\,\text{Å}$ its energy is about $10\,\text{kcal mol}^{-1}$; however, this energy is sufficient to distort the electron envelopes of H-atoms by H to O attraction. Hydrogen atoms are most sensitive in this respect: their single-electron envelope is drawn towards O and therefore undergoes the distortion most easily. It takes much more energy to distort, say, the electron envelope of oxygen that has eight "own" electrons and a share of the single electron of both hydrogens of the water molecule.

It is the ease of distortion of the electron cloud of a hydrogen atom that turns a normal electrostatic interaction into a hydrogen bond. This is true for all hydrogen bonds among which those of interest for us are O—H ::: O, N—H ::: O, N—H ::: N.

Thus, a hydrogen atom has the thinnest cloud whose significant distortion results from attraction between the partial positive charge of hydrogen and the partial negative charge of oxygen (or nitrogen). This gives distances between H and O (or N) nuclei as small as 1.8–$2.1\,\text{Å}$ (reported for crystals of small molecules) instead of the 2.35–$2.75\,\text{Å}$ typical for van der Waals interactions discussed in the previous lecture.

$\leftarrow 1.8\,\text{Å} \rightarrow$	$\leftarrow 1.9\,\text{Å} \rightarrow$	$\leftarrow 2.1\,\text{Å} \rightarrow$
O—H : : : O	N—H : : : O	N—H : : : N
$\leftarrow 2.8\,\text{Å} \rightarrow$	$\leftarrow 2.9\,\text{Å} \rightarrow$	$\leftarrow 3.1\,\text{Å} \rightarrow$

This close approach yields a *hydrogen bond* (or *H-bond*). The H-atom (to be more exact, the O—H or N—H group) is called the *donor* of the hydrogen bond, and the O- or N-atom towards which the hydrogen moves is called the *acceptor* of the hydrogen bond.

Note that in crystals H-bonding occurs when the distance between O- and/or N-atoms of the donor and acceptor is about $3\,\text{Å}$ (for example, in ice it is $2.8\,\text{Å}$). This is similar to the optimal van der Waals distance between O- and/or N-atoms, i.e., the presence of the mediating H-atom does not increase the distance between these atoms of the donor and acceptor, as it pushes them not apart but together.

Each H-bond has one donor and one acceptor. The hydrogen atom almost always acts as a donor of only one H-bond, while the oxygen atom may participate as an acceptor in two H-bonds simultaneously, thereby causing "fork-like" H-bonding:

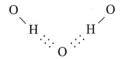

Since the "fork-like" H-bonding implies a short distance between two H atoms (with $+1/3$ charge on each), its total energy is less than double the energy of a single (see previous diagram) H-bond.

Unlike van der Waals interactions, H-bonding is rather orientation-sensitive, especially as concerns orientation of the donor group. Usually, a valence bond of the donor is directed at the acceptor atom (O or N) to be involved in the hydrogen bond.

Orientation of the acceptor group is of considerably less importance for the H-bond:

The H-bond energy is about $5\,\text{kcal mol}^{-1}$. This estimate results from comparison of experimental evaporation heats of similar compounds, some of which are capable of H-bonding while others are not. For example, the evaporation heat of dimethyl ether, $H_3C\!-\!O\!-\!CH_3$, is about $5\,\text{kcal mol}^{-1}$, and that of ethanol, $CH_3\!-\!CH_2\!-\!OH$, is about $10\,\text{kcal mol}^{-1}$. These compounds consist of the same atoms (i.e., their van der Waals interactions are nearly the same), but ethanol is capable of H-bonding while dimethyl ether is not (as it lacks the O—H group). Each O–H group can participate as donor in only one H-bond, and its O-atom can accept this bond. Since each H-bond is supposed to have one donor and one acceptor, there is only one H-bond per ethanol molecule, that is, an H-bond "costs" about $10\,\text{kcal mol}^{-1}(ethanol) - 5\,\text{kcal mol}^{-1}(ether) = 5\,\text{kcal mol}^{-1}$.

The same estimate follows from the value of ice evaporation heat ($680\,\text{cal g}^{-1} = 680\,\text{cal}\,(1/18\text{ mole})^{-1} = 12\,\text{kcal mol}^{-1}$). Here a couple of kilocalories per mole

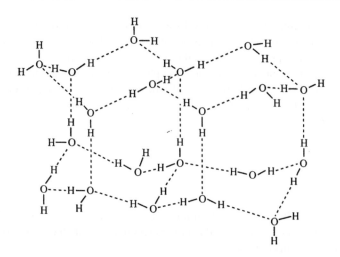

Figure 4.1. Normal ice. The dashed lines show H-bonding. As shown, the openwork structure of ice has small cavities surrounded by H_2O molecules. Adapted from [4] with minor modifications.

are the share of van der Waals interactions, as seen from evaporation of small molecules like methane (CH_4) or O_2. The remaining $10\,\text{kcal mol}^{-1}$ are for H-bonding. In ice, there are two H-bonds (Fig. 4.1) per molecule of H_2O, since its two O—H groups can serve as donors for two H-bonds (and as many can be accepted by its oxygen). So, again, the "cost" of one H-bond appears to be about $(10\,\text{kcal mol}^{-1})/2 = 5\,\text{kcal mol}^{-1}$.

The structure of normal ice is determined by H-bonds (Fig. 4.1): it is good for their geometry (O—H is directed at O), although not so good for close van der Waals contacts between water molecules. In ice, water molecules envelop tiny (smaller than H_2O molecules) pores, thereby giving it an openwork structure. This results in two well-known phenomena: (1) ice is not as dense as water, it floats, and (2) under strong pressure (e.g., caused by skate blades) ice melts.

The majority of H-bonds existing in ice (Fig. 4.1) persist in liquid water. This follows from the low melting heat of ice ($80\,\text{cal g}^{-1}$) as compared with water boiling heat ($600\,\text{cal g}^{-1}$ at $0\,°C$). We might think that only as many as $80/(600+80) = 12\%$ of all H-bonds existing in ice break down in liquid water. However, this picture – with some H-bonds broken while others persist – is not quite true: rather, in liquid water all H-bonds become slightly loose. This is well illustrated by the following experimental results.

In Fig. 4.2, Curve 1 shows the maximum of the IR absorption spectrum for O—H groups in ice (where all H-bonds are saturated); Curve 2 illustrates this maximum for the O—H groups of separate water molecules dissolved in CCl_4 (where no H-bonding occurs because of the extreme dilution); and Curve 3 shows the absorption spectrum for liquid water. Suppose liquid water contained two types of O—H groups: those participating and those not participating in H-bonds. Then the former would vibrate

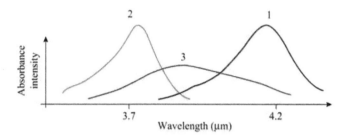

Figure 4.2. Typical IR absorption spectra for O—H groups in ice (Curve 1), in CCl₄ solution (Curve 2) and in liquid water (Curve 3).

with frequencies typical for ice (where they have been involved into H-bonding), while the latter would vibrate like those in water molecules dissolved in CCl₄ (where no H-bonding has occurred). Then the IR absorption spectrum for liquid water would be double-peaked to reflect the two types of O—H groups and their two typical frequencies, since the vibrational frequency of a group is equal to its light absorption frequency. However, no such "double-peaked" picture is actually observed. Instead, Curve 3 presents one broad peak that goes from the peak on Curve 1 to that of Curve 2. This means that in liquid water all O—H groups are involved in H-bonding and all resulting H-bonds are loose, though in different ways.

So, strictly speaking, the model with some H-bonds persisting in liquid water and others broken is incorrect. However, people often use it owing to its simplicity and convenience in describing the thermodynamic properties of water – and we may use it as well, although its drawbacks must be kept in mind.

Now let us concentrate on interactions between the protein chain and water molecules.

Like water molecules, the backbone of a protein chain is polar. To be more exact, it is its peptide groups that are polar. The net charge of the protein chain backbone is 0, and the distribution of charges on its atoms (again, the charge on each atom is expressed in fractions of the proton charge) is as follows:

$$
\begin{array}{cc}
 & 0 \\
\frac{-2}{5}\ O & C^{\alpha}- \\
\quad \backslash\!\backslash & \quad / \\
\frac{+2}{5}\ C' == N\ \frac{-1}{5} \\
\quad / & \quad \backslash \\
-C^{\alpha} & H\ \frac{+1}{5} \\
0 &
\end{array}
$$

Partial charges are acquired by some side-groups too, e.g., by that of Ser (its side-group, —CH₂—OH, is similar to ethanol). Charged amino acid residues are even more polarized: a charge of -1 is typical for the acidic side-groups of Asp and Glu when they are ionized (at about neutral pH), while $+1$ is characteristic of the ionized basic side-groups of Arg, Lys and His.

Both main-chain peptide groups and polar side-groups participate as donors and acceptors in hydrogen bonds. They can be – and mostly are – involved in H-bonds formed among themselves or to water molecules: since the H-bond energy ($5\,\mathrm{kcal\,mol^{-1}}$) is about an order of magnitude higher than that of thermal movement, H-bonds are mostly preserved by these movements.

When an H-bond between donor (D) and acceptor (A) is formed within a protein molecule in the aqueous environment, it replaces two hydrogen bonds between the protein and water molecules (that existed earlier, since an H-bond is too expensive to be neglected), and an additional bond between the two freed water molecules is formed (for the same reason):

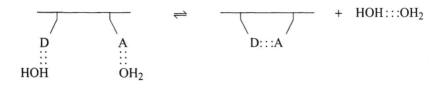

The energy balance of this reaction is close to zero: two bonds yield two bonds. However, the *entropy* of the water molecules increases since they are no longer bonded to the protein chain but only mutually H-bonded and free to go anywhere (and the entropy results just from the free movement). This entropy increase caused by the water dimer release is approximately equal to an entropy increase resulting from the molecule H_2O transition from ice to liquid water (in both cases one particle becomes free in its movements).

The entropy difference resulting from one molecule H_2O transition from ice to liquid water can be estimated as follows. At the melting-point (for ice, at $0\,^\circ\mathrm{C}$, i.e., 273 K), the increase in entropy caused by melting is known to fully compensate for the corresponding decrease in energy. And we know the value of this decrease – it is $80\,\mathrm{cal\,g^{-1}} \times 18\,\mathrm{g\,mol^{-1}} = 1440\,\mathrm{cal\,mol^{-1}}$ ($18\,\mathrm{g\,mol^{-1}}$ is the molecular weight of H_2O).

So, owing to this increase of water entropy, the free energy of the "protein in water" system decreases by about 1.5 kilocalories per mole of emerging intra-protein hydrogen bonds D:::A.

This free energy released by waters may fully or partially compensate the free energy increase that results from decreasing conformational entropy of the chain during the D:::A bonding. As we will see later, a decrease in free energy of the water molecules nearly compensates for the entropy of fixation of the amino acid residue conformation required for H-bonding (N—H:::O) in secondary structures of the polypeptide chains. As a result, in water, the regular secondary structure of polypeptides is just at the edge of stability.

So, the awful words "entropy" and "free energy" have been articulated... My experience shows that a normal biologist has heard about entropy (that it is a measure of the multitude of possible states, a measure of disorder) but is most uncertain about what the free energy is... As we will often appeal to the concept of free energy (which is a measure of stability), I would like to devote a few minutes to it just now, and later, when a need arises, to discuss it (and entropy too) in more detail.

Let us start with a simple example: suppose a molecule can have two states, "a" and "b"; "a", when it is here, in this room, 200 m above sea level; "b", when it is in the Dalai Lama's monastery, in The Himalayas, 5 km above sea level. What is the relationship of the probabilities of these two states provided that (1) temperature T is the same in both places, and (2) we watch the molecule long enough for it to visit both places?

As stated by the usually well-remembered Boltzmann formula,

$$\text{Probability of being in the state of energy } E \propto \exp(-E/kT). \qquad (4.2)$$

Physically, the sense of this formula is that the heat of the medium (that is, collisions with other molecules) excites our molecule to a certain extent (proportional to the medium temperature T, on average) and thereby enables it to enter the region of more or less high energies. In more detail all this will be discussed later, and at the moment I take the liberty of believing that you remember this formula. Still let me remind you that k is Boltzmann's constant, and T is the absolute temperature in Kelvin (counted from the "absolute zero", $0\,K = -273.16\,°C$), and "$\propto$" means "is proportional to".

No, we'd better deduce the Boltzmann formula not relying on your memory. We will deduce it for at least the case that is of interest to us, that is, for distribution of gas molecules over height, when their energy is described as $E(h) = mgh$, where m is mass of a gas molecule, g is acceleration of gravity, and h is height.

As stated by Clapeyron–Mendeleev law, the pressure of an ideal gas $P = nkT$, where n is the number of gas molecules per unit volume. If T remains unchanged at any h, then $dP/dh = (dn/dh)kT$. On the other hand, when considering a gas column of unit cross-section, we see that $dP = (mgn)(-dh)$, since the weight of the gas pressing down on the unit cross-section decreases by $(mgn)dh$ when the height grows by dh. Therefore, $dP/dh = (dn/dh)kT = -mgn$. Hence, $dn/dh = -(mg/kT)n$, or $d[\ln(n)]/dh = -mg/kT$, that is, $n \propto \exp(-mgh/kT) = \exp(-E(h)/kT)$.

Thus, as applied to the problem in question (What is the relationship of probabilities for a molecule to be at different heights?), the Boltzmann formula reduces to the barometrical relationship

[*Probability of being at a height b*] relates to [*Probability of being at a height a*]

$$\text{as } \exp(-E_b/kT) \text{ relates to } \exp(-E_a/kT) \qquad (4.3)$$

where E_a, energy of the molecule in the state "a" (i.e., "here"); E_b, in the state "b" ("at a height of 5 km"); T, absolute temperature (for simplicity, as mentioned above, is assumed to be invariable with height).

Because of gravity, "here" the energy of the molecule is lower than "at a height of 5 km"; so, according to Boltzmann, the molecule will stay "at a height of 5 km" for a shorter time than "here" (the time will be 1.5–2 times shorter).

Make the calculations yourselves by taking $T = 300\,\mathrm{K}$ and recalling that $E = mgh$, where $m =$ average mass of an air molecule (≈ 30 dalton $= 30\,\mathrm{g\,mol^{-1}}$), $g \approx 10\,\mathrm{m\,s^{-2}}$ (gravitational acceleration), $h \approx 5\,\mathrm{km}$ (height difference), and Boltzmann's constant $k = 1.38 \times 10^{-23}$ Joule degree^{-1} particle$^{-1} = 0.33 \times 10^{-23}$ cal degree^{-1} particle^{-1} (since J $= \mathrm{kg\,m^2\,s^{-2}} = 0.24\,\mathrm{cal}$), i.e. $2\,\mathrm{cal\,deg^{-1}\,mol^{-1}}$. ($1\,\mathrm{mol} = 6 \times 10^{23}$ particles.) As you may remember, $R = 2\,\mathrm{cal\,mol^{-1}\,K^{-1}}$ is the "gas constant".

In other words, at a height of 5 km the molecules will be about two times less numerous than here. Or rather, this will be the case *for equal volumes*, for example, your lungs (as you may easily ascertain by breathing at different heights). However, *in total*, the molecules in the Dalai Lama's monastery are more numerous than they are here just because the monastery is much larger than our room. That is, over there, the molecule may have more positions, since for a freely flying molecule the number of positions is proportional to the room volume. In this case physicists would say that in the monastery the number of *microstates* of the molecule is far greater than in our room. So, the probability that our molecule is *somewhere* in the Dalai Lama's monastery relates to the probability that it is *somewhere* in this room as

[*Probability of being somewhere in volume* "b"] :

[*Probability of being somewhere in volume* "a"]

$$= [V_b \exp(-E_b/kT)] : [V_a \exp(-E_a/kT)], \qquad (4.4)$$

where V_a is the volume of "a" ("our room"), and V_b is the volume of "b" ("his monastery"). From elementary college math you remember that V can be presented as $\exp(\ln V)$, so the above formula can be written as

[*Probability of being somewhere in volume* "b"] :

[*Probability of being somewhere in volume* "a"]

$$= [\exp(-E_b/kT + \ln V_b)] : [\exp(-E_a/kT + \ln V_a)]$$
$$= [\exp(-(E_b - T \times k \ln V_b)/kT)] : [\exp(-(E_a - T \times k \ln V_a)/kT)] \quad (4.5)$$

The last expression looks very much like eq. (4.3), the Boltzmann equation, but it is applicable not to a volume *unit* but to the *total* volume of a system, and – note carefully – it has $E - T \times k \ln V$ instead of E.

It is the value $F = E - T \times k \ln V$ that is called *free energy* of our molecule of air in a given volume V at temperature T. And the value $S = k \ln V$ is called the *entropy* of our molecule in the volume V (which, in our case, is proportional to the "number of accessible states" of our molecule).

In the general case, entropy S is simply equal to $k \times$ [*logarithm of the number of accessible states*]. And free energy F relates to energy E, entropy S and temperature T according to the general equation

$$F = E - TS \qquad (4.6)$$

Of two states, the more stable (i.e., more *probable*) is the one having a lower free energy:

[*Probability of being somewhere in volume* "b"] :

[*Probability of being somewhere in volume* "a"]

$$= \exp(-F_b/kT) : \exp(-F_a/kT) = \exp[-(F_b - F_a)/kT] \qquad (4.7)$$

In other words, a more probable, that is, a more stable, state of the system is that with a lower F, while the *most* stable state of a system (at a given temperature and volume of this system) corresponds to *the minimum free energy F*.

Thus, the "free energy" is a natural generalization of the regular "energy" for the case when *the system exchanges heat* with the environment. Let me remind you that if a body is *not* excited by environmental heat, its stable state corresponds to its minimum energy (or simply, everything that can fall down – eventually falls down). When excited by environmental heat, molecules of the system start acquiring numerous states of a higher energy (i.e., the entropy of this system increases), and as a result, air molecules fly and do not drop onto the ground.

This can also be put as follows.

A change in the energy, $E_b - E_a$, is the work required to transfer a body from state "a" into state "b" when there is no heat exchange with the environment. And a change in the free energy $F_b - F_a$ is the work required to transfer a body from state "a" into state "b" *when the body keeps exchanging heat* with the environment.

Well, let us now descend from The Himalayas to proteins.

So, what is the balance of energy, entropy and free energy in the previous example of H-bonding in the protein chain? To visualize this, let us compare how the process goes under different conditions.

Here $E_H < 0$ is the H-bond energy, and $S_H > 0$ is the entropy of movements and rotation of a free body, i.e., of the free molecule H_2O or the free dimer $HOH:::OH_2$. The H-bonds – between water molecules or between water and protein molecules – are stable when $E_H - TS_H < 0$ (and if $E_H - TS_H > 0$, i.e., when the H-bonding is unstable, what we deal with is not liquid water but water vapor).

A comparison of the diagrams given below shows that H-bonds within the protein chain surrounded by water display a lower stability than when in a vacuum. Indeed, in water, the H-bond free energy is $F_b - F_a = -TS_H$, i.e., it is smaller than in a vacuum where $F_b - F_a = E_H$.

I would like to emphasize again that the reason for this decrease in the H-bond stability is that *in water* an H-bond formed within the protein chain *replaces* the H-bond between the chain and water. For the same reason, the hydrogen bonds

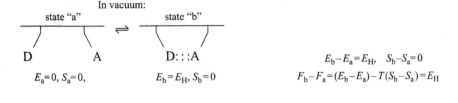

$$E_b - E_a = E_H, \quad S_b - S_a = 0$$
$$F_b - F_a = (E_b - E_a) - T(S_b - S_a) = E_H$$

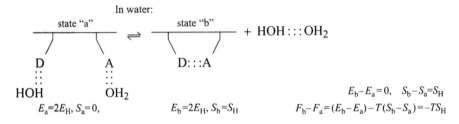

$$E_b - E_a = 0, \quad S_b - S_a = S_H$$
$$F_b - F_a = (E_b - E_a) - T(S_b - S_a) = -TS_H$$

that stabilize protein structure *in water* are entropic but not energetic in nature: the energies of the two states of the chain (with and without the intra-chain H-bond) are approximately equal, and of these two, the state with a higher entropy (with a greater number of microstates) is more stable. And a free water molecule has a greater number of microstates (i.e., a greater number of positions in space) than a bound molecule.

I call your attention to the following: in the protein chain (surrounded by water molecules) hydrogen bonds are entropic and *not* energetic in nature just *because* the energy of H-bonding is extremely high. Consequently, donors and acceptors that are "free" of bonding *within the protein* are not really free of *any* bonding, as they participate in H-bonds to water molecules. The water molecules released from the protein during H-bonding inside the chain immediately bind to one another, thereby compensating for the energy, such that the free energy gain of intra-protein H-bonds occurs only because of the increasing number of possible microstates of the released water molecules. While it is true that to bond to one another the water molecules have to sacrifice a part of their gained freedom (entropy), it is better to lose a small entropy than a large energy.

Two facts determine the behavior of water as a specific solvent: (1) water molecules are strongly H-bonded to one another; (2) this H-bonding occurs only at a certain mutual orientation of water molecules. A variety of interesting effects are thereby caused. These will be the subject of our next two lectures.

LECTURE 5

In the initial part of this lecture I would like to talk on thermodynamics. This will be useful for further consideration of water as a specific solvent. The focus of this consideration will be the free energy of immersing various molecules in water.

To study the free energy of immersion of a molecule in water, we take a closed tube half of which is filled with water and the other half with vapor, and watch how introduced molecules are distributed between these phases.

As we have learned, the difference between free energies F defines a more favorable state of the system (in this case, a more favorable position of the studied molecule) according to the formula

[*Probability of being somewhere in "b"*] :

[*Probability of being somewhere in "a"*] $= \exp[-(F_b - F_a)/kT]$ (5.1)

The free energy $F = E - TS$ comprises the energy E and the entropy S. I believe you know what energy is. As to temperature, your knowledge, I think, is rather *intuitive*, so we will return to it later on. The problem of entropy is still more complex, so let us discuss it once again.

In the simplest case of a particle in the container (that we discussed in Lecture 4), $S = k \ln V$, where V is the volume accessible to the particle.

In what sense are we incorporating this entropy into the free energy in the form of $-TS$? When considering $\exp[-F/kT]$ irrespective of E, we get $\exp[-(-TS)/kT] = \exp[-(-Tk \ln V)/kT] = V$, that is just the accessible volume defining *the number of accessible states* of the particle in space. The greater the entropy, the larger this number, and the higher the probability that the particle is somewhere in this volume.

In the general case (when the accessible states of the particles are limited not only by the walls surrounding the volume V but also by collisions between the particles), the entropy S is given by

$S = k \times$ [*logarithm of the number of accessible states of the studied particle*] (5.2)

In molecular physics, biology and chemistry not a single particle but a mole (6×10^{23}) of particles are usually considered. Then the entropy of a mole is given by

$S = R \times$ [*logarithm of the number of accessible states of one particle*] (5.3)

43

where $R = k \times (6 \times 10^{23} \text{ mol}^{-1})$. The only difference between k and R is that k refers to a single particle and R to a mole of particles.

Strictly speaking, the value

$$F = E - TS \tag{5.4}$$

is known as the Helmholtz free energy. It is easy to describe and convenient to calculate this value since it refers to a system that (like the molecule we have just discussed) is enclosed in some fixed volume.

However, a normal experiment deals not with a fixed volume V but with a constant pressure P (e.g., atmospheric pressure). In this case, we measure not a change in the energy E of the studied body but a change in its *enthalpy* $H = E + PV$. The change in enthalpy H includes, apart from the change in energy (E), the work against the external pressure P in changing the body volume V.

The value of PV will be negligible for all objects considered in these lectures, since we will deal with liquids or solids (where the volume per molecule is small) at a rather low (e.g., atmospheric) pressure. Under these conditions, the value of PV is many times less than the thermal energy of the body.

Indeed, even for a gas (where the volume per molecule is particularly large), $PV \approx RT \times$ [number of moles] (remember the Clapeyron–Mendeleev law?). So, per mole of gas, the correction $PV = 1RT \approx 0.55 \div 0.75 \text{ kcal mol}^{-1}$ at temperatures ranging from 0 to 100 °C, i.e., at $T = 273$ to 373 K. In other words, even in gases this value is commonly low compared with the magnitudes of the effects of interest, which usually amount to a few kilocalories per mole. Under a pressure close to atmospheric, for liquids and solids, the effect on H of correcting for PV is still hundreds or thousands of times less: here the volume of one mole is a minor fraction of a liter [$\approx 1/55$ liter for H_2O, $\approx 1/10$ liter for $(CH_2)_6$, etc.], while for a gas at a pressure of 1 atmosphere and room temperature this volume amounts to about 25 liters.

That is why, further on, I will neglect the difference between H and E and refer to them both simply as "energy".

Similarly there is little difference (for us) between *Helmholtz free energy* $F = E - TS$ and *Gibbs free energy*

$$G = H - TS = (E + PV) - TS = F + PV \tag{5.5}$$

Making no difference between G and F we will refer to them both simply as "free energy".

However, it is worth remembering that for processes occurring in a fixed volume we should use E and F, while for processes occurring under constant pressure we should use H and G.

Remark. If we add particles one by one to a system having a constant volume, it takes more and more effort to drive more particles in (it is easy to inject one drop into a sealed bottle, but what about another? One more? Still more?). However, *under constant pressure* the effort remains the same for each successive particle. That is, at a constant volume, E and F are *not* proportional to the number of particles in the system, while *under constant pressure H and G are* proportional to it. The latter allows the introduction of a convenient quantity, "Gibbs free energy per particle" (this quantity is called "chemical potential", see below).

You should also remember that at a given temperature (T = const), any system adopts the equilibrium *stable* state at the minimum $F = E - TS$ if the *volume* is fixed, and at the minimum $G = H - TS$ if the *pressure* is constant.

With a minor change in state of the system its free energy varies as:

$$F \rightarrow F + dF = F + dE - T \, dS - S \, dT$$

or

$$G \rightarrow G + dG = G + dH - T \, dS - S \, dT. \tag{5.6}$$

This means that all possible rearrangements of the system about its stable state are described by the following equations characterizing the free energy minimum (as we know, a peculiarity of the point of minimum (and maximum) is that minor deviations from it result in almost no change in the function's value):

1. at V = const, the stable state (at a given T) is gained with F at a minimum, where

$$dF|_{V=\text{const}} = dE|_{V=\text{const}} - T \, dS|_{V=\text{const}} = 0 \tag{5.7}$$

(taking into account that $dT = 0$ at T = const, i.e., $S \, dT = 0$ in eq. (5.6));
2. at P = const, the stable state (at a given T) is gained with G at a minimum, where

$$dG|_{P=\text{const}} = dH|_{P=\text{const}} - T \, dS|_{P=\text{const}} = 0. \tag{5.8}$$

These equations yield *the thermodynamic definition of absolute temperature*:

$$T = \frac{dE}{dS}\bigg|_{V=\text{const}} = \frac{dH}{dS}\bigg|_{P=\text{const}} \tag{5.9}$$

I realize that here this definition has been obtained rather formally, and its physical sense is not obvious. Therefore, I will return to it soon in my lectures.

Now let us consider the chemical potential. This quantity describes the thermodynamic characteristics of one molecule in a system rather than those of the system as a whole.

If molecules are added to a system one by one *under constant pressure*, identical efforts are required for driving in each of them; in this process the volume will grow, while the density of the system and the intensity of interactions in its interior will remain unchanged. In this case, the thermodynamic state of a single molecule is adequately described by the Gibbs free energy G divided by the number of molecules N,

$$\mu = \frac{G}{N} \tag{5.10}$$

where μ is known as the *chemical potential* (and since in liquids or solids $F \approx G$ at low pressures, here $\mu \approx F/N$). (If N is not the number of molecules but, as usual in physical chemistry, the number of moles of molecules, then μ refers not to one molecule but to a mole of molecules.)

The chemical potential, or, which is the same, the Gibbs free energy per molecule, will be of use later in this lecture for considering the distribution of molecules between phases. The thing is, molecules pass from the phase where their chemical potential is higher to a phase where it is lower, thereby lowering the total free energy of the system and shifting it to equilibrium. And the equilibrium state is characterized by identical chemical potentials of molecules in both phases.

For further considerations, we will need two more equations.

First, the definition of heat capacity reflecting a temperature-dependent increase of energy:

$$C_P = \frac{dH}{dT}\Big|_{P=\text{const}} \tag{5.11}$$

(this is for a constant pressure; one can also calculate heat capacity at constant volume, but we do not need such details).

Second, the relationship between the entropy and the free energy:

$$S = -\frac{dG}{dT}\Big|_{P=\text{const}} = -\frac{dF}{dT}\Big|_{V=\text{const}}. \tag{5.12}$$

Equation (5.12) is one of the most important in thermodynamics. It results directly from the fact that the small increment of free energy $dG = d(H - TS) = dH - T\,dS - S\,dT$ (and $dF = d(E - TS) = dE - T\,dS - S\,dT$)), whereas, in equilibrium, $dH - T\,dS = 0$ (and $dE - T\,dS = 0$) in accordance with the thermodynamic definition of temperature: $T = dH/dS = dE/dS$ (see eq. (5.9)).

Equation (5.12) shows that the free energy has its minimum (or maximum) value at such temperature T where $S(T) = 0$.

It is also useful to know that G/RT is at a minimum (maximum) when $H(T) = 0$, since

$$\frac{d(G/RT)}{dT} = \frac{(dG/dT)}{RT} - \frac{G}{RT^2} = \frac{-S}{RT} - \frac{(H - TS)}{RT^2} = \frac{-H}{RT^2}. \tag{5.13}$$

Remarks

1. Using equations (5.5) and (5.12) one can show that $H = G + TS = G - T(dG/dT)$, and therefore C_P (see eq. (5.11)) can be obtained in the form

$$C_P = -T\frac{d^2G}{dT^2}\Big|_{P=\text{const}} \tag{5.14}$$

2. We are never interested in values of energy, entropy and free energy themselves. We are interested only in *changes* of these values. Indeed, speaking on energy: we can count off the gravitational energy of particles from the sea level, from the floor level, from the center of the Earth, etc. The values will be all different, but this does not matter. The only important thing is the *difference* between gravitational energies of a particle in two states. Just the

same, when we define the particle's entropy as $k \ln V$, we can measure the volume V in liters, in cubic feet, etc. The values will be all different, but this does not matter: the only important thing is the *difference* between particle's entropies in two volumes (V and V'. And this difference, $\Delta S = k \ln(V/V')$, is independent of the unit of measurements.

With this introduction we can now go on to discuss water as a specific solvent.

First of all, let us consider the so-called *hydrophobic effect*.

"Hydro-phobicity" is "fear of water". Who is "afraid" of water? – all nonpolar molecules such as inert gases (argon, xenon), hydrogen and all purely hydrocarbon molecules (methane, ethane, benzene, cyclohexane, etc.). We will focus on hydrocarbons in water, since proteins have many hydrocarbon side chains. It is these water-fearing and water-escaping side chains that form the *hydrophobic core of a protein globule*.

So, what does hydrophobicity mean in terms of experiment?

Methane's (CH_4) concentration in water is about an order *lower* than in gas above this water at a temperature of 20–40 °C (the exact value depends on temperature). For H_2 and for propane $CH_3CH_2CH_3$, the difference is nearly the same. All of them are water-fearing (phobic) molecules: they are *more numerous* in a vapor than in the water. In contrast, ethanol CH_3CH_2OH and water are easily miscible, and as you know, their separation is quite laborious; but the ethanol molecule is polar (its polar O—H group is capable of H-bonding). As to the purely nonpolar molecules, they would prefer a vacuum (vapor is nearly a vacuum) rather than water... And this happens *in spite* of van der Waals attraction between *any* molecules, even H_2 or CH_4, and water.

Let us consider (using one example shown in Fig. 5.1) some typical thermodynamic effects for hydrocarbons in water. The thermodynamic parameters presented in Fig. 5.1 (free energies, energies and entropies of transfer from one phase to another) resulted from studies of the equilibrium distribution of molecules of cyclohexane, $(CH_2)_6$, among three phases: vapor, aqueous solution and liquid cyclohexane.

The free energy of transfer of a molecule from the non-polar liquid to water is defined as

$$\Delta G_{\text{liquid} \to \text{aqueous solution}} = -RT \ln(X_{aq}/X_{liq}). \qquad (5.15)$$

Here X_{aq} and X_{liq} are equilibrium concentrations of studied molecules in the aqueous solution and in the non-polar liquid that contacts with the former. The value $\Delta G_{\text{liquid} \to \text{aqueous solution}}$ is a difference between the mean force potentials affecting our molecule in water and in the non-polar liquid. The mean force potential is created by all interactions of our molecule with its molecular surrounding; it includes both energy and entropy terms that arise from these interactions. Since concentration is the number of molecules in a given volume, equation (5.15) follows from the Boltzmann distribution over two phases with different values of the mean force potential.

At 25°C and low (~ 1 atm.) pressure, $X_{liq} = 9.25 \, \text{mol}^{-1}$ for pure liquid $(CH_2)_6$ and $X_{aq} = 0.0001 \, \text{mol} \, l^{-1}$ for its saturated solution in water. Accordingly, $\Delta G_{\text{liquid} \to \text{aqueous solution}} = +6.7 \text{kcal} \, \text{mol}^{-1}$.

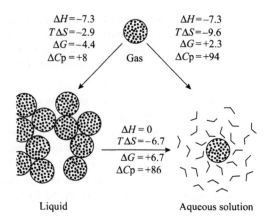

Figure 5.1. The thermodynamics of transfer of a typical nonpolar molecule, cyclohexane $(CH_2)_6$, from vapor (top) to water (right), and from liquid cyclohexane (left) to water. The numerical values are for $25\,°C$ (i.e., $T \approx 300\,K$, $RT \approx 0.6\,kcal\,mol^{-1}$). ΔH (like energy, it is measured in kilocalories per mole) is the enthalpy change per mole of molecules; ΔS is the corresponding entropy change ($T\Delta S$, the contribution of the entropy change to the free energy, is also measured in kilocalories per mole); $\Delta G = \Delta H - T\Delta S$ ($kcal\,mol^{-1}$) is the change in Gibbs free energy per mole of transferred molecules; ΔC_p [$cal\,mol^{-1}\,K^{-1}$] is the change in heat capacity per mole of transferred molecules. Adapted from [4]. All values are recalculated using equations (5.15)–(5.17) and experimental values from reference books.

The free energy of transfer of our molecule from gas to the non-polar liquid and to the aqueous solution can be defined in the same way:

$$\Delta G_{gas \to liquid} = RT \ln(X_{liq}/X_{gas}). \tag{5.16}$$

$$\Delta G_{gas \to aqueous\ solution} = -RT \ln(X_{aq}/X_{gas}). \tag{5.17}$$

where X_{gas} is the equilibrium concentration of our molecules in gas above the liquid(s). The pressure of the saturated $(CH_2)_6$ vapor at $25°C$ is about 0.05 atm., which means that $X_{gas} = 0.002\,mol\,l^{-1}$ (i.e., X_{gas} is 50 times higher than X_{aq}). The resulting ΔG values are presented in Fig. 5.1.

Since the mean force potential of interactions affecting a molecule in a rarified gas $(G_{in\ gas})$ is virtually zero, $\Delta G_{gas \to liquid} = G_{in\ liquid} - G_{in\ gas}$ is very close to $G_{in\ liquid}$ that is the mean force potential of interactions of the molecule in the non-polar liquid; and the mean force potentials of the molecule's interactions in water, $\Delta G_{in\ aqueous\ solution}$ equals to $\Delta G_{gas \to aqueous\ solution}$.

When ΔG and its temperature dependence is known, the values of ΔS, ΔH, ΔC_P are derived from this dependence according to equations (5.12), (5.13) and (5.14).

Figure 5.1 shows that the *energy* of attraction of $(CH_2)_6$ molecules to water is as high as that to liquid cyclohexane ($\Delta H = -7.3\,kcal\,mol^{-1}$), but $(CH_2)_6$ molecules do not want to go into water, though they go into liquid cyclohexane readily. As is clear from Fig. 5.1, it is entropy that causes this hydrophobicity: it becomes too low when a cyclohexane molecule comes into water.

Why are nonpolar molecules like CH_4 or $(CH_2)_6$ hydrophobic? The reason is that, unlike water molecules, nonpolar H_2 or Ar, CH_4 or $(CH_2)_6$ are incapable of H-bonding. This is confirmed by the well-known fact that polar ethanol molecule, CH_3CH_2OH (which consists mainly of hydrocarbon groups, like $(CH_2)_6$, but is capable of H-bonding, like H_2O) is not hydrophobic.

A naive suggestion could be made that upon coming into water, CH_4 or $(CH_2)_6$ molecules disrupt H-bonding in water. But it is not that simple. If it were so, the solution energy would have *increased* drastically with incoming $(CH_2)_6$, while in fact it actually *decreases* (see Fig. 5.1) by eight kilocalories per mole of incoming $(CH_2)_6$. Actually, instead of increasing energy, we have *decreasing entropy* (and this decrease is substantial: $T\Delta S = -9.6\,kcal\,mol^{-1}$).

This entropy decrease prevents cyclohexane from dissolving in water. The free energy $G = H - TS$ increases not only with increasing energy H but also with *decreasing* entropy S (S contributes to G in the form of $-TS$). Thus, a large decrease of the entropy S, even with a simultaneous decrease of energy H, results in increasing free energy G, and hence (see eq. (5.1)) in a *decreasing probability* of molecules remaining in the current state (or more simply, in their decreasing concentration in this state).

Now the main physical question arises as to why the entropy of water molecules decreases as a result of their contact with a nonpolar surface.

It decreases because an H_2O molecule must not point at a hydrophobic surface with its H atom. Otherwise, its hydrogen bonds will be lost (see Fig. 5.2; as you may remember, H-bonds are orientation-dependent and emerge only when an O—H group is directed towards the O atom of another water molecule).

Let me remind you that H_2O molecules are almost fully hydrogen-bonded in water, so it is impossible to sacrifice some H-bonds without a great loss of the free energy. To avoid the loss of H-bonds (that is, to avoid O—H groups being directed towards the hydrophobic surface), water molecules seek favorable positions (see the upper molecule in Fig. 5.2) and partially freeze their thermal motions. Thereby, they

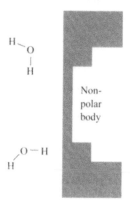

Figure 5.2. Water molecules near the surface of a nonpolar body. The upper molecule can make all its H-bonds, but this "favorable" position about the body surface is restricted and therefore entropy-expensive. The lower molecule loses one energy-expensive H-bond to water because its O—H group is directed towards the nonpolar obstacle.

preserve their valuable H-bonds at the expense of some of their entropy. The entropy-driven loss of free energy is about 0.2 kcal per mole of mean-surface waters, which is an order of magnitude less than the price of lost H-bonds would be.

Low temperature (up to 20 °C, and for some hydrocarbons, up to 60 °C) even allows the hydrogen bonds near a hydrophobic surface (see Fig. 5.3) to gain a little energy, see Fig. 5.4 (because now these bonds are less damaged by movements of half-frozen near-surface waters), but this gain does not fully compensate for the entropy loss resulting from freezing the near-surface waters.

Note that again the net effect is entropic rather than energetic in nature *just because* the energy of H-bonds is extremely high: since it is so, waters would prefer to become frozen (although this is also thermodynamically bad) and sacrifice a part of their freedom (entropy) than to lose the large energy of a hydrogen bond.

I would like to emphasize that the resultant entropic effect on the free energy value is of *the same sign* as the energetic effect expected by naivete, but *less* in magnitude.

The suggestion that waters close to a nonpolar surface are, in a way, frozen is additionally supported by the anomalously high heat capacity of cyclohexane (and other hydrocarbons) in water. The excessive heat capacity of a $(CH_2)_6$ molecule in aqueous surrounding is 10 times as high as that amidst its fellow cyclohexanes. To be more exact, a high heat capacity in water is typical not of the hydrocarbon itself but of its ice-like water shell; with increasing temperature this "iceberg" tends to melt out, and this explains the anomalous heat capacity.

In considering frozen hydrogen-bonded surface waters, bear in mind that their relative orientation *differs* from that in normal ice. In ice, water molecules must have regular space positions because they have to form a huge three-dimensional crystal. At the surface they can adopt any position they like, provided that it is favorable for H-bonding. The water molecules do not observe translational symmetry of the three-dimensional lattice of ice because the resultant "microiceberg" is not going to grow infinitely: it only tends to coat the introduced hydrophobic molecule (Fig. 5.3) or a group of such molecules.

In the latter case, these hydrophobic molecules form a "hydrophobic bond".

The so-called *clathrates* represent an extreme ordering of waters caused by hydrophobic molecules. Clathrates are crystals built up by water and nonpolar molecules. From your chemistry course you may know that they are far less stable than

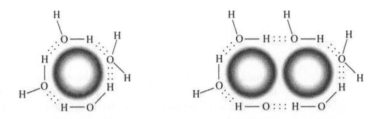

Figure 5.3. Irregular packing of H-bonded waters around a nonpolar molecule (left) and around a pair of such molecules. A hydrophobic bond is formed in the latter case.

the crystalline hydrates built up by water and polar molecules. Clathrates emerge only at low temperatures (about 0 °C) and high pressures, which cause many molecules of nonpolar gas to penetrate into water. In clathrates, as in ice, water molecules have their hydrogen bonds saturated, although their geometry is different from that in normal ice. In the resultant crystal, quasi-ice keeps nonpolar molecules in its pores. Incidentally, clathrates are thought to contain much more natural gas than ordinary gas fields, and gas production from clathrates (from a great depth where high pressure ensures their existence) is perhaps a project of tomorrow.

The hydrophobic effect is rather temperature-dependent (Fig. 5.4). The temperature affects ΔG, but even more it affects the magnitude (and even sign) of its constituents, ΔH and ΔS.

As the temperature is increased, the surface hydrogen bonds tend to melt out. Up to ≈ 140 °C, this is accompanied by an increasing hydrophobic effect, since the thermodynamically unfavorable ordering of surface waters persists, while favorable hydrogen bonds are destroyed.

While at low and room temperature the hydrophobic effect results from entropy only, at elevated temperatures the energy of the lost near-surface hydrogen bonds becomes more and more important. But ΔG keeps increasing until $\Delta S < 0$, i.e., up to ≈ 140 °C (see Fig. 5.4).

However, at still higher temperatures, there are too many disrupted H-bonds in water (which remains liquid only under high pressure); now the hydrophobic surface interferes with hydrogen bonding less and less, and the hydrophobic effect begins to diminish.

Fig. 5.4 illustrates the transfer of a nonpolar molecule to water from a nonpolar solvent rather than from a vapor. This is done deliberately: we are going to consider

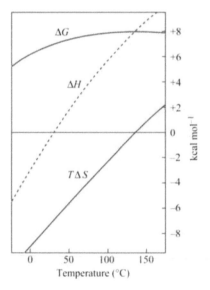

Figure 5.4. The thermodynamics of transferring a typical nonpolar molecule, pentane C_5H_{12}, from liquid pentane to water at various temperatures. The transfer free energy, $\Delta G = \Delta H - T\Delta S$, and its enthalpic ($\Delta H$) and entropic ($T\Delta S$) components are measured in kilocalories per mole of transferred molecules. ΔG is at a maximum when $\Delta S = 0$; the proportion of pentane distribution between the water and the liquid pentane phases (which is proportional to $\exp(-\Delta G/RT)$) is at a minimum when $\Delta H = 0$. Adapted from Privalov P.L., Gill S.J., *Adv. Protein Chem.* (1988) **39**: 191–234.

the hydrophobic effect in proteins where amino acid residues are transferred from water to the protein core, which is similar to a nonpolar solvent rather than to a vapor.

The hydrophobicity of amino acids will be in focus a little later, and now it would be useful to consider Table 5.1: it contains some characteristics measured at room temperature for nonpolar groups similar to protein ones.

It is immediately apparent from Table 5.1 that ΔG (unlike ΔH and ΔS) increases with increasing size of a hydrophobic molecule. How exactly does it increase? A more detailed analysis of a variety of nonpolar molecules shows that their hydrophobic free energy ΔG increases almost proportionally to the accessible surface area of the nonpolar molecules. The physical sense and mode of construction of the accessible surface is demonstrated in Fig. 5.5 (see also Fig. 5.3).

Table 5.1. Typical thermodynamic parameters of hydrophobic group transfer from a nonpolar liquid to an aqueous solution at 25°C.

Molecule	Transfer from →to	ΔG (kcal mol^{-1})	ΔH (kcal mol^{-1})	$T\Delta S$ (kcal mol^{-1})	ΔC_p (kcal mol^{-1} K^{-1})
Ethane $(CH_3)_2$ (compare with Ala side group: $-CH_3$)	benzene→ water	+3.6	−2.2	−5.8	+59
	CCl_4→ water	+3.8	−1.8	−5.4	+59
Benzene C_6H_6 (compare with Phe side group: $-CH_2-C_6H_5$)	benzene→ water	+4.6	+0.5	−4.1	+54
Toluene $(C_6H_5)_2CH_3$ (compare with Phe side group)	toluene→ water	+5.4	+0.4	−5.8	+63

Values taken from Tanford C. *The hydrophobic effect*, 2nd edn, New York: Wiley-Interscience, 1980.

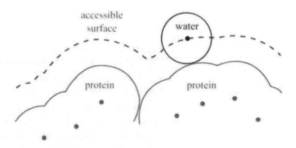

Figure 5.5. The "accessible surface" of a molecule in water. Dots indicate the centers of atoms exposed to water; the solid line denotes their van der Waals envelopes. A water molecule is shown as a sphere of radius 1.4 Å. The "accessible" (to water) surface is defined as the area described by the center of a 1.4 Å sphere that rolls over the van der Waals envelope of the protein. Adapted from [12].

The hydrophobic free energy is about $+0.02 - +0.025 \, \text{kcal mol}^{-1} \, \text{Å}^{-2}$ of the accessible nonpolar area of a molecule transferred from a nonpolar solvent to water. Specifically, for benzene, the accessible area is about $200 \, \text{Å}^2$, and $\Delta G \approx 4.6 \, \text{kcal mol}^{-1}$; for cyclohexane, the accessible area is about $300 \, \text{Å}^2$, and $\Delta G \approx 6.7 \, \text{kcal mol}^{-1}$.

The same regularity is observed for hydrophobic amino acid residues (Fig. 5.6).

The hydrophobicities of amino acids are derived experimentally from equilibrium distributions of amino acids between water and a nonpolar or slightly polar solvent; the latter is usually a high-molecular-weight alcohol (for example, octanol) or dioxane. These experiments are far from simple because some amino acids are hardly soluble in water while others are hardly soluble in organic solvents (for instance, polar amino acids are virtually insoluble in purely nonpolar cyclohexane or benzene). Therefore, a nonpolar solvent is replaced by a specially selected slightly polar one (like octanol, which is satisfactory for both strongly and weakly polar amino acids), and other tricks are used too. According to the various solvents used, the results differ, specifically for charged and strongly polar amino acids. However, the qualitative agreement of these results is not bad.

The hydrophobic effect is smaller for the side chains with polar atoms than for completely nonpolar groups (provided their total accessible surface area is the same). However, if we consider only the nonpolar-atom-produced portion of their accessible surface area (i.e., the total accessible surface area minus approximately $50 \, \text{Å}^2$ for each polar atom), essentially the same surface dependence of hydrophobicity can be revealed for all the groups (see Fig. 5.6).

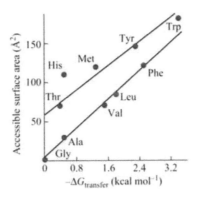

Figure 5.6. The accessible surface area of amino acid side chains and their hydrophobicity. (The accessible surface of a side chain X is equal to the accessible surface of amino acid X minus that of Gly having no side chain; the hydrophobicity of a side chain X is equal to the experimentally measured hydrophobicity of amino acid X minus that of Gly.) The side chains of Ala, Val, Leu, Phe consist of hydrocarbons only. Those of Thr and Tyr additionally have one O-atom each, Met – S-atom, Tpr – N-atom, and His – two N-atoms; therefore, their accessible *nonpolar* area is smaller than their total accessible surface. Adapted from [12] with minor modifications.

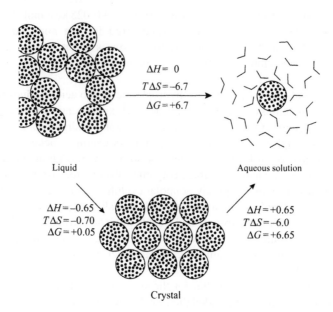

Figure 5.7. The thermodynamics of transfer of a typical nonpolar molecule, cyclohexane, from the liquid to the solid phase and to aqueous solution. The numerical values are for 25 °C. ΔH, $T\Delta S$ and ΔG are measured in kilocalories per mole. Adapted from [4]. All values are recalculated using equation (5.15) and experimental data from reference books.

The hydrophobic effect is of major importance in maintaining the stability of the protein structure. It is this effect that is responsible for the compact globule formation of the protein chain. As seen from Fig. 5.7, the free energy of transfer of a hydrophobic group from water to hydrophobic media is high and amounts to a few kilocalories per mole, while the free energy of hardening of the nonpolar liquid is close to zero at physiological temperatures. (To be more accurate, I should say that the free energy of hardening of comparatively small hydrocarbons (like cyclohexane shown in Fig. 5.7) is actually positive, which prevents their hardening at room temperature. The hardening is prevented by entropy of rotations and movements of a molecule in the liquid, where each molecule is more or less free, in contrast to a solid in which the crystal lattice keeps the molecules fixed. However, the entropy of these motions of a molecule does not depend on its size, while the enthalpy of a molecule increases with increasing number of intermolecular contacts, i.e., with the size of the molecular surface. In a side chain the entropy of motions is lower, since amino acid residues are linked into a chain and cannot move freely, and this facilitates hardening.)

However, even if the entire entropic component of cyclohexane crystallization, $\Delta G_{\text{liquid} \rightarrow \text{crystal}}$, was neglected (i.e., if we assume that $\Delta G_{\text{liquid} \rightarrow \text{crystal}} \approx \Delta H_{\text{liquid} \rightarrow \text{crystal}} = -0.65\,\text{kcal}\,\text{mol}^{-1}$), the thermodynamic effect of crystallization would be much weaker than the hydrophobic effect condensing the water-dissolved cyclohexane molecules into a liquid drop ($\Delta G_{\text{aqueous solution} \to \text{liquid}}$ = $-6.7 \, \text{kcal mol}^{-1}$).

Thus, the hydrophobic effect is responsible for approximately 90% of the effort required to make a compact globule. However, by itself it cannot provide a native solid protein. It creates only the molten globule that is yet to be discussed. The hardening of a protein, like that of all compounds, results from van der Waals interactions, as well as from hydrogen and ionic bonds, which are far more specific and more sensitive to peculiarities in the atomic structure than simple hydrophobicity (which is, actually, an effect operating in liquid). But what they perform is the final "polishing" of the native protein, whereas the bulk of the basic work is done by the hydrophobic effect.

LECTURE 6

In this lecture we will discuss electrostatic interactions and, specifically, their features induced by the protein globule and its aqueous environment.

It may seem that there's nothing to discuss: undoubtedly, you remember that in a medium with permittivity (dielectric constant) ε the charge q_1 creates an electric field whose potential at the distance r is

$$\varphi = \frac{q_1}{\varepsilon r} \qquad (6.1)$$

and q_1 interacts with the charge q_2 at this distance with the energy

$$U = \varphi \, q_2 = \frac{q_1 q_2}{\varepsilon r} \qquad (6.2)$$

You may also remember that in a vacuum (or air) $\varepsilon = 1$, in water ε is close to 80, and in media like plastics (and dry protein as well) ε is somewhere between 2 and 4.

Inner voice: All this is true, or, better to say, almost true. First, strictly speaking, U is not the energy but the *free* energy when our charges are in some medium instead of a vacuum (it is the free energy that under conditions of heat exchange between our system (charges in a medium) and its environment tends to a minimum at equilibrium, i.e., it is the free energy that governs the attraction between charges of opposite signs and of repulsion between those of the same sign). Besides, the energy of elementary interaction must be temperature-independent, while U depends on permittivity ε of the medium and hence varies with temperature.

Lecturer: This comment is absolutely right, but inasmuch as we are considering how charge interactions affect protein stability, these are minor items, a matter of purism. For the sake of simplicity, let me use the term "energy" for a while . . .

What is more important, eq. (6.2) is valid only for homogeneous media. And when studying charge interactions in proteins, we are dealing with a most heterogeneous medium. The permittivity of a protein itself, like that of plastics, is not high and amounts to about 2–4, while that of water is 80. And the charged groups of the protein are mostly located on its surface, close to the water (we will see why later). So, what value of ε should be chosen to make estimates of electrostatic interactions in the protein? If we take $\varepsilon \approx 80$, the energy of interaction between two elementary (proton) charges at a 3 Å distance will be approximately 1.5 kcal mol^{-1}; with $\varepsilon \approx 3$, this energy will be about 40 kcal mol^{-1}. The difference is too large: the additional 40 kcal mol^{-1} can destroy any protein structure . . .

By the way, how can we estimate the extent to which this or that free energy affects the protein? Here, two magnitudes are helpful: (a) $kT \approx 0.6$ kcal mol^{-1}; any effect below this value (per protein molecule) can be neglected as being "washed away" by thermal fluctuations; (b) the typical "reserve of stability" of a protein structure (the difference in free energies between the native and denatured states of the protein) is about 10 kcal mol^{-1}; any effect exceeding this value causes an "explosion" of the protein.

The other problem is as follows. Equations (6.1) and (6.2) are valid when the distance r between charges much exceeds the size of surrounding molecules. However, in proteins, charges are often in immediate contact with as little as 3–4 Å distance between them, which does not allow even a water molecule, not to mention a side chain, to take an intervening position. How can we estimate electrostatic interactions in this case? Should we assume that $\varepsilon = 1$, as in a vacuum? Or should we take $\varepsilon \approx 80$? Or rather . . . ?

A brief philosophical digression. Why are these rough estimates wanted at all? Indeed, it is often believed that with the powerful computers available now, one can input "all as it is in reality": water molecules, the coordinates of protein atoms including the coordinates of the charges, the temperature (i.e. the energy of thermal motion) and obtain "the precise result". As a matter of fact, this is a rather utopian picture. The calculation – I mean the detailed one (made using so-called "molecular dynamics") – will take hours or days because you will have to follow both thermal motions and polarization of many thousands of interacting atoms (and, by the way, will not be absolutely accurate either: if nothing else, remember that atoms are "nonspherical" owing to their p-orbitals and other quantum effects). And what really interests you is most likely a simple quick estimate such as whether it is possible to introduce a charge into the protein at this or that site without a risk of protein explosion. That's why my aim is to teach you how to make such estimates.

First of all, let us estimate the change in energy of a charge upon its transfer from water ($\varepsilon \approx 80$) into the middle of the protein (where $\varepsilon \approx 3$). For the time being, let us consider water and protein as continuous media and disregard their corpuscular (i.e., atomic) structures; or rather, let us postpone considering these details.

According to classical electrostatics, a sphere of charge q and radius R in a medium of permittivity ε has the energy

$$U = \frac{q_1^2}{2\varepsilon R} \tag{6.3}$$

This expression directly follows from eq. (6.2): when we charge up the sphere (from zero to q_1) by bringing small charges dq onto its surface, each small charge dq

increases the energy of the sphere by $dU = q\,dq/\varepsilon R$ (according to eq. (6.2)), and the integral of $q\,dq$ from 0 to q_1 is $q_1^2/2$.

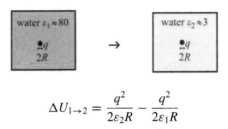

$$\Delta U_{1 \to 2} = \frac{q^2}{2\varepsilon_2 R} - \frac{q^2}{2\varepsilon_1 R}$$

If the radius of the charged atom is about 1.5 Å, then its (free) energy is nearly 1.5 kcal mol^{-1} at $\varepsilon \approx 80$ (in water) and nearly 40 kcal mol^{-1} at $\varepsilon \approx 3$ (in protein). This great difference explains why inside the protein, in contrast to its surface, charged groups are virtually absent (it is easy to estimate that even immersion of a close pair of oppositely charged particles into the medium with $\varepsilon \approx 3$ increases the free energy by about the same magnitude of 40 kcal mol^{-1}). Therefore, an ionizable group is virtually always uncharged when it is deeply involved in the protein globule: that is, a positively charged side group donates its surplus H$^+$ to water, and a negatively charged side group takes its missing H$^+$ from water. True, this discharging costs some additional free energy – but "only" a few kilocalories mol^{-1}, as you will see in Lecture 10, and not a few dozens of kilocalories mol^{-1}. To be more exact, there are some charges in the interior of a protein; but these are almost always functional, and the protein has to put up with their presence just to keep functioning.

Now let us learn how to estimate the interaction of charges taking into account the interface between the protein ($\varepsilon \approx 3$) and water ($\varepsilon \approx 80$).

Consider the following school problem. Let water (medium with $\varepsilon_1 \approx 80$) occupy half of the space, while "protein" (a medium with $\varepsilon_2 \approx 3$) occupies the other half, the two being separated by a flat interface. Let a charge q be in water, at the point "1" above the flat surface of the "protein" (Fig. 6.1).

Find the electric field φ induced by this charge at an arbitrary point "2".

If the protein was far away or absent, this field for the arbitrary point "2" would be calculated from eq. (6.1) with $\varepsilon = 80$ (the dielectric constant of water) and with $r = r_{12}$ (the distance between points "1" and "2").

What difference is caused by the nearby protein, i.e., by a medium of different permittivity?

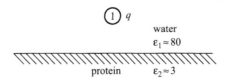

Figure 6.1.

As a hint, recall a similar problem that you must have considered in general physics. There, the specification is the same, except that "medium 2" is a metal (conductor) where ε_2 is infinity (Fig. 6.2a). Its solution is as follows.

$$\varphi = \begin{cases} 0 & \text{there is no field within metal} \\ q/\varepsilon_1 r_{12} + (-q)/\varepsilon_1 r_{02} & \text{above the metal surface} \end{cases} \tag{6.4}$$

There is no field *inside* the metal: otherwise, free charges would keep moving in the field until it was reduced to zero.

The resultant field *above* the face of the metal (at the point "2" where the field must be calculated, see Fig. 6.2b) is created by the charge q *and* its "mirror reflection" in the metal. This reflection "0" sums up the effect resulting from the charge "1"-induced shift of electrons. The distribution of these electrons over the face of the metal results in the "reflection effect". The charge of this "reflection" is $-q$, and its position "0" has the same location *below* the surface as the point "1" *above* it (i.e., r_{02} is the distance between the reflected charge "0" and the point "2").

I will not prove this solution (or rather, its last part, which is not obvious and refers to the "field above the metal"); I'll only give a brief idea of the proof.

Since in equilibrium, i.e., in the absence of current, there is no electric field inside the metal, the force lines of the electric field (which show the direction of the charge-affecting force, see arrows in Fig. 6.3) must be strictly perpendicular to the face of the metal (otherwise, they would have had a surface-parallel component inducing current in the metal). However, the lines of force of a field induced (in the absence of any metal) by two equal and opposite charges are also perpendicular to the plane positioned precisely between the charges. And similar pictures of force lines reflect similar behaviors of the potentials.

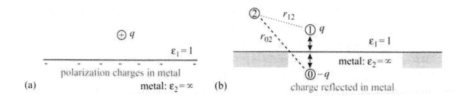

(a) polarization charges in metal metal: $\varepsilon_2 = \infty$

(b) charge reflected in metal

Figure 6.2.

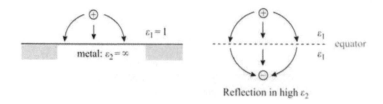

metal: $\varepsilon_2 = \infty$

Reflection in high ε_2

Figure 6.3.

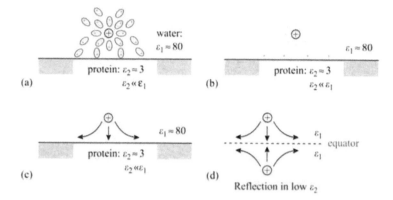

Figure 6.4. (a) Each water dipole is turned so as to point with its "minus" at our charge ⊕ and with its "plus" in the opposite direction. (b) As a result, polarized charges appear at water boundaries (while inside the bulk water the "pluses" and "minuses" of water dipoles are mutually compensated). The polarization charges adjacent to ⊕ decrease its field, thereby creating a high permittivity of water (this effect is independent of the nearby protein). The water dipoles positioned near the water–protein interface are responsible for the polarization charge induced here; note that it is of the same sign as our charge ⊕. (c) The resultant force lines of the electrostatic field near the surface of the medium with a low permittivity (compared with that of the medium hosting our charge ⊕). (d) The equivalent picture of the lines of force produced by the reflection charge.

The solution of the problem on "the field of charge above the metal" shows that it is easy to solve such problems using "reflections" and analysis of the force lines. Note also, that these lines *tend to be attracted by a medium with higher permittivity*.

Now let us return to our problem that also deals with a charge at the interface between two media. Similarly, here the key role is played by polarization of the medium with a higher permittivity, that is, of water (Fig. 6.4a and b). The small polarized charges of the protein (at the interface between the protein and water) are negligible compared with the large polarized charges of water. Polarization of the medium results in the tendency for the force lines to stay in water (Fig. 6.4c), since water permittivity is higher than that of the protein. A field with force lines of the same kind (when only the part *above* the interface is considered) is induced by two charges of the same sign (Fig. 6.4d). This is suggestive as to the position and sign of the reflection charge: it must be placed below the interface at the point where we see the mirror reflection, and its sign must be the same as that of our charge "1" (that is above the interface).

I will not bore you with all the necessary calculations, but give you the solution right away. *The field above the protein* is induced by both the charge q and the reflection charge $q' = q(\varepsilon_1 - \varepsilon_2)/(\varepsilon_1 + \varepsilon_2) \approx +q$ (close to $+q$ at $\varepsilon_1 \gg \varepsilon_2$) that is positioned as far below the interface as q is above it (Fig. 6.5).

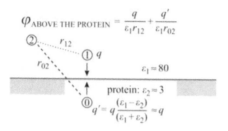

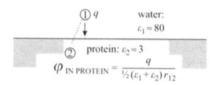

Figure 6.5.

Figure 6.6.

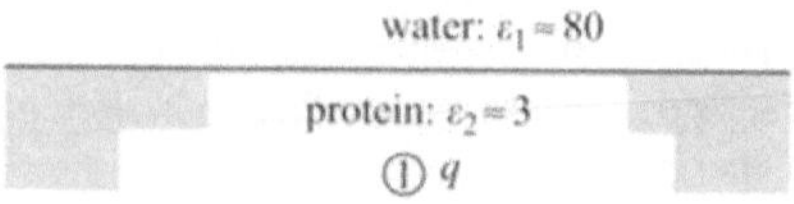

Figure 6.7.

As a result, above the protein, the effective permittivity ε_{eff} (that is the magnitude to be used in the formula $\varphi = q/\varepsilon_{\text{eff}} r_{12}$) is close to ε_1, i.e., to 80, *only* at a minor distance from the charge "1" (when r_{12} is much less than the distance between "1" and the face of the protein; then, $r_{12} \ll r_{02} \approx r_{01}$).

But everywhere (in water) at large distances from "1", the ε_{eff} value is close to 40. Indeed, $r_{12} \approx r_{02}$ when $r_{12} \gg r_{10}$, and since $q' \approx q$ at $\varepsilon_1 \gg \varepsilon_2$, we have $\varphi \approx 2q/[\varepsilon r_{12}] = q/[(\varepsilon_1/2) r_{12}]$.

Close to the interface we have always $r_{02} \approx r_{12}$, see Fig. 6.5. Then $\varepsilon_{\text{eff}} \approx 40$ for the field not only above but also immediately below the protein surface: the potentials of the "interior field" and the "exterior field" must coincide on the interface.

Moreover, we can show that the total "interior" field is induced by the charge "1" (without any mirror images, see Fig. 6.6), and for this field (I take the liberty of omitting the calculations again), $\varepsilon_{\text{eff}} = (\varepsilon_1 + \varepsilon_2)/2$, i.e., it is close to 40 for the entire area below the face.

Thus, the effective permittivity ε_{eff} is close to $(\varepsilon_1 + \varepsilon_2)/2$ (i.e., to 40) *everywhere* at large distances from the charge q, and it is close to ε_1 (i.e., to 80) only when r_{12} is much less than the distance between the charge and the protein surface.

We obtain a similar result for the complementary problem of the field of a charge located *below* the protein surface where the permittivity is $\varepsilon_2 \approx 3$ (Fig. 6.7).

In this case the effective permittivity ε_{eff} is also close to $(\varepsilon_1 + \varepsilon_2)/2$ (i.e., to 40) *everywhere* at large distances from the charge q, and it is close to ε_2 (i.e., to 3) only when r_{12} is much less than the distance between the charge and the protein surface.

We obtain a still more peculiar result for the situation when a charge located at one side of the protein induces a field on the other side (Fig. 6.8).

In this case of interaction via the protein, that is, via the medium of low permittivity ($\varepsilon_2 \approx 3$), we might expect ε_{eff} to be close to 3 or at least between 3 and 80 because of the aqueous environment. But in fact, it gets beyond 200 (!!). How does this happen?

For explanation, let us consider how polar water molecules are oriented in this case around the protein (and around "our" charge "1"), and how this changes the field of the charge (see Fig. 6.9; to make it more clear, I have used a scale different from that of Fig. 6.8, although the event illustrated is the same).

The water molecules are oriented in conformity with the field: their "−" are directed mostly at "our" charge $\oplus$, while their "+" are directed oppositely. This results, first, in partial compensation of the charge $\oplus$ by adjacent "−" of waters; this event is trivial and simply causes a large water permittivity ε_1. But, second, this also produces the event that is of interest to us. Namely, the water "pluses" turn to

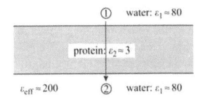

Figure 6.8.

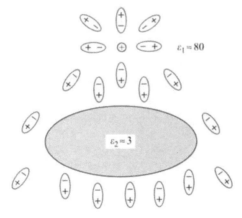

Figure 6.9. The orientation of water molecules (shown as dipoles ⊕⊖) around the protein ▢ and the charge $\oplus$ (which is shown as positive for the sake of simplicity).

the protein side facing the charge ⊕, and this polarization induces a positive charge at this side of the water/protein interface, while the water "minuses" coming near the protein on the other side induce there an opposite (negative) polarized charge (polarization of the protein itself can be neglected, since it is small as compared with water polarization because $\varepsilon_2 \ll \varepsilon_1$).

The induced charges emerge only at the interfaces ("charge–water" and "protein–water"): as stated by an electrostatics theorem, inside homogeneous (and initially uncharged) media the induced charges are absent. According to the same theorem, the sum of the polarized charges induced by an external charge near the surface of a body (our protein) surrounded by a homogeneous medium (here, by water) is equal to zero.

So, the net polarized charge "+" induced by oriented waters near the protein surface is directed at ⊕, while an equal polarized "−" is induced by waters at the other side of the protein.

As a result, on the ⊕-facing protein side, the field potential becomes higher compared with what would have been in the absence of the protein: here, the induced "pluses" add to the potential of the charge ⊕ (and the induced "minuses" are far away and their effect is minor). That is why here $\varepsilon_{\text{eff}} \approx 40$, as has been stated.

At the same time, on the ⊕-opposite side of the protein, the field potential becomes *lower* than it would have been in the absence of the protein: here, the potential of the charge ⊕ is diminished by the opposite in sign potential of "minuses" induced at the adjacent protein surface (and the induced "pluses" are far away and their effect is minor). And since here (on the ⊕-opposite side of the protein) the field potential is *lower* than it would have been in the absence of the protein, ε_{eff} is much *higher* than ε_1 (the value in the absence of the protein), i.e., it considerably exceeds $\varepsilon_1 = 80$.

In general, the distribution of numerical values of ε_{eff} "in and around the protein" for the field produced by the charge "1" and the induced polarized charges looks as shown in Fig. 6.10a if "1" is in water near the protein surface, and as in Fig. 6.10b if it is deep in the protein. Let me remind you that ε_{eff} is the effective value of permittivity for the point r to be used in the formula $\varphi(r) = q_1 / [\varepsilon_{\text{eff}} \mid r - r_1 \mid]$ to calculate the potential of the charge "1" (located at r_1) at the point r.

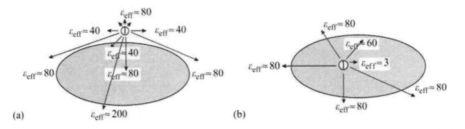

(a)   (b)

Figure 6.10.

It is worth considering one more consequence of the interface. This is the effect of the charge on itself (see also the discussion of eq. (6.3)): a charge located outside the protein is *repelled* from the protein (i.e., from the charge's reflection in the protein), while a charge inside the protein is *strongly attracted* to water. In both cases the medium of higher permittivity attracts the charge, and that with a lower permittivity repels it. The values of these effects can be easily estimated as interactions between the charge and its reflection, and I leave these calculations to you.

Now let us focus on the effects connected with the particulate nature, i.e., the atomic structure of the medium.

Actually, the value of permittivity ε is determined by the atomic structure of the medium. If the medium consists of nonpolar molecules, an electric field only shifts electrons in the molecules, which is rather difficult; therefore, the polarization is insignificant, and ε is low. If the medium consists of polar molecules (as exemplified by water), an electric field turns these molecules, which is easier, and, hence, ε of such medium is high. In both cases, with electrons shifted or molecules turned, polarization of the medium partially screens the immersed charges ($\oplus$ and $\ominus$, see Fig. 6.11) and thereby diminishes the electric field in the medium compared to what it would have been in a vacuum.

It would be only natural to expect that the particulate effects must strongly affect the interaction of charges at short distances, since the classical equations (6.1)–(6.4) are valid, strictly speaking, only when the charges are separated by many medium molecules. And if the charges are 3–4 Å apart (as often happens in proteins), no other atom can get between them to change their interaction.

In the case of such close contact, the permittivity might be believed to approach 1 even in the aqueous environment. This viewpoint, or rather this misapprehension, may still be encountered in the literature.

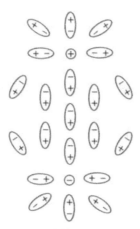

Figure 6.11.

However, strange as it may seem, the medium's particulate nature makes no drastic changes in the "macroscopic" (i.e., derived for large distances between the charges) permittivity, even if the distance is as short as 3 Å. In other words, even at the smallest distances the value of water permittivity is much closer to 80 or 40 than to 1 or 3.

This is well supported by the fact that salt easily dissolves (and dissociates) in water, which is possible only with a weak attraction between counter-ions even at very short distances of about 3 Å.

Indeed, the distance between Na^+ and Cl^- ions can be as short as 3 Å in a direct van der Waals contact. Then their free energy of attraction would amount to $-1.5\,kcal\,mol^{-1}$ at $\varepsilon = 80$, $-3\,kcal\,mol^{-1}$ at $\varepsilon = 40$, and $-6\,kcal\,mol^{-1}$ at $\varepsilon = 20$. The latter ($6\,kcal\,mol^{-1}$) exceeds the energy of a hydrogen bond. Such an energy would make the counter-ions stick together more tightly than water molecules do, and then the concentration of a saturated salt solution would be about $10^{-4}\,mol\,l^{-1}$, like the concentration of saturated water vapor. But this obviously cannot be true: it is no problem to dissolve one mole of NaCl (58 g) in one liter of water (this will be an ordinary, perhaps a bit too salty brine). Consequently, water permittivity is considerably higher than 20 even at a distance of about 3 Å.

The value of ε at the closest distances inside the molecule can be estimated more precisely using the first and second constants of dissociation of dihydric acids in water. For example, dissociation of oxalic acid occurs as follows:

$$
\begin{array}{ccccc}
\mathrm{HO}\quad\mathrm{OH} & & \mathrm{HO}\quad\mathrm{O-} & & \mathrm{-O}\quad\mathrm{O-} \\
\backslash\ \ / & & \backslash\ \ / & & \backslash\ \ / \\
\mathrm{C-C} & \longrightarrow & \mathrm{C-C} & \longrightarrow & \mathrm{C-C} \\
/\!/\ \ \backslash\!\backslash & & /\!/\ \ \backslash\!\backslash & & /\!/\ \ \backslash\!\backslash \\
\mathrm{O}\quad\mathrm{O} & & \mathrm{O}\quad\mathrm{O} & & \mathrm{O}\quad\mathrm{O} \\
\mathrm{pH} < 1.5 & & 2 < \mathrm{pH} < 4 & & 4.5 < \mathrm{pH}
\end{array}
$$

The second dissociation is shifted from the first one by approximately 2.5 pH units, i.e., it occurs when the H^+ concentration is $10^{2.5} = e^{2.3 \times 2.5}$ times lower. This shows that the free energy of interaction of the first charge with the second one is $2.5 \times 2.3RT \approx 3.5\,kcal\,mol^{-1}$ when the distance between the charges is about 3 Å. This value of interaction energy corresponds to $\varepsilon \approx 40$ at a distance of 3 Å. A similar result ($\varepsilon \approx 30$–40 at a distance of about 2–2.5 Å) was obtained for carbonic acid dissociation, $H_2CO_3 \rightarrow HCO_3^- \rightarrow CO_3^{2-}$.

Hence, even a salt bridge between oppositely charged side-chains on the protein surface must "cost" 2–3 $kcal\,mol^{-1}$ only. In the interior of the protein its "price" is higher, but immersion of charged side-chains into the protein would have been still more expensive, so it is no wonder that such bonds are rather rare in native proteins.

Thus, our conclusion is that the particulate nature makes no drastic changes in the "macroscopic" (derived for large distances between the charges) permittivity of water even at a distance of about 3 Å, which is too small for any other molecule to get between the interacting charges. The reason is that the charges are quite well shielded

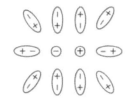

Figure 6.12.

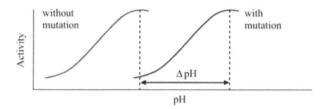

even by solute molecules coming from other sides and from the flanks (Fig. 6.12): these molecules become polarized (in the case of water, they simply turn), so that their "+"s shift towards the charge ⊖, and their "−"s towards the charge ⊕.

Here we see again (cf. Figs. 6.10a and b) that the electrostatic interaction between the charges seemingly occurs mainly via the medium of a higher permittivity and nearly ignores the medium of weak polarization.

All our previous discussions referred to "micromolecular" systems. But are the conclusions drawn valid for proteins (where the particulate effects are coupled with the huge difference in permittivity between the water and the protein)?

Experiments reported by the research team headed by A. Fersht, the founder of protein engineering, show that the above estimates are valid for proteins.

The basis of these experiments is as follows. There are proteins (enzymes) that exhibit a particularly high activity at a certain value of pH (they are said to have a pH-optimum).

This pH-optimum can be shifted (Fig. 6.13) using a charged residue introduced to the protein by mutating its gene, and the electric field induced at the active site by the charge of the mutated residue can be estimated by the shift of the pH-optimum.

The pH-optimum is caused by the fact that, to keep the enzyme functioning, a group at its active site must have a certain charge that, in turn, depends on the concentration of hydrogen ions in the medium. The H^+ concentration, i.e., $[H^+]$ (mol l^{-1}), is equal to 10^{-pH} by definition, and the OH^- concentration in water is about 10^{-14+pH}.

Let the active site (AS) accept the ion H^+: $AS + H^+ = ASH^+$. Then, according to the active mass law, the ratio between these two forms of the active site (with and without H^+) is

$$\frac{[ASH^+]}{[AS]} = \exp(-\Delta F_{ASH^+}/RT) \times [H^+]$$

$$= \exp(-\Delta F_{ASH^+}/RT) \times 10^{-pH}$$

$$= \exp\{-(\Delta F_{ASH^+}/RT + 2.3 \times pH)\}, \qquad (6.5)$$

where ΔF_{ASH^+} is the free energy of H^+ binding (at $[H^+] = 1 \, mol\, l^{-1}$) to the active site, and the symbol [ ] denotes concentration.

If the mutation-introduced charge induces a potential φ at the protein active site, then ΔF_{ASH^+} changes as $\Delta F_{ASH^+}|_{\text{with mutation}} = \Delta F_{ASH^+}|_{\text{without mutation}} + \varphi e$, where e is the charge of H^+. Since at the pH-optimum the magnitude of $[ASH^+]/[AS]$ (and the magnitude of $\Delta F_{ASH^+}/RT + 2.3 \times pH$) must remain the same both with and without the mutation, then

$$\frac{\Delta F_{ASH^+}|_{\text{without mutation}}}{RT} + 2.3 \times pH|_{\text{opt. without mutation}}$$

$$= \frac{\Delta F_{ASH^+}|_{\text{with mutation}}}{RT} + 2.3 \times pH|_{\text{opt. with mutation}}. \qquad (6.6)$$

That is,

$$\varphi e = \Delta F_{ASH^+}|_{\text{with mutation}} - \Delta F_{ASH^+}|_{\text{without mutation}}$$

$$= 2.3RT \times (pH|_{\text{opt. without mutation}} - pH|_{\text{opt. with mutation}})$$

$$= 2.3RT \times (-\Delta pH). \qquad (6.7)$$

Thus, having learned the shift of the pH-optimum, ΔpH, we can estimate the potential induced at the active site by the mutated protein residue. Then, using the known three-dimensional structure of the protein, and, hence, the distance r from the mutated residue to the active site, we can estimate the effective permittivity ε_{eff} (a term in the equation $\varphi = q/\varepsilon_{\text{eff}}r$) for the interaction between the mutation-introduced charge q and the active site.

In Fersht's experiments, the mutations were performed at the surface of the protein in order not to damage its structure (as we have already learned, the energy of a charge deeply immersed in the protein is high and can literally explode the protein globule).

The experimental result reported: the effective permittivity ε_{eff} ranges from ~ 40 to ~ 120, the former being typical of mutations at short distances from the active site, and the latter for remote ones. The fact that ε_{eff} can reach a value of 120 appeared to be not a little surprising to those believing that ε_{eff} must lie between 3 (as inside the protein) and 80 (as in water), and probably closer to 3 due to the particulate nature of water. However, for us, these values are not surprising, since they are in good agreement with what follows from Fig. 6.10a.

A brief digression on protein engineering. Its major advantage is that by changing a codon in a protein gene, we can perform a mutation at an exact site of the protein globule, since the gene in question, the amino acid sequence of the protein and the protein 3D structure are known. Besides, the mutation effect on the structure can also be monitored by X-ray analysis or NMR. Thus, the entire work is performed with one's eyes open.

In experiments that we discussed above the protein served as a microscopic (or rather, nanoscopic) electrometer. And protein engineering allows us to use such instruments as well as jumping from the physical theory to genetic manipulations, which is great fun.

Here I would like to make some additions concerning electrostatic interactions.

First: So far I have discussed only the interaction of separate charges. However, electrostatics also covers the interactions of dipoles (for example, the dipoles $H^{(+)}$—$O^{(-)}$ and $H^{(+)}$—$N^{(-)}$ involved in hydrogen bonding) as well as quadrupoles; the latter are present, for example, in aromatic rings (Fig. 6.14).

The reason for my considering interactions of ions is simply the strength of these interactions: even in immediate contact they are a few times stronger than interactions between dipoles (also, their decrease with increasing distance is slower), while interactions of dipoles are stronger than those between quadrupoles.

Second: With free charges (e.g., salt) available in the water, the electrostatic interactions diminish with distance r as $(1/r) \times \exp(-r/D)$ rather than as $1/r$. Here D, the Debye–Hückel radius, corresponds to the typical size of the counter-ion cloud around the charge. The value of D is independent of the charge itself but depends on the charges of the salt ions, on their concentration in the medium, on its permittivity and on temperature. In water, at room temperature

$$D \approx \frac{3}{I^{1/2}} \text{ Å},\qquad(6.8)$$

where

$$I = \frac{1}{2}\sum_i c_i z_i^2\qquad(6.9)$$

is the *ionic strength* of the solution given in moles per liter. In eq. (6.9), the sum is taken over all kinds of ions present in the solution, z_i is the ion's charge (in proton charge units), c_i is the concentration (in moles per liter) of ion i. Under ordinary physiological conditions, $I \approx 0.1$–0.15; then $D \approx 8$ Å. However, some microorganisms live at

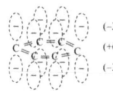

Figure 6.14. The electric quadrupole of an aromatic ring: the layer of "halves" of six p-electrons (charge -3) – the layer of cores (charge $+6$) – the layer of the last "halves" of p-electrons (charge -3).

$I \approx 1 \, \mathrm{mol} \, l^{-1}$ and more; then the persisting electrostatic interactions of the charged groups of the protein, though much weakened, are only those corresponding to "salt bridges", i.e., to the immediate contact of the charges.

In general, with an ionic atmosphere present in the solution, the energy of interaction of the two charges is

$$U = \frac{q_1 q_2}{\varepsilon_{\mathrm{eff}} r} \times \exp(-r/D). \qquad (6.10)$$

Third: The electrostatic interaction is a striking example of the *non*-pairwise interaction of particles (unlike, say, van der Waals interaction). It depends not only on the distance r between the charges q_1 and q_2 but also on the medium properties (those change both ε and D), and specifically, on the distance between the charges and other bodies and on the shape of these bodies (these affect $\varepsilon_{\mathrm{eff}}$), as well as on the concentration of free ions in the solution (which affects D).

And one more addition. So far, I have used the term "the energy of electrostatic interactions". This was done for the sake of simplicity; as I mentioned at the beginning of this lecture, the strict term would be "free energy". This is because we focused only on the attraction and repulsion of charges, without preventing their *heat exchange* with the environment. And with heat exchange allowed, what we dealt with is the free energy by definition.

Moreover, the temperature dependence of electrostatic effects in aqueous environment may be used to show the predominance of the entropic constituent over the energetic (enthalpic) one; the latter, by the way, is close to zero. This follows from the fact that water permittivity decreases from 88 to 55 (i.e., the electrostatic interactions increase by 40%) with increasing absolute temperature T from 273 to 373 K (i.e., by 35%). This means that the electrostatic interactions increase approximately proportionally to the absolute temperature. And the interaction increasing proportionally to the absolute temperature implies an exclusively entropic constituent. Hence, in water, the entire electrostatic effect is caused not by energy, but by the ordering of water molecules around the charges and by its variation with variation distance between the charges.

Hence, strange as it may seem, electrostatics in water originates not from energy but from entropy, just like hydrophobic interactions or hydrogen bonding in an aqueous environment.

In concluding this section on "Elementary interactions in and around proteins", I would like also to mention disulfide and coordinate bonds. Although not as abundant in proteins as hydrogen bonds, these may often be of great importance.

Disulfide (S—S) bonds are formed by cysteine (Cys) amino acid residues (the Cys side-chain is $-\mathrm{C}^{\beta}\mathrm{H}_2-\mathrm{SH}$). No direct oxidation of cysteines accompanied by hydrogen release (according to the scheme $-\mathrm{CH}_2-\mathrm{SH} + \mathrm{HS}-\mathrm{CH}_2- \rightarrow -\mathrm{CH}_2-\mathrm{S}-\mathrm{S}-\mathrm{CH}_2- + \mathrm{H}_2$) occurs in proteins because at room temperature this process is too slow. However, in proteins, S—S-bonding can be rapid when assisted by thiol-disulfide exchange. In the cell, the exchange is thought to involve *glutathione* that has both the monomeric thiol (GSH) and dimeric disulfide (GSSG) form, and to follow the scheme

$$
\begin{array}{ccccc}
\text{PROTEIN} + \text{GS–SG} & \rightleftharpoons & \text{PROTEIN} + & \text{GSH} & \rightleftharpoons & \text{PROTEIN} + & 2\text{GSH} \\
\diagup \quad \diagdown & & \diagup \quad \diagdown & & & \diagdown \quad \diagup \\
\text{SH} \quad \text{HS} & & \text{SH} \quad \text{S–SG} & & & \text{S–S}
\end{array}
$$

Both breakdown and formation of S—S-bonds in cells are catalyzed (i.e., accelerated but not directed) by a special enzyme, disulfide isomerase.

In the cell, S—S-bonding is reversible, since the energetic equilibrium of this reaction (thiol-disulfide exchange) is close to zero (there were two covalent S—H bonds and one S—S bond, and now there are as many; this very much resembles the energy balance of hydrogen bonding in proteins, does it not?). Moreover, the available (rather high) concentration of GSH in the cell shifts the equilibrium towards breaking the bonds that might be produced by an "occasional" cysteine approach. Therefore, the only S—S bonds that can be formed and persist are between cysteines brought close to one another by other interactions.

S—S bonds are of particular importance for proteins that have to reside and function out of the cell. On the one hand, the absence of disulfide isomerase and glutathione is favorable for bonds that have been already formed (either in the cell or when leaving it) as they become "frozen" and run no risk of breaking or rearranging. On the other hand, the external conditions may be different, and the extended margin of safety provided by the stable, "frozen" S—S bonds would not be out of place for the protein. That is why S—S bonds are typical of secreted proteins rather than cellular ones. Usually, in secreted proteins, all available cysteines (less one, if their number is odd) participate in S—S bonds.

Coordinate bonds are formed by N, O, and S atoms of the protein (as well as by O atoms of water) to di- and trivalent ions of Fe, Zn, Co, Ca, Mg, and other metals.

The ions of these metals have vacant orbitals lying low (as to energy), only slightly above their filled electron orbitals. Each of the vacant orbitals is capable of bonding an electron pair. And O, N, and S atoms ("electron donors") have electron pairs that can occupy the vacant orbitals of the ions. The resultant bonds are identical to ordinary chemical bonds, except that ordinary bonds comprise electrons from both parent atoms, while coordinate bonds are formed by electrons coming from one bonded atom only.

During coordinate bonding the metal ion binds to several donors of electrons. Then a small (radius ~ 0.7 Å) di- or trivalent ion is surrounded by large-atom donors (radius ~ 1.5 Å). Mostly, there are six of these coordinating donor atoms located at the apices of a regular octahedron (Fig. 6.15).

Since the ion can be bonded to both electron donors of the protein and to oxygens of water, it passes (despite the high energy of each bond) from water to the protein and back without dramatic gains or losses in energy. What is of greater importance: if the positions of the donor atoms in the protein are "proper" for coordinate bonding, the ion

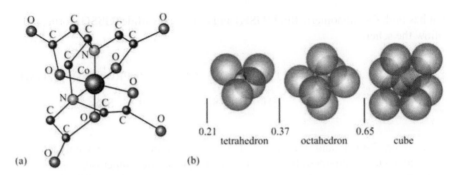

Figure 6.15. (a) The structure of the octahedral complex between Co^{3+}-atom and EDTA. (b) Typical coordination of the central ion at various ratios between its radius and the radii of the surrounding electron donors. Adapted from [9].

can release the water molecules of its previous environment and bind to the protein; then a strong bond occurs owing to the gained entropy of motion of the released water molecules (very much like the energy and entropy balance of hydrogen bonding, is it not?). On average, each coordinate bond costs several kilocalories per mole, i.e., it is a little more expensive than a hydrogen bond in a water environment.

Such bonds formed by several atoms of one molecule are called *chelate* ("claw-shaped"). The role of these bonds in proteins, specifically at their active sites, will be considered later on. Also, we will see that chelate complexes coating ions completely can be members of the hydrophobic protein core. And at the moment I would like to draw your attention again to Fig. 6.15 presenting the widely used reagent EDTA (ethylenediaminetetraacetic acid) that participates in a chelate bond to the metal. For EDTA, this bond is particularly strong because the negatively charged COO^- groups of EDTA are bound to the positively charged metal ion.

Part III

SECONDARY STRUCTURES OF POLYPEPTIDE CHAINS

LECTURE 7

Having dealt with elementary interactions, in this lecture we will consider the secondary structure of proteins.

First of all we will discuss regular secondary structures, that is, α-helices and β-structures.

These secondary structures are distinguished by regular arrangements of the main chain with side chains of a variety of conformations. The tertiary structure of a protein is determined by the arrangement of these structures in the globule (Fig. 7.1).

We shall consider helices first. They can be right-handed or left-handed (Fig. 7.2) and have different periods and pitches. Right-handed (R) helices come closer to the viewer as they move counterclockwise (which corresponds to positive angle counting in trigonometry), while left-handed (L) helices approach the viewer as they move clockwise.

In the polypeptide chain, major helices are stabilized by hydrogen bonds. The bonds are formed between C=O and H—N groups of the polypeptide backbone,

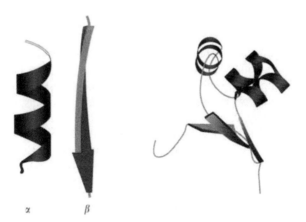

α β

Figure 7.1. The secondary structures of a polypeptide chain (α-helix and a strand of β-sheet) and the tertiary structure of a protein globule. Usually, taken together, α- and β-structures make up about a half of the chain in a globular protein.

75

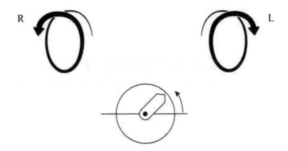

Figure 7.2. Right-handed (R) and left-handed (L) helices. The bottom picture shows positive angle counting in trigonometry; the arrow that is "close" to the viewer moves *counterclockwise*.

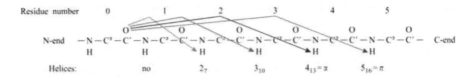

Figure 7.3. Hydrogen bonds (shown with arrows) typical of different helices. The chain residues are numbered from the N- to the C-end of the chain.

the latter being closer to the C-terminus of the chain. In principle, the following H-bonded helices can exist (Fig. 7.3): 2_7, 3_{10}, 4_{13} (usually called α), 5_{16} (called π), etc. The 2_7-helix derives its name from the 2nd residue participating in the H-bond (see Fig. 7.3) and **7** atoms in the ring (O:::H—N—C'—C$^\alpha$—N—C') closed by this bond. Other helices (3_{10}, 4_{13}, etc.) are named accordingly.

Which of these helical structures are most abundant in proteins? α-Helices are. Why? This question is answered by the Ramachandran map for a typical amino acid residue, alanine (Fig. 7.4), where I marked the conformations that, being repeated periodically, cause formation of the H-bonds shown in Fig. 7.3.

As seen, only the α_R-helix (right-handed α-helix) is deep inside the region allowed for alanine (and for all other "normal", i.e., L amino acid residues). Other helices are either at the very edge of this allowed region (e.g., the left-handed α_L-helix or the right-handed 3_{10}-helix), which gives rise to conformational strains, or in the region allowed for glycine only.

Therefore, it may be expected that the right-handed α-helix is the most stable and hence most abundant in proteins – and this is really so. In the right-handed α-helix (Fig. 7.5) the arrangement of all atoms is optimal, i.e., tight but not strained. Therefore, it is no wonder that in proteins α-helices are numerous, and in fibrous proteins they are extremely extended and incorporate hundreds of residues.

Left-handed α-helices are not (or hardly ever) observed in proteins. This is also true for 2_7-helices that not only lie at the very edge of the allowed region but also have a large angle between their N—H and O=C groups, that is energetically disadvantageous for hydrogen bonding.

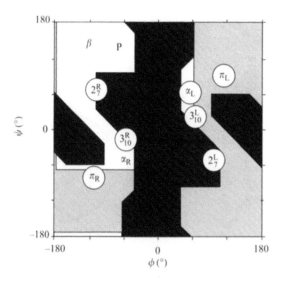

Figure 7.4. The conformations of various regular secondary structures against the map of allowed and disallowed conformations of amino acid residues. 2_7^R, 2_7^L: the right-handed and left-handed 2_7-helix; 3_{10}^R, 3_{10}^L: the right-handed and left-handed 3_{10}-helix; α_R, α_L: the right-handed and left-handed α-helix; π_R, π_L: the right-handed and left-handed π-helix. β, the β-structure (for details, see Fig. 7.8b). P, the poly(Pro)II-helix. $\square$, conformations allowed for alanine (Ala); ▨, regions allowed for glycine only, but not for alanine and other residues; ■, regions disallowed for all residues; φ and ψ, dihedral angles of rotation in the main chain.

π-Helices are absent from proteins too. They also occur at the very edge of the allowed region and their turns are far too wide, which results in an energetically unfavorable axial "hole". In contrast, 3_{10}-helices (mainly right-handed; left-handed ones are good for glycines only) are present in proteins, although only as short (3–4 residues) and distorted fragments: the 3_{10}-helix is too tight and gives rise to steric strains; its conformation lies close to the edge of the allowed region.

Pay attention to the feature clearly seen in Fig. 7.5a: the helical N-terminus is occupied by "free" H atoms of N—H groups uninvolved in intrahelical H-bonds, while the C-terminus is occupied by H-bond-free O atoms of C=O groups. Since the electron cloud of the H-atom is partially pulled off by the electronegative N-atom, and the electronegative O-atom attracts the electron of the C' atom, the N-terminus assumes a positive and the C-terminus a negative partial charge. That is, the helix is a long dipole where the N-terminal partial "+" charge (coming from three "free" H-atoms) amounts to about half of the proton charge, while the C-terminal "−" charge amounts to about half of the electron charge.

Now let us consider the regular main-chain structures lacking hydrogen bonds inside each of them but periodically H-bonded with one another.

The extended (all angles in the main chain are nearly *trans*), slightly twisted chains ("β-strands") form the sheet of the β-structure. A β-sheet can be (Fig. 7.6)

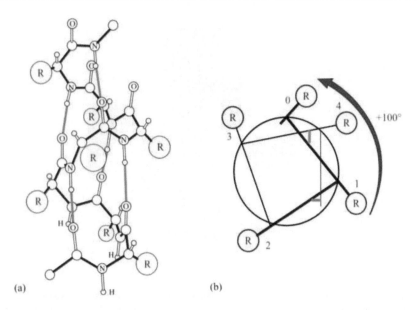

(a)   (b)

Figure 7.5. The right-handed α-helix. (a) Atomic structure; R = side-chains. Hydrogen bonds are shown as light-blue lines. (b) Axial view of one turn of this α-helix. The arrow shows the turn of the helix (per residue) when it approaches the viewer (the closer to the viewer, the smaller the chain residue number). The circle depicts the cylindrical surface enveloping the C^α atoms of the helix. Adapted from [12].

parallel ($\beta\uparrow\uparrow$), antiparallel ($\beta\downarrow\uparrow$), and mixed (comprising $\beta\uparrow\uparrow$ and $\beta\downarrow\uparrow$). The β-structure is stabilized by H-bonds (shown in light-blue lines in Fig. 7.6). Since the surface of β-structure sheets is pleated (Fig. 7.7), this structure is also called the "pleated β-structure".

As a whole, the β-sheet is usually somewhat twisted (Fig. 7.8a) because each separate β-strand is twisted by itself (Fig. 7.8b), thereby slightly altering the direction of H-bonds along the strand and thus favoring a certain angle between the H-bonded strands. In its turn, as explained by C. Chothia, this twist of a strand results from shifting the energetically advantageous conformation of all side-chain-possessing residues towards the center of the sterically allowed region (Fig. 7.8c). The twist of an individual β-strand is left-handed (for "normal", i.e., L amino acids; for D, it would be opposite): as seen from Fig. 7.8b, the strand's side-chains turn clockwise (by about $-165°$ per residue) as the strand comes closer to the viewer.

Because the strands twist, H-bonds turn as well (by about $-165°$ per residue, i.e., by $-330° = +30°$ per residue pair, a regular element of the β-structure). As a result, the angle between the neighboring β-strands (viewed from the edge of the sheet, Fig. 7.8a) usually amounts to about $-25°$ (as always, "$-$" means a clockwise turn of the nearby β-strand about the remote one). Thus, the β-sheet has a left-handed twist if viewed from its edge (and *right*-handed if viewed along the β-strands, as is usually done).

$\beta\uparrow\uparrow$

N-end $-\mathrm{N}-\mathrm{C}^3-\mathrm{C}'-\mathrm{N}-\mathrm{C}^3-\mathrm{C}'-\mathrm{N}-\mathrm{C}^3-\mathrm{C}'-\mathrm{N}-\mathrm{C}^3-\mathrm{C}'-\mathrm{N}-\mathrm{C}^3-\mathrm{C}'-\mathrm{N}-\mathrm{C}^3-\mathrm{C}'-$ C-end

N-end $-\mathrm{N}-\mathrm{C}^3-\mathrm{C}'-\mathrm{N}-\mathrm{C}^3-\mathrm{C}'-\mathrm{N}-\mathrm{C}^3-\mathrm{C}'-\mathrm{N}-\mathrm{C}^3-\mathrm{C}'-\mathrm{N}-\mathrm{C}^3-\mathrm{C}'-\mathrm{N}-\mathrm{C}^3-\mathrm{C}'-$ C-end

$\beta\downarrow\uparrow$

N-end $-\mathrm{N}-\mathrm{C}^3-\mathrm{C}'-\mathrm{N}-\mathrm{C}^3-\mathrm{C}'-\mathrm{N}-\mathrm{C}^3-\mathrm{C}'-\mathrm{N}-\mathrm{C}^3-\mathrm{C}'-\mathrm{N}-\mathrm{C}^3-\mathrm{C}'-\mathrm{N}-\mathrm{C}^3-\mathrm{C}'-$ C-end

C-end $-\mathrm{C}'-\mathrm{C}^3-\mathrm{N}-\mathrm{C}'-\mathrm{C}^3-\mathrm{N}-\mathrm{C}'-\mathrm{C}^3-\mathrm{N}-\mathrm{C}'-\mathrm{C}^3-\mathrm{N}-\mathrm{C}'-\mathrm{C}^3-\mathrm{N}-\mathrm{C}'-\mathrm{C}^3-\mathrm{N}-$ N-end

N-end $-\mathrm{N}-\mathrm{C}^3-\mathrm{C}'-\mathrm{N}-\mathrm{C}^3-\mathrm{C}'-\mathrm{N}-\mathrm{C}^3-\mathrm{C}'-\mathrm{N}-\mathrm{C}^3-\mathrm{C}'-\mathrm{N}-\mathrm{C}^3-\mathrm{C}'-\mathrm{N}-\mathrm{C}^3-\mathrm{C}'-$ C-end

$\beta\uparrow\uparrow\downarrow$ N-end $-\mathrm{N}-\mathrm{C}^3-\mathrm{C}'-\mathrm{N}-\mathrm{C}^3-\mathrm{C}'-\mathrm{N}-\mathrm{C}^3-\mathrm{C}'-\mathrm{N}-\mathrm{C}^3-\mathrm{C}'-\mathrm{N}-\mathrm{C}^3-\mathrm{C}'-\mathrm{N}-\mathrm{C}^3-\mathrm{C}'-$ C-end

C-end $-\mathrm{C}'-\mathrm{C}^3-\mathrm{N}-\mathrm{C}'-\mathrm{C}^3-\mathrm{N}-\mathrm{C}'-\mathrm{C}^3-\mathrm{N}-\mathrm{C}'-\mathrm{C}^3-\mathrm{N}-\mathrm{C}'-\mathrm{C}^3-\mathrm{N}-\mathrm{C}'-\mathrm{C}^3-\mathrm{N}-$ N-end

Figure 7.6. Chain pathway and location of hydrogen bonds in the parallel ($\beta\uparrow\uparrow$), antiparallel ($\beta\uparrow\downarrow$) and mixed ($\beta\uparrow\uparrow\downarrow$) β-structures. As shown, in each β-strand, the H-bonds (light blue) of one residue are directed oppositely to those of its neighbor in the chain.

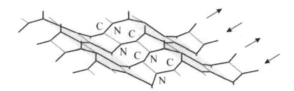

Figure 7.7. The β-sheet surface is pleated. The side-chains (shown as short red rods) are at the pleats and directed accordingly; i.e., the upward and downward side-chains alternate along the β-strand. The H-bonds are shown in light-blue. Adapted from [12].

There are also helices without any hydrogen bonds. Their tight (and hence, energetically advantageous) arrangement is stabilized by van der Waals interactions only. This is exemplified by a polyproline helix consisting of three chains; each chain forms a rather extended left-handed helix. Winding these three chains together forms a *right-handed superhelix*. Of two possible types of polyproline helix, that of interest to us is poly(Pro)II, since a helix of this kind is realized in collagen. In this helix, the Pro peptide groups are in the usual (*trans*) conformation. Let us postpone a detailed consideration of the collagen helix till due time and at the moment restrict ourselves to its overall view (Fig. 7.9) and mark in Fig. 7.4 the region corresponding to its conformation: one can see that it is close to the β-structural conformation.

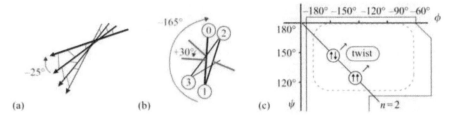

Figure 7.8. (a) The twist of the β-sheet. β-strands are shown as arrows and hydrogen bonds between them as light-blue lines. (b) Axial view of one turn of the β-strand. Side-chains are shown as circles; their numbers increase with increasing distance from the viewer. Blue lines indicate the direction of C=O groups involved in H-bonding in the sheet. The large arrow shows the turn of the β-strand as it comes closer to the viewer by one residue, and the small arrow shows the turn of similarly directed H-bonds when the β-strand comes closer to the viewer by two residues. (c) Conformation of the ideal (non-twisted) parallel ($\uparrow\uparrow$) and antiparallel ($\uparrow\downarrow$) β-structure for poly(Gly), and the averaged conformation of a real twisted β-structure (composed of L amino acids). The dashed line encircles the energy minimum for a separate Ala; the allowed region for its conformations is contoured by the red line. The diagonal of the $\phi\psi$-map corresponds to the flat regular structure with two residues per turn. Left-handed (L) helices are above the diagonal, and right-handed (R) ones are below it. Parts (a) and (c) are adapted from [12].

Figure 7.9. The overall view of poly(Pro)II, the right-handed superhelix composed of three left-handed helices.

The features of major regular secondary structures of protein chains are summed up in Table 7.1.

Apart from the regular secondary structures, in polypeptide chains there are irregular ones, i.e., standard structures that do not form long periodic systems.

The most important are the so-called β-turns (or β-bends; "β" in the name shows that they often bridge the neighboring β-strands in antiparallel β-hairpins). The appearances of most typical β-bends and conformations of their constituent residues are presented in Fig. 7.10. Compare Fig. 7.10c with Fig. 7.4 and Table 7.1 and pay attention to the fact that conformations of turn I (and especially of turn III) are close to that of the turn of a 3_{10}-helix.

Usually, bends comprise about half of the residues uninvolved in the regular secondary structures of the protein.

Another kind of irregular secondary structure is the β-bulge (Fig. 7.11). It is formed by a residue (or sometimes by a few residues) "inserted" in the β-strand and having a non-β-structural conformation. The bulge is typical only for edge strands of

Table 7.1. The main geometric parameters of the most abundant secondary structures in proteins.

Structure	H-bonding	Residues per turn	Shift per residue (Å)	ϕ	ψ
Helix α_R	$CO_0 - HN_{+4}$	+3.6	1.5	$-60°$	$-45°$
Helix $(3_{10})_R$	$CO_0 - HN_{+3}$	+3	2.0	$-50°$	$-25°$
Sheet $\beta\uparrow\downarrow$	Between chains[a]	-2.3	3.4	$-135°$	$+150°$
Sheet $\beta\uparrow\uparrow$	Between chains[a]	-2.3	3.2	$-120°$	$+135°$
Helix poly(Pro)II	No	-3	3.0	$-80°$	$+155°$

[a]The distance between strands in the β-sheet is 4.8 Å.

Data are from [4] and [12]. All values are approximate. In the "Residues per turn" column "+" denotes the right-handed helix, and "$-$" the left-handed.

a β-sheet (usually of an antiparallel β-sheet); it increases the twist of this sheet and bends it.

In conclusion, a few words on how the secondary structure is determined by experiment.

Of course, with X-ray (or accurate multi-dimensional NMR) protein structures available, the secondary structure can be derived from atomic coordinates. However, by detecting closely positioned H-atom nuclei (with less than 4–5 Å between them), NMR (nuclear magnetic resonance) reveals the secondary structure (mainly, α and β) even before the full atomic structure of the protein has become known.

NMR spectroscopy is based on applying radio-waves to excite the magnetic moments of nuclei aligned in a strong magnetic field. These nuclei must have an odd number of nucleons (protons and neutrons): then they have a magnetic moment, or spin. In proteins, these are natural "light" hydrogens (^1H), as well as introduced isotopes (^{13}C, ^{15}N, etc.). The magnetic resonance occurs at a radio frequency typical (in the given magnetic field) of the nucleus in question and slightly modified by its neighbors in chemical bonds and in space (which helps us to understand which atom of which residue is excited). The excitation can be propagated from the initial nucleus to a neighboring one (if it has a magnetic moment); the recipient will report on its excitation at its own frequency, thereby demonstrating the closeness of the two nuclei.

The characteristic feature of α-helices is the closeness of the H-atom of the $C^\alpha H$ group to that of the NH-group of the fourth residue down the chain (towards the C-terminus), while the typical feature of the β-structure is closeness between H-atoms of NH- and $C^\alpha H$-groups pertaining to immediate neighbors in the chain and to H-bonded residues in the β-sheet (Fig. 7.12).

However, the most important role in determining the secondary structure (mainly, α and β) is played by circular dichroism (CD).

For CD, a knowledge of the overall spatial structure of the protein is not required. On the contrary, structural studies of a protein are usually started with CD. This method is based on differing absorptions of clockwise and counterclockwise polarized light, which are caused specifically by helices of different handedness. Owing to this difference, plane-polarized light turns into elliptically polarized light.

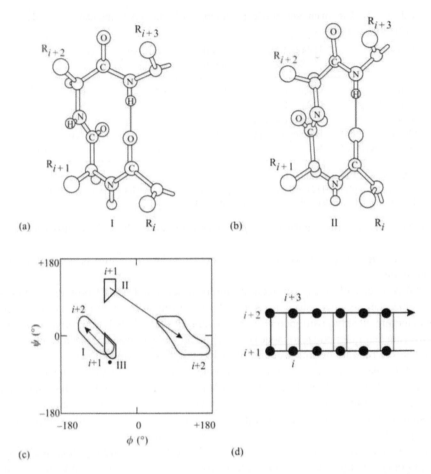

Figure 7.10. β-turns. (a) The β-turn of type I (type III looks very similar and therefore is not given here). (b) The β-turn of type II. It differs from the β-turn I mainly by the inverted peptide group between the residues $i + 1$ and $i + 2$. (c) Conformations of the residues $i + 1$ and $i + 2$ fixed by the H-bond closing the β-turns. In the β-turn III both these residues have the same conformation (denoted as bold dot). The conformations of residues i and $i + 3$ are not fixed in β-turns; they are fixed by the β-structure, when this structure extends the turn as shown in (d), which sketches a β-hairpin with the β-turn at its top. H-bonds are shown in light-blue. Parts (a), (b) and (c) are adapted from [12].

The typical ellipticity spectra for the "far UV region" (190–240 nm) are given in Fig. 7.13. These spectra depend on the asymmetry of the peptide group environment, and therefore report on the secondary structure.

The optical excitation of peptide groups in the far UV region occurs at a wavelength of about 200 nm. This wavelength is approximately twice as large as that required

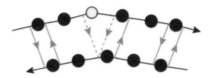

Figure 7.11. The β-bulge in an antiparallel β-structure. Hydrogen bonds are shown as blue arrows directed from N—H to O=C groups. All residues (shown as circles) have β-structural conformations, except for that shown as a yellow circle, whose conformation is often nearly α-helical. One or even both H-bonds adjacent to this residue (shown by broken arrows) may be broken.

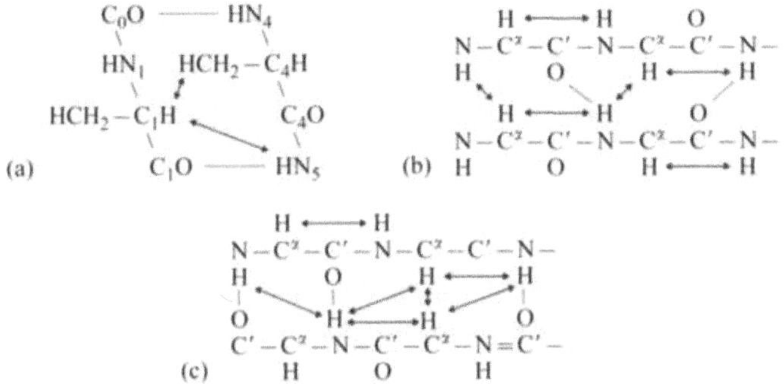

Figure 7.12. Approach ($\leftrightarrow$) of the nuclei of H-atoms characteristic of the α-helix (a), and of the parallel (b) and antiparallel (c) β-structure. In (a), indices at atoms of the main chain indicate the relative location of residues in the chain.

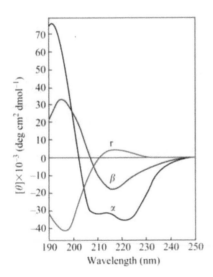

Figure 7.13. Typical far UV CD spectra for polylysine as the α-helix (α), β-structure (β), and random coil (r). Adapted from Greenfeld N.J., et al., *Biochemistry* (1969) **8**: 4104–4116.

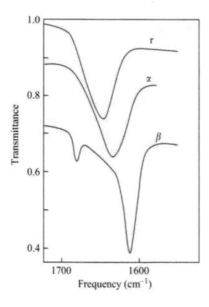

Figure 7.14. Typical infra-red (IR) transmittance spectra measured in heavy water (D_2O) for polylysine as the α-helix (α), β-structure (β), and random coil (r). The measurements were taken in the "amide I" region, reflecting vibrations of the C=O bond. Adapted from Susi H., *Methods Enzymol.* (1972) **26**: 455–472.

for excitation of separate atoms. The possibility to excite a peptide group with light of lower frequency is explained by delocalization of electrons over the group, which has been discussed previously.

In aromatic side groups, delocalization of electrons is still greater: here, they are "spread over" six atoms (while in peptide groups there are three atoms). The CD spectra for aromatic groups fall within the wavelengths of about 250–280 nm (although their "tails" can reach even 220 nm). In this range of 250–280 nm (in the "near-UV region") the asymmetry of aromatic side-chain environments is studied, i.e., the effects characteristic of not the secondary but the tertiary structure of the protein.

In passing, I could add that when the electron is delocalized still more (i.e., in larger molecules with double partial bonds) its excitation moves from UV to visible (400–600 nm) light, and such molecules become dyes.

Apart from UV spectra, IR spectra can be exploited to reveal the secondary structure of polypeptides and proteins. They reflect the difference in vibration of peptide groups involved and uninvolved in various secondary structures (Fig. 7.14). These measurements are more complicated than study of UV spectra, since the absorption range of ordinary water (H_2O) is nearly the same; therefore, heavy water (D_2O) is used. Also, these measurements, compared with UV ones, require more protein to be used and a higher protein concentration in solution.

LECTURE 8

At this point we could pass on to the formation and decay of the secondary structure. However, prior to that I would like to talk about the fundamentals of statistical physics, thermodynamics, and kinetics "in general", inasmuch as otherwise it is difficult to discuss the stability of the secondary structure, the stability of proteins, cooperative transitions in polypeptides and proteins and the kinetics of these transitions.

Thermodynamics provides an idea of the types of possible cooperative transitions in systems incorporating a great many particles. Statistical physics advises on when and what transitions may occur in the system of particles in question and gives details of these transitions based on the properties of the studied particles and their interactions.

First of all we will consider the main concepts of statistical physics and thermodynamics, namely, entropy, temperature, free energy and partition function.

Systems with numerous degrees of freedom (i.e., comprising a lot of molecules or even just one large and flexible molecule) are described by means of statistical physics. It is "statistical" because such a large system has zillions of configurations. Here is an illustrative example. If each of N links of a chain may have only two configurations (e.g., "helical" and "extended" ones), then the whole N-link chain has 2^N possible configurations. In other words, a "normal" 100-link protein chain may have at least 2^{100} configurations, that is, about 1 000 000 000 000 000 000 000 000 000 000 of them. This is an enormous number. Consideration of all of them at a rate of 1 nanosecond (10^{-9} s) per configuration would take 3×10^{13} years, that is more than one thousand lifetimes of the Universe ... And in the experimental tube there are billions of such chains, not to mention the solvent. If we were going to consider all their configurations we would have been lost forever. Certainly, we are interested in much more simple and reasonable things, like the average (i.e., *statistically averaged*) helicity of chains and its change upon heating. A peculiar feature of statistical averaging (i.e., neglecting all minor details) is a crucial simplification of events.

In statistical averaging the major role is played by *entropy*. It shows how many configurations (in other words, *microstates*) of the system correspond to its observed *macrostate* (averaged by the observation time and the number of molecules studied). We considered a similar example earlier, though it was only a special case with the number of *microstates* of a molecule limited by a volume, and the entropy of the molecule proportional to the logarithm of this volume (and the molecule's being in the given volume, e.g., "in this room", was the "macrostate" of the molecule).

Here, it would be timely to answer the question that is only natural: why do physicists prefer to consider the *logarithm* of microstate number but not the number itself? The answer is that in considering many separate systems (e.g., separate molecules) we have to sum up their energies and degrees of freedom, while the numbers of accessible states are to be multiplied (if one molecule has 10 microstates and another as many, these two molecules have 100 different combinations of microstates). This makes the calculations inconvenient and too bulky. But *logarithms* of the numbers of accessible states are to be summed up (you remember that $\ln(AB) = \ln(A) + \ln(B)$), like energies or volumes. Therefore, the logarithms are convenient for calculations. Moreover, the additivity of logarithms allows us to use the potent differential calculus.

Now let us talk about *temperature*. It is closely connected with entropy: *where there's no numerous states (no entropy), there's no temperature either.*

To clarify this connection, let us consider a closed system (with no energy exchange with the environment). Let its total energy be E, and let this system be in equilibrium, that is, all its microstates with the energy E are equally probable (and those of different energy $E' \neq E$ have zero probability).

Let us pick out an "observed small part" of the system (e.g., a molecule in gas or a macromolecule with its liquid environment). Then the rest of the system may be regarded as a thermostat in which the "small part" is immersed.

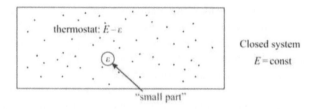

Let us divide all the system's microstates having the total energy E into such classes that each of them corresponds to one microstate of the chosen "small part". The more microstates incorporated in a given class, the higher probability of observation of this class, i.e., the higher the probability of observation of a certain state of our "small part".

Let a microstate of our "small part" be given (e.g., let a molecule in gas have a certain position in space and a certain speed). Let its energy be ε. Since the system ("small part" + "thermostat") is closed, *its total energy is conserved* (as stated by the energy conservation law), and thus the thermostat energy is $E - \varepsilon$. Let as many as $M_{\text{therm}}(E - \varepsilon)$ of thermostat microstates correspond to this energy. Then the probability of observation of this state of our "small part" is simply proportional to $M_{\text{therm}}(E - \varepsilon)$.

Remark. Here an implicit assumption has been made that a certain microstate of the system produces no effect on thermostat microstates. Strictly speaking, this is not quite true (or rather, this is true only for an ideal gas as the thermostat), but taking into consideration all events at the border of "our system" and the thermostat would obscure the entire narration. So, to combine strictness and clearness, let us assume for the time being that our "observed part" is encapsulated and thereby separated from thermostat molecules; if required (not in these lectures), the interaction between the "observed part" and the thermostat can be considered separately.

With the number of the thermostat states equal to $M_{therm}(E - \varepsilon)$, its logarithm *by definition* is proportional to the thermostat entropy:

$$S_{therm}(E - \varepsilon) = \kappa \ln [M_{therm}(E - \varepsilon)] \tag{8.1}$$

The coefficient κ is used here only to have the entropy measured in cal K^{-1}, as usual; as you will see later, it appears to be simply the Boltzmann constant.

The energy of the "small part" ε must be relatively small as well. Therefore, we can use an ordinary differential expansion of $S_{therm}(E - \varepsilon)$ over the small parameter ε (you may remember that $f(x_0 + dx) = f(x_0) + (df/dx)|_0 \, dx + \frac{1}{2}(d^2f/dx^2)|_0(dx)^2 + \cdots = f(x_0) + (df/dx)|_0 \, dx$ at a small dx with $(df/dx)|_0$ meaning that the derivative df/dx is taken at the point x_0). So,

$$S_{therm}(E - \varepsilon) = S_{therm}(E) - (dS_{therm}/dE)|_E \varepsilon \tag{8.2}$$

Note. Since both S and E are proportional to the number of particles, dS_{therm}/dE is independent of the number of particles in the thermostat, while d^2S_{therm}/dE^2 is *inversely* proportional to this number, i.e., $d^2S_{therm}/dE^2 \to 0$ in a very large thermostat; this allows us to neglect members of the order of ε^2 (as well as ε^3, etc.) in eq. (8.2).

Thus, the number of accessible thermostat microstates depends on the energy ε of our "small part" as

$$M(E - \varepsilon) = \exp \left[\frac{S_{therm}(E - \varepsilon)}{\kappa} \right]$$

$$= \exp \left[\frac{S_{therm}(E)}{\kappa} \right] \times \exp \left\{ -\varepsilon \left[\frac{(dS_{therm}/dE)|_E}{\kappa} \right] \right\} \tag{8.3}$$

Here, neither the common multiplier $\exp[S_{therm}(E)/\kappa] = M(E)$ nor the number $(dS_{therm}/dE)|_E$ depends on ε or on a concrete microstate of our "small part" in general.

Since the number of microstates must increase with increasing energy (the higher the energy, the greater the number of ways it can be divided), eq. (8.3) expresses the following simple idea: the more energy taken from the thermostat by our "small part", the less energy kept by the thermostat, and hence, the smaller the number of ways it can be divided. Moreover, this equation shows that the decrease of the number of accessible thermostat microstates (the number of ways to divide its energy) depends *exponentially* on the energy of our "small part".

Conclusion: The probability of observation of a certain microstate of our "small part" (molecule, etc.) is proportional to $\exp[-\varepsilon\{(dS_{therm}/dE)|_E/\kappa\}]$, where ε is the energy of this "small part", and the magnitude $(dS_{therm}/dE)|_E/\kappa$ depends not on the "small part" but only on averaged features of its environment.

But according to Boltzmann, the probability of a molecule keeping a certain state with energy ε is proportional to $\exp(-\varepsilon/k_B T)$ (where T is temperature, and k_B is the Boltzmann constant). A comparison of the identical expressions $\exp[-\varepsilon\{(dS_{therm}/dE)|_E/\kappa\}]$ and $\exp(-\varepsilon/k_B T)$ yields

$$(dS_{therm}/dE)|_E = \frac{1}{T}, \tag{8.4}$$

and κ in eqs (8.1) and (8.3) turns out to be the Boltzmann constant (k_B), provided the energy is measured (as usual) in Joules (or in calories), temperature in K, and entropy in $J\,K^{-1}$ (or in cal K^{-1}).

Equations (8.1) and (8.4) are the main equations of statistical physics and thermodynamics: they define the temperature as the reciprocal of the rate of entropy (or of the logarithm of the number of microstates) change with the system energy E.

In particular, they show that $\ln[M(E + k_B T)] = S(E + k_B T)/k_B = [S(E) + (k_B T)(1/T)]/k_B = \ln[M(E)] + 1$, i.e., an energy increase by $k_B T$ results in an "e" $(= 2.72)$-fold (approximately three-fold) increase of the number of microstates, *independently* of the system size, its inner forces, etc.

They also allow us to find the corresponding temperature value for each energy of any large system ("thermostat"), provided we know the number of its microstates at different energies (also called the "energy spectrum density"), or rather, the dependence of the logarithm of the energy spectrum density on this energy. The schematic diagram is given in Fig. 8.1.

It is essential that the microstates are highly numerous because the derivative dS/dE can be taken only when a small (as compared with $k_B T$) energy interval

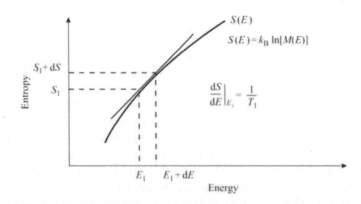

Figure 8.1. Determination of the temperature of a large system. The bold curve shows the dependence of entropy S on the system energy E. The curve slope, dS/dE, determines the temperature T corresponding to this energy E. $M(E)$ is the number of microstates with the energy E, i.e. the density of the energy spectrum of the system.

houses many microstates. That is why temperature appears only in sciences dealing with enormous numbers of accessible states.

Let us continue considering a small system in the thermostat whose temperature is equal to T. Equations (8.1) and (8.3) show that for our system the probability of being in a given state i with the energy ε_i at temperature T is

$$w_i(T) = \frac{\exp(-\varepsilon_i/k_B T)}{Z(T)} \tag{8.5}$$

where

$$Z(T) = \sum_j \exp(-\varepsilon_j/k_B T) \tag{8.6}$$

is the normalization factor which takes into account that the sum of probabilities of all states, $\sum_j w_j$, is necessarily equal to 1 (here and above the sum $\sum_j$ is taken over all microstates j of the studied "small system").

The value Z is called *partition function* for the studied system. Provided Z is known, eq. (8.5) allows us to calculate the probability of each microstate of this system at a given temperature. Then the average energy of the system at this temperature

$$E(T) = \sum_j w_j \varepsilon_i, \tag{8.7}$$

and its average entropy

$$S(T) = k_B \sum_j w_j \ln(1/w_j). \tag{8.8}$$

Note that eq. (8.8) averages $\ln(1/w_j)$ over all microstates j of the system, allowing for their probabilities w_j. This equation provides a direct generalization of the determination of entropy $S = k_B \ln[M(E)]$ already familiar to us (see eq. (8.1)). This is a generalization of averaging for the case when all states of the system have several probabilities, and not just two probabilities, and follows from the energy conservation law: $w_j = 1/M(E)$ for all $M(E)$ states where $E_j = E$ and $w_j = 0$ when $E_j \neq E$.

Inner voice: I feel that the meaning of eq. (8.8) must be explained better, and that the term $S(T)/k_B$ should be proved to be the logarithm of the average number of system states.

Lecturer: Please be tolerant and we will go into some math ...

Let us consider a large number N of systems each of which may be in the state "1" with the probability w_1, in the state "2" with the probability $w_2, \ldots$, in the state "J" with the probability w_J. Then on average, among N systems considered, those in the state "1" amount to $n_1 = w_1 N$, those in the state "2" amount to $n_2 = w_2 N$, and so on (while $\sum_j w_j N = N$).

In how many ways can these N systems be distributed over the J states so that n_1 of them are in the state "1", n_2 in the state "2" and so on?

This is how this can be calculated. Let each system have its number: 1, 2, $\ldots$, N. Then let us select those n_1 systems that are in the state "1". How many ways are there to do so? Since the first selected system is to be picked up from N systems, the second from the remaining $N - 1$ systems, ..., the n_1th from $N - (n_1 - 1)$, *the given order of selection* allows us to select n_1 systems out of N by $N(N-1)\cdots[N-(n_1-1)]$ ways. But it is of no importance for us in what order the systems were picked up (first the 5th system and then the 10th one, or vice versa): these selections yield *the same result*. And since n_1 systems can be selected using $n_1(n_1 - 1)\cdots 1$ orders of actions (first we can pick up the first one out of all n_1 systems, then the second one out of the rest $(n_1 - 1)$ of them, and so on), selection of n_1 from N systems gives $\{N(N - 1)\cdots[N - (n_1 - 1)]\}/\{n_1(n_1 - 1)\cdots 1\}$ *different* results. Having denoted $n(n - 1)\cdots 1$ as n! (it reads as "n factorial"), we see that selection of n_1 systems out of N gives $N!/[(N - n_1)!n_1!]$ different results.

Then we select n_2 systems out of the remaining $N - n_1$ systems. According to the above, this can be done by $(N - n_1)!/[(N - n_1 - n_2)!n_2!]$ ways for each way of selecting n_1 out of N systems as the first step. In total, there are $\{N!/[(N - n_1)!n_1!]\}\{(N - n_1)!/[(N - n_1 - n_2)!n_2!]\} = N!/[(N - n_1 - n_2)!n_2!n_1!]$ different ways of selecting n_1 systems in the state "1" and n_2 systems in the state "2" from N systems. Similarly, we can conclude that there are $N!/[n_J!\cdots n_2!n_1!]$ different ways to have n_1 out of N systems in the state "1", n_2 in the state "2", ..., n_J in the state "J" (let me remind you that $n_J = N - n_1 - n_2 - \cdots - n_{J-1}$).

Now we can use Stirling's approximation, which states that $n! \approx (n/e)^n$. Then

$$N!/[n_J!\cdots n_2!n_1!] \approx (N/e)^{(n_J+\cdots+n_1)}/[(n_J/e)^{n_J}\cdots(n_1/e)^{n_1}]$$

$$= (N/n_J)^{n_J}\cdots(N/n_1)^{n_1} = (1/w_J)^{Nw_J}\cdots(1/w_1)^{Nw_1} \quad (8.9)$$

In the last equation, we used the relationship $n_1 = w_1 N$, etc.

Finally, we obtain the number in question (that is, the number of distributions of N systems over J states yielding n_1 of them in the state "1", n_2 in the state "2", etc.) from eq. (8.9) as $[1/(w_J^{w_J}\cdots w_1^{w_1})]^N$. And since this number (for N independent systems) is simply the average number of states of one system to the Nth power, the average number of states of one system is equal to $1/(w_J^{w_J}\cdots w_1^{w_1})]$, and its logarithm is (as it has been expected to be) the term $S(T)/k_B$ of eq. (8.8). That's all.

The partition function plays a most important role in statistical physics because the quantity $Z(T)$ allows direct calculation of the free energy of a system enclosed in a fixed volume:

$$F(T) = E(T) - TS(T)$$

$$= \sum_j w_j \left\{\varepsilon_j - T\left[-k_B \ln(w_j)\right]\right\}$$

$$= -k_B T \ln[Z(T)] \quad (8.10)$$

(in the last but one transformation we used eq. (8.5)).

Then the system entropy can be found directly as $S(T) = -dF/dT$ (we have already seen this equation, and I want you to check it up by yourselves using definitions derived from eqs (8.10), (8.6), (8.5) and (8.8).

Now we can find the energy of the system: $E(T) \equiv F(T) + TS(T) = F(T) - T(dF/dT)$.

Important though secondary notes:

1. If the "small system" has many degrees of freedom, it has its own entropy and therefore its own (internal) temperature. The internal temperature of the "small system", T_{in}, is equal to the thermostat temperature T because, as follows from the above definitions and equations, $T_{in} \equiv dE(T)/dS(T) = d[F(T) - T(dF/dT)]/d(-dF/dT) = [dF/dT - T(d^2F/dT^2) - dF/dT] dT/(-d^2F/dT^2) dT = T$.

2. The total energy incorporates kinetic and potential energies. The former depends on particle speeds only, while the latter on their positions in space, and not on the speeds. The "microstate" of each particle is determined by its coordinate in space and by its speed. In classical (not quantum) mechanics, any combinations of speeds and coordinates are allowed (as we know, Heisenberg's quantum ratio, $\Delta v \Delta x \approx \hbar/m$, imposes restrictions on the speed–coordinate combinations, but at room temperature this is important for very light particles, i.e., virtually electrons and protons only). This means that probabilities of coordinates and speeds can be "uncoupled", i.e., $w(\varepsilon_{kinet} + \varepsilon_{coord}) \sim \exp(-\varepsilon_{kinet}/k_B T) \times \exp(-\varepsilon_{coord}/k_B T)$. Further simple calculations (you can make them yourselves) will show that free energies, energies and entropies can also fall into kinetic and coordinate parts, i.e., $F = F^{kinet} + F^{coord}$, etc. It is important that kinetic parts are *independent* of the system configuration (actually, they only contribute constantly to heat capacity) and can be neglected when considering conformational changes. **Therefore, further on we will discuss *only* configurational (or "conformational") energy spectra, energies, entropies, etc**.

3. Above, we summed over microstates, whereas in the frame of classical mechanics we can equally well integrate over coordinates and speeds that determine the microstate of each particle.

4. Equilibrium temperature must be *positive*. Otherwise, probability integration over speeds, i.e., $\int \exp(-mv^2/2kT) \, dv$, turns into infinity at great speeds, and the system "explodes".

   Therefore, the stable state cannot be observed in those conformational energy spectrum regions where the spectrum density (and, hence, the entropy of the system) decreases with increasing energy: for these regions, $T < 0$ (see eqs (8.4) and (8.1), and Fig. 8.2a).

5. The quantity $k_B T$ is measured in units like "energy per particle" or "energy per mole ($= 6.02 \times 10^{23}$) of particles". Had temperature been expressed in energy units from the very outset, Boltzmann's constant k_B would never have been used at all. However, historically it happened that the "degree" was first introduced as a temperature unit, and then it became evident that it could be easily converted into something like "energy per particle" through multiplying by a certain (Boltzmann) constant. Accordingly, k_B is measured in the units "energy per particle per degree". Its numerical value depends on the energy unit used: "joule per particle", "calorie per mole of particles", etc. In accordance with the measured energy cost of a "degree" (K), $k_B = 1.38 \times 10^{-23}$ joule particle^{-1} K^{-1}. However, apart from "per particle", k_B can be calculated per mole ($= 6.02 \times 10^{23}$) of particles. To do this we can multiply and divide k_B by 6.02×10^{23}, and, since [6.02×10^{23} particles] = one mole of particles, we have: $k_B = 1.38 \times 10^{-23}$ (joule/particle)/degree

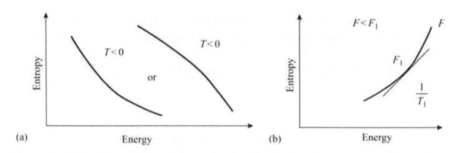

Figure 8.2. Regions of the $S(E)$ plot that do not correspond with any stable state of the system. (a) Regions where entropy $S(E)$ decreases with increasing energy (E): here $T = 1/(dS/dE) < 0$. (b) A "concave" region of the $S(E)$ plot: here at the point of contact (where $dS/dE = 1/T_1$) the free energy $F = E - TS$ is not lower but higher than that of neighboring sections of the plot $S(E)$; cf. Fig. 8.3.

   $= 1.38 \times 10^{-23}$(joule $\times |6.02 \times 10^{23}|/[6.02 \times 10^{23}$particles$|$)degree^{-1} $= [1.38 \times 10^{-23} \times 6.02 \times 10^{23}|$(joule/mol of particles)degree^{-1} $= 8.31$(J/mol)degree^{-1} $= 1.99$ (cal/mol)degree^{-1} (since 1 cal $= 4.18$ J). The last value, $1.99\,\mathrm{cal\,mol^{-1}\,K^{-1}}$, is traditionally referred to as the "gas constant" R.

6. Specific entropy is often measured in entropy units, "e.u." : e.u. $=$ cal mol^{-1} K^{-1} $= k_B/1.99 \approx k_B/2$. Hence, 1 e.u. corresponds to $e^{1/2} \approx 1.65$ states per particle of the system.

Now at last we can start considering ***conformational changes***. It was in order to give them competent consideration that we reviewed the fundamentals of statistical physics and thermodynamics.

Conformational changes can be gradual or sharp (the latter are called "phase transitions"). Let us try and distinguish one kind from the other using energy spectrum density plots like those presented in Fig. 8.1. First of all we have to learn how to locate the stable state(s) of our system at a given temperature of the medium.

Let the number of states of our system ("macromolecule") be $M(E)$ when its energy is E, and let the medium ("thermostat") temperature be T_1. Given that we have the plot of entropy $S(E) = k_B \ln[M(E)]$ and know T_1, how can we find the plot's point corresponding to the stable state of our system? The temperature T_1 determines the plot's slope dS/dE at the sought point: here $dS/dE = 1/T_1$ since the temperature of our macromolecule is equal to the medium's temperature. Thus, the corresponding tangent to the curve $S(E)$ gives us the possibility of finding the point we are looking for.

However, there may be several such points with the given plot slope of $1/T_1$ (see below, Fig. 8.5). Which of them corresponds to the stable state? Let us consider the tangent at the point E_1 where $dS/dE|_{E_1} = 1/T_1$. The equation describing such tangent is $S - S(E_1) = (E - E_1)/T_1$, or, what is the same, $E - T_1S = E_1 - T_1S(E_1)$. The value $F_1 = E_1 - T_1S(E_1)$ is simply the system's free energy at temperature T_1. As seen, along the tangent the value $E - T_1S$ is constant. Everywhere to the left of the

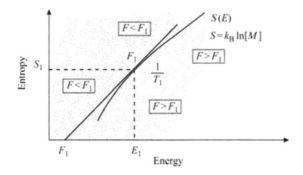

Figure 8.3. Graphical definition of temperature and free energy. The bold curve $S(E)$ shows the dependence of entropy $S = k_B \ln(M)$ on the energy E. $M(E)$ is the energy spectrum density of the system's states. The slope of $S(E)$ determines the temperature T : $dS/dE = 1/T$. Physically possible temperatures $T > 0$ correspond to a rise in $S(E)$. F_1 is the free energy (corresponding to the given energy spectrum) at temperature T_1 (that determines the tangent to the $S(E)$ curve). The magnitude $F = E - T_1 S$ is constant along the tangent and equal to $F_1 = E_1 - T_1 S_1$. At the left and above the tangent, $F = E - T_1 S < F_1$, while at its right and below, $F = E - T_1 S > F_1$. Since here the curve $S(E)$ is convex, i.e., it is below the tangent, the contact point corresponds to the minimum free energy at temperature T_1. For other explanations, see text.

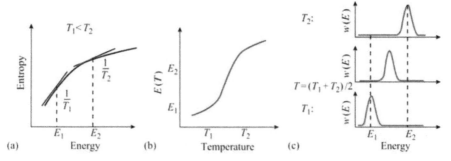

Figure 8.4. A gradual change of the system state with changing temperature (from T_1 to T_2). The requirement is convexity of the curve $S(E)$. With $S(E)$ known (a), $T(E)$ and then $E(T)$ can be found (b). $w(E)$ (c) is the probability of having energy E at the given temperature T; $w(E)$ is proportional to $\exp[-(E - TS(E))/k_B T]$.

tangent the value of $E - T_1 S$ is lower and everywhere to its right higher than on the tangent itself (Fig. 8.3).

Specifically, the latter means that concave regions of $S(E)$ (Fig. 8.2b) cannot correspond to a stable state, since at the contact point, F is not lower but higher than that in the neighboring sections of the curve $S(E)$. The latter means that the system can decrease its free energy (i.e., shift towards a more stable state) by moving from the contact point along the concave curve.

If the curve $S(E)$ is convex along its entire length, then its slope decreases with increasing energy E. Hence, each value of slope $1/T$ corresponds to only one point of the curve (Figs 8.1 and 8.4), that is, this point corresponds to the sought point for the stable state at the given temperature T. As the temperature changes, this point gradually moves and the system gradually changes its thermodynamic state, i.e., its entropy and energy (Fig. 8.4).

If the $S(E)$ slope alternatively decreases and increases with increasing energy E (Fig. 8.5), then there may be several tangents with the same slope, and the contact point of the extreme left of these tangents (with the lowest F) reflects the most stable state. Up to a certain temperature T^*, the "best" tangent (with the lowest F) will be one in the region of small energies, whereas beyond T^* the "best" tangent will be found in the region of greater energies (and greater entropies).

At temperature T^*, structures with low and high energies will have equal free energies and equal probabilities of existence. This means, that among many identical systems at transition temperature T, half of them are in the low-energy state while the other half is in the high-energy state (see Fig. 8.5, right). In other words, half of the time each system is in the high-energy state, and for the other half it is in the low-energy state. This "co-existence" of two systems, equally probable but utterly different in energy, will occur within a certain very narrow (specifically, for large "macroscopic" systems) temperature range around T^*; we will estimate it soon.

It is of major importance, that the states with "medium" energies will not be displayed as probable states of the system, since, owing to the $S(E)$ plot concavity, the points for "intermediate" states lie on the right of the tangent corresponding to the transition temperature T^*. In other words, at temperature T^*, the free energy of these

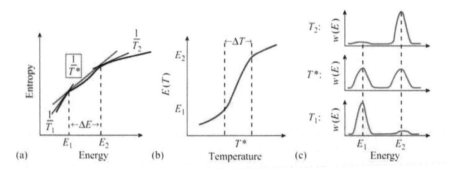

Figure 8.5. Phase transition of the "all-or-none" type (in macroscopic bodies it is called a "first-order phase transition") is characterized by a sharp change of the system's state with changing temperature. The requirement is a concave region on the curve $S(E)$ in (a): at the transition temperature T^*, the free energy at the center of this region is higher than at its flanks (this region corresponds to unstable states of the system, see Fig. 8.2b). The transition occurs within a narrow temperature range (ΔT) corresponding to the co-existence (i.e., approximately equal probabilities) of low- and high-energy states. The tangents correspond to the mid-transition temperature T^*, as well as to $T_1 < T^*$ and $T_2 > T^*$. Note that temperature T^* of the "all-or-none" transition (i.e., of the first-order phase transition) exactly coincides with the middle of the sharp energy change (b) and with maximum splitting of the distribution of probability $w(E)$ over the states (c).

"intermediate" states is higher than that of structures corresponding to both contact points, and the probability of manifestation of the intermediates is nearly zero. Then the two stable states are said to be separated by a "free energy barrier".

These are conditions for the "all-or-none" transition.

In microscopic systems (in proteins, in particular) this transition can occur as a jump over the free energy barrier at the transition temperature T^*. We will discuss this at the end of the lecture.

In macroscopic systems (e.g., in a tube with freezing water) the barrier (at temperature T^*) is so high that it would take almost infinity to overcome it. Therefore, macroscopic systems possess *hysteresis*, i.e., they preserve their state up to slight overcooling (when freezing) or overheating (when melting) as compared with temperature T^*; after that the transition runs through a temporary (i.e., unstable) phase separation in the system (e.g., into liquid and solid). Such a transition in macroscopic systems is called a *first-order phase transition*.

Now I would like to estimate the temperature interval corresponding to co-existence of low- and high-energy states of the system. In the middle of the transition, at temperature T^*, the free energy of the low-energy phase, $F_1(T^*) = E_1 - T^*S_1$, is equal to the free energy of the high-energy phase, $F_2(T^*) = E_2 - T^*S_2$. That is, in the middle of transition

$$(E_2 - E_1) = T^*(S_2 - S_1) \tag{8.11}$$

At a small (δT) temperature deviation from T^*, the free energies of the phases change slightly. The difference between them amounts to

$$\delta F = F_1(T^* + \delta T) - F_2(T^* + \delta T)$$
$$= [F_1(T^*) + (dF_1/dT)\delta T] - [F_1(T^*) + (dF_1/dT)\delta T]$$
$$= -S_1\delta T + S_2\delta T = \delta T(S_2 - S_1) \tag{8.12}$$

The phases co-exist (i.e., their probabilities are nearly equal: say, the probability ratio varies from $10 : 1$ to $1 : 10$) as long as $\exp(-\delta F/k_B T^*)$ is between about 10 and 1/10, i.e., as long as $\delta F/k_B T^*$ is somewhere between $\ln(10) \approx +2$ and $\ln(1/10) \approx -2$. In this region δT is somewhere between $+2k_B T^*/(S_2 - S_1)$ and $-2k_B T^*/(S_2 - S_1)$. So, the temperature interval of phase co-existence is

$$\Delta T \approx \frac{2k_B T^*}{S_2 - S_1} - \left(\frac{-2k_B T^*}{S_2 - S_1}\right)$$
$$= \frac{4k_B T^*}{S_2 - S_1}$$
$$= \frac{4k_B (T^*)^2}{E_2 - E_1} \tag{8.13}$$

Let us consider an instructive numerical example. When $T^* \sim 300\,\mathrm{K}$ (i.e., $k_B T^* \sim 0.6\,\mathrm{kcal\,mol^{-1}}$) and $E_2 - E_1$ amounts to about $50\,\mathrm{kcal\,mol^{-1}} \approx 100\,k_B T^*$, which is typical of protein melting (i.e., $E_2 - E_1 \approx 50/(6 \times 10^{23}) \sim 10^{-22}$ kilocalories per protein particle), then ΔT is about ten degrees. This means that molten and intact protein molecules co-exist in the range of a few degrees around the melting mid-point. However, when $E_2 - E_1$ is about 50 kilocalories per system (equivalent to melting a piece of ice as big as a bottle), then the co-existence range, ΔT, is about 10^{-23} degrees only.

In other words, "all-or-none" phase transitions of small systems are characterized by an energy jump within a narrow temperature range. This is true to a much greater extent for first-order phase transitions in macroscopic systems. In these systems the range of the jump is almost infinitely narrow, while for macromolecules it covers several degrees (and thus still remains narrow as compared with the usual "observation range" from $0\,°C$ to $100\,°C$). And in small oligopeptides no jump is observed: here the range of energy change can cover the entire "experimental window" of the temperatures studied.

A few words should be added concerning *second-order* phase transitions. Whereas first-order phase transitions are characterized by a *jump in the energy* of the system (along with its entropy, volume and density), the typical feature of a second-order phase transitions is *an abrupt change* in the slope of the function $E(T)$, that is *a jump in the heat capacity.*

It should be stressed that a second-order transition occurs at the abrupt change in the $E(T)$ curve (Fig. 8.6) and *not* in the middle of the subsequent more-or-less S-shaped dependence of energy (or other observed parameter) on the temperature. Such a transition is typical, for example, for the situation when the system changes up to a certain temperature, and then this change stops (e.g., in ferromagnetism, at a temperature below the Curie point, the heat is used to destroy the spontaneously magnetized state; and at a temperature above the Curie point, with the spontaneous magnetization already removed, the added heat is spent to increase the fluctuations only).

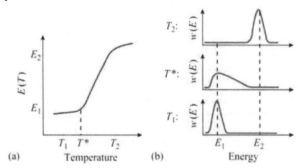

Figure 8.6. Typical appearance of (a) the energy, $E(T)$, and (b) the energy distribution function, $w(E)$, for a second-order phase transition. T^* is the temperature of this phase transition. At this temperature, the distribution of energies, $w(E)$, becomes dramatically expanded, and the energy *begins* (or *stops*, depending on the direction of temperature change) its rapid change. Note that the second-order phase transition temperature T^* coincides with the *beginning* of this dramatic energy change (and not with the middle, which is typical for "all-or-none" transitions shown in Fig. 8.5).

Figures 8.4–8.6 show (deliberately) the cases when the system first changes slowly with increasing temperature, then rapidly, and then slowly again. These are selected to show that such "S-shaped" behavior of $E(T)$ is compatible with both a first-order phase transition and a gradual change of the system in the absence of any phase transition at all, as well as with a gradual change triggered by a second-order phase transition.

I ask you to pay attention to these examples because for unknown reasons some non-physicists are prejudiced that a "transition" always corresponds exactly in the middle of any S-shaped profile, and that any "transition" that is not first order is second order. This is not true.

Note that although the curves $E(T)$ in Figs 8.4–8.6 are alike, the behavior of the distribution curves, $w(E)$, for first-order ("all-or-none") transitions is *utterly different* from the others. It is *only* "all-or-none" transitions that are characterized by two peaks on $w(E)$ curves reflecting the *co-existence* of two phases. Therefore, to see whether the transition between two extreme states is jump-like or gradual, it is *insufficient* to register a rapid change of energy or any other parameter in a narrow temperature range. Some additional measurements are required, which will be the subject of discussion in a future lecture.

Now let us talk about the *kinetics* of conformational changes, or rather, about why their rate is sometimes extremely slow. What is "slow" here? Suppose you know the rate of one elementary step of the process. For example, it takes a residue 1 ns to join the secondary structure. Also, you know that the chain contains 1000 residues. But the entire process takes not 1000 ns but 1000 s. That's what is "slow": orders of magnitude slower than expected from the rate of the steps and their necessary number. We have to understand why there should be this difference.

In some processes, the slow rate is caused by slow diffusion at high viscosity. However, the slow rate often – not always but often – results from the necessity of overcoming a high *free-energy barrier*. This is a very characteristic feature of "all-or-none" transitions, where the free-energy barrier separates two phases (Fig. 8.5); its value is always considerable here.

This barrier is very similar to the activation energy barrier of chemical reactions, although here it has both energy and entropy constituents. Let me remind you how to estimate the rate of such "barrier-overcoming" reactions using the classical theory of transition states.

At first, let us consider the simplest process of a system transition from state 0 to state 1. On the pathway $0 \rightarrow 1$ let there be one barrier # (Fig. 8.7) and no "traps" (Fig. 8.8; the presence of "traps" complicates the process, but its nature remains unchanged).

Then the process rate is determined (a) by the population of the barrier ("transition") state, and (b) by the rate of transition from the barrier to the post-barrier state.

If $\Delta F^{\#}$ is the free energy of the barrier relative to the free energy of the initial state, with $\Delta F^{\#} \gg k_{B}T$, if there are no "traps" (Fig. 8.8), and if n "particles" (molecules,

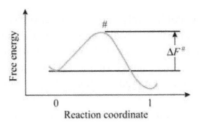

Figure 8.7. The free-energy (activation) barrier # in the transition from the state "0" to state "1". $\Delta F^{\#}$ is the free energy of the barrier (i.e., of the "transition state").

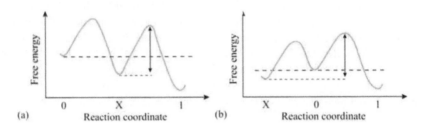

Figure 8.8. A "trap": this may be either the on-pathway intermediate "X" (a) or an out-of-pathway (for the reaction $0 \rightarrow 1$) state "X" (b). The kinetic trap occurs provided it is more stable than the initial state "0" but less stable than the final state "1", and there is a higher free energy barrier (shown with an arrow) between "X" and "1" than between "0" and "X".

systems, etc.) are in the initial state, then due to fluctuations, $n^{\#} \approx n \, \exp(-\Delta F^{\#}/k_B T)$ particles are on the barrier. Let it take each of these the time τ to jump from the barrier (τ being the time of "an elementary reaction step"). Then, during a time interval of about τ, all $n^{\#}$ "on-barrier" particles will cross the barrier. For all of the n particles, the time of $n/n^{\#}$ elementary steps is required, and this time amounts to about

$$t_{0 \rightarrow 1} \approx \tau(n/n^{\#}) \approx \tau \, \exp(+\Delta F^{\#}/k_B T) \tag{8.14}$$

The reciprocal $k_{0 \rightarrow 1} \equiv 1/t_{0 \rightarrow 1}$ is called the rate of transition from 0 to 1.

With the "trap" X present, the $0 \rightarrow X$ and $X \rightarrow 1$ transition time is estimated in the same way. Then the total time of transition from 0 to 1 is the sum of the times of $0 \rightarrow X$ and $X \rightarrow 1$ transitions.

Note two further points.

First, if there are several transition pathways used *in parallel* (Fig. 8.9a), their *transition rates* must be *summed*:

$$k_{1+2+\cdots} = k'_{0 \rightarrow 1} + k''_{0 \rightarrow 2} + \cdots . \tag{8.15}$$

Here $k'_{0 \rightarrow 1} \equiv 1/t'_{0 \rightarrow 1} = (1/\tau) \exp(-\Delta F^{\#1}/k_B T)$ is the transition rate for the first pathway, $k''_{0 \rightarrow 2}$ is that for the second pathway, etc. So, if the parallel pathways are not

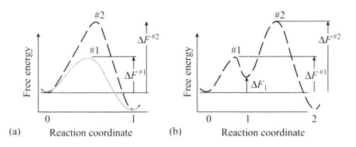

Figure 8.9. The free energy barriers in parallel (a) and consecutive (b) processes.

too numerous, the process time is determined by the most rapid of them, i.e., by the one overcoming the *lowest* barrier.

Second, if there are several *consecutive* barriers on the transition path, then the individual barrier-overcoming *times* must be *summed*.

This statement is evident for the process shown in Fig. 8.8a where the stable intermediate X is first accumulated before giving rise to the final state 1. However, it is far less evident for the process shown in Fig. 8.9b where the intermediate "1" is *un*stable. Moreover, in this case it should be specified that here "the individual barrier-overcoming time" implies the time required for a particle to reach the on-barrier position from the *deepest* prior minimum of the free energy, and *not* from the *immediately* preceding one (e.g., in Fig. 8.9b, the height of barrier #2 must be taken relative to state 0 not to state 1).

To prove this (I will provide the idea only without boring you with calculations), let us consider the process

$$\boxed{\begin{array}{c} 0 \\ \text{start} \end{array}} \underset{k_{1 \to 0}}{\overset{k_{0 \to 1}}{\rightleftarrows}} \boxed{1} \underset{k_{2 \to 1}}{\overset{k_{1 \to 2}}{\rightleftarrows}} \cdots\cdots \underset{k_{M \to M-1}}{\overset{k_{M-1 \to M}}{\rightleftarrows}} \boxed{M} \underset{k_{M+1 \to M}}{\overset{k_{M \to M+1}}{\rightleftarrows}} \boxed{\begin{array}{c} M+1 \\ \text{finish} \end{array}}$$

(where $k_{i \to i+1}$ is the rate of transition from i to $i+1$, and $k_{i+1 \to i}$ is the rate of inverse transition from $i + 1$ to i). Here the free energies of all intermediate states $(1, 2, \ldots, M)$ are higher by many times $k_B T$ than the free energies of both initial (0) and final $(M + 1)$ states.

Because of their high free energies, all intermediate states accumulate only a few molecules (as compared with the total number of the molecules in the initial and final states). Therefore, the rate at which the numbers of molecules change is also very low for the intermediate states as compared to that for the initial and final states. In other words, it can be assumed that the flow rate is approximately constant over the entire reaction pathway. Thus:

$$-dn_0/dt \equiv k_{0 \to 1} n_0 - k_{1 \to 0} n_1 = k_{1 \to 2} n_1 - k_{2 \to 1} n_2 = \ldots.$$

$$= k_{M \to M+1} n_M - k_{M+1 \to M} n_{M+1} \equiv dn_{M+1}/dt \tag{8.16}$$

(This assumption is called the "*quasi-stationary approximation*", and is widely used in chemical kinetics). Solution of the above equations gives the result:

$$t_{0\to\cdots\to M+1} = (1/k_{0\to 1}) + (1/k_{1\to 2})\exp(\Delta F_1/k_B T)$$
$$+ \cdots + (1/k_{M\to M+1})exp(\Delta F_M/k_B T) \tag{8.17}$$

where $1/k_{i\to i+1}$ is the barrier i overcoming time, provided the starting point was the immediately preceding state i, and ΔF_i is the free energy of the intermediate state i relative to the free energy of the initial state.

It should also be mentioned that to obtain this equation I used the well-known equation

$$k_{i\to j}/k_{j\to i} = \exp[(F_i - F_j)/k_B T] \tag{8.18}$$

which follows from the fact that the equilibrium population of these (i and j) states, n_i^0 and n_j^0, must satisfy both the kinetic equation $n_i^0 k_{i\to j} = n_j^0 k_{j\to i}$ (accounting for zero flow of particles from one state to the other in equilibrium) and the thermodynamic equation $n_i^0/n_j^0 = \exp[-(F_i - F_j)/k_B T]$ (connecting the equilibrium probability of the particle being in either state with its free energy).

And one last point: since in accordance with eq. (8.14) $1/k_{i-1\to i} = \tau_i \exp[(\Delta F^{\#i} - \Delta F_{i-1})/k_B T]$, then $(1/k_{i-1\to i})\exp(\Delta F_{i-1}/k_B T) = \tau_i \exp(\Delta F^{\#i}/k_B T)$. This is just the time required to overcome the individual barrier #i, if it were the only one on the pathway of the process $0 \to M + 1$; let us refer to this time as $t_{0\to \#i}$. Then

$$t_{0\to\cdots\to M+1} = t_{0\to \#1\to} + t_{0\to \#2\to} + \cdots + t_{0\to \#M+1} \tag{8.19}$$

which proves the statement that the time of a *consecutive* reaction is the *sum of the times* required to overcome individual barriers the heights of which are taken relative to *the deepest* of the prior free energy minima.

Incidentally, eq. (8.19) shows that if the sequential barriers are not too numerous, the time of the process is determined simply by the *highest* one among them.

Finally, let us consider the following. As I have already said, the existence of a free energy barrier is suggested by a reaction rate that is much lower than the rate of diffusion. And what is the typical diffusion time? To have some idea, let as talk a little about diffusion.

Before we do this, it is useful to estimate how long a molecule needs to forget the direction of its movement and to start diffusing. That is, we have to know how long it takes for the molecule's kinetic energy to dissipate because of friction against a viscous fluid.

One can show that this occurs in picoseconds.

Indeed, the particle's movement in a viscous fluid is described by the simple Newton equation

$$m(dv/dt) = F_{\text{frict}} \tag{8.20}$$

where m is the particle's mass, dv/dt is acceleration and F_{frict} is the force of friction. The mass can be estimated as $m = \rho V$, where ρ is the particle's density and V its volume. The friction force $F_{frict} = -3\pi D\eta v$ according to Stokes' law, where D and v are the particle's diameter and speed and η is the viscosity of the fluid. The equation

$$m(dv/dt) = -3\pi D\eta v \tag{8.21}$$

determines the time

$$t_{kinet} \approx \frac{m}{(3\pi D\eta)} \tag{8.22}$$

typical of the friction-caused movement damping. In fact, $t_{kinet} \approx 0.1\rho D^2/\eta$, since $m \approx \rho D^3$, and $3\pi \approx 10$. Since $\rho \approx 1\,g\,cm^{-3}$ for all the molecules we deal with, and $\eta \approx 0.01\,g\,cm^{-1}\,sec^{-1}$ for water (see any data-book), we have:

$$t_{kinet} \approx 10^{-13}\,sec\,(D/nm)^2 \tag{8.23}$$

where (D/nm) is the particle's diameter expressed in nanometers.

This means that the kinetic energy of a small $(D \approx 0.3\,nm)$ molecule dissipates within $\sim 10^{-14}\,sec$, of a small protein $(D \approx 3\,nm)$ within $\sim 10^{-12}\,sec$, and of a large protein $(D \approx 10\,nm)$ within $\sim 10^{-11}\,sec$. Thus, for aqueous solutions, the typical time is a picosecond. For a more viscous environment, a membrane for example, the kinetic energy dissipation is proportionally faster.

Of course, collisions with other molecules compensate for this energy loss. But the direction of the initial movement is forgotten within a picosecond . . .

Now we can turn to the diffusion movement of a molecule.

The heat-maintained kinetic energy of each particle, $mv^2/2$, amounts, on average, to about $k_B T$. The particle "memorizes" the direction of its movement for a time t_{kinet}. During this time, the time of one step, it covers a distance $\Delta l \approx vt_{kinet}$. Then the direction of its movement changes, and it covers approximately the same distance Δl in some new direction. That is, at each step its displacement is about $\pm\Delta l$. The mean square displacement of the molecule from the initial point grows proportionally with time. Indeed, if, after n steps, the particle is moved by a distance l_n in some direction, then its displacement after $n + 1$ steps is $l_{n+1} = l_n \pm \Delta l$, and $(l_{n+1})^2 = (l_n \pm \Delta l)^2 = (l_n)^2 \pm 2l_n\Delta l + \Delta l^2$. That is, on average, $(l_{n+1})^2 = (l_n)^2 + \Delta l^2$, since the mean value of the term $\pm 2l_n\Delta l$ is zero. The particle makes t/t_{kinet} steps within time t; thus, its mean square displacement after time t is

$$l_t^2 = (t/t_{kinet})\Delta l^2 \tag{8.24}$$

Since $\Delta l \sim vt_{kinet}$,

$$l_t^2 \sim (t/t_{kinet})(vt_{kinet})^2 = t(v^2 t_{kinet}) \tag{8.25}$$

and since $t_{kinet} \approx m/(3\pi D\eta)$, and $mv^2/2 \approx k_B T$,

$$l_t^2 \sim t[k_B T/(1.5\pi D\eta)] \tag{8.26}$$

This answer is only approximate, since we have used the symbols "$\approx$" ("approximately equal") and "$\sim$" ("equal in the order of magnitude") many times. However, as usually happens in such cases, the approximate answer is close to the precise answer (which requires much more refined calculations). The precise answer is:

$$l_t^2 = t[2k_B T/(\pi D\eta)] \tag{8.27}$$

The characteristic diffusion time is the time spent by a molecule in diffusing for a distance of its diameter D. It is easy to estimate this time from eq. (8.27):

$$t_{\text{diff}} = (\pi D^3 \eta)/(2k_B T) \tag{8.28}$$

Since water viscosity $\eta \approx 0.01\,\mathrm{g\,(cm^{-1}\,sec^{-1})}$, and $k_B T \approx 600\,\mathrm{cal\,mol^{-1}}$ at room temperature,

$$t_{\text{diff}} \approx 0.4 \times 10^{-9}\,\sec\,(D/\mathrm{nm})^3 \tag{8.29}$$

where (D/nm) is again the particle's diameter expressed in nanometers.

It is possible to show that a particle's inversion takes the same time (the inversion can be regarded as displacement of the particle's pole by a distance of $\sim D$).

Thus, we can conclude that in water, the typical diffusion time of a molecule falls within a nanosecond range: within this time a molecule driven by collisions with its fellow molecules covers a distance equal to its size and/or is overturned.

Any process that takes much more time than diffusion allows us to suggest the existence of a free energy barrier on its pathway. For inter-molecular reactions (or reactions between remote chain regions), this barrier is created, in part, by the entropy loss required to bring together the reacting pieces.

LECTURE 9

With basic physics learned, let us move on to discuss the stability of the secondary structure and the kinetics of its formation. Today we will consider only *homo*polypeptides, i.e., chains formed by identical amino acid residues.

We start by considering an α-helix. The conformations of its first three residues (1, 2, 3) in this helix are fixed with its first hydrogen bond, $(CO)_0$—$(HN)_4$; the next hydrogen bond, $(CO)_1$—$(HN)_5$, additionally stabilizes the conformation of only one residue (residue 4); the hydrogen bond $(CO)_2$—$(HN)_6$ provides additional binding for residue 5, and so on.

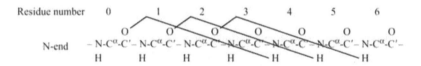

Thus, n residues are fixed by $n-2$ hydrogen bonds. Let us consider the free energy of formation of such a helix from a coil in aqueous solution (a coil is a polymer without any fixed structure and without interactions between non-neighboring residues). This free energy is given by

$$\Delta F_\alpha = F_\alpha - F_{\text{coil}}$$
$$= (n - 2)f_H - nTS_\alpha$$
$$= -2f_H + n(f_H - TS_\alpha) \qquad (9.1)$$

Here f_H is the free energy of formation of a hydrogen bond in the α-helix. Apart from the free energy of the H-bond *per se* (which, as you remember, is not just the energy as would be the case in a vacuum, but includes both the energy and entropy of the subsequent H-bond rearrangement in the aqueous environment), it also includes the free energy of other interactions accompanying formation of the H-bond in the helix. S_α is the entropy loss caused by fixation of one residue in the helix.

As you see, ΔF_α has two terms. One of them $(-2f_H)$ is independent of the helix length; the quantity

$$f_{INIT} = -2f_H \tag{9.2}$$

is known as the free energy of helix initiation (actually, f_{INIT} reflects both helix initiation and termination). The other term, $n(f_H - TS_\alpha)$, is directly proportional to the helix length; the quantity

$$f_{EL} = (f_H - TS_\alpha) \tag{9.3}$$

is known as the free energy of helix elongation per residue. Generally, we have

$$\Delta F_\alpha = f_{INIT} + n f_{EL} \tag{9.4}$$

The relationship between the probabilities of the purely helical state of an n-residue-long sequence and its purely coil (free of any helical admixtures) state is expressed as

$$\exp(-\Delta F_\alpha/kT) = \exp(-f_{INIT}/kT)[\exp(-f_{EL}/kT)]^n = \sigma s^n \tag{9.5}$$

Here I have used the conventional notation:

$s = \exp(-f_{EL}/kT)$, the helix elongation parameter;

$\sigma = \exp(-f_{INIT}/kT)$, the helix initiation parameter.

It is obvious that $\sigma \ll 1$, since $\sigma = \exp(-f_{INIT}/kT) = \exp(+2f_H/kT)$; and the free energy of a hydrogen bond is a large negative value of about several kT.

The quantity $\exp(-\Delta F_\alpha/kT) = \sigma s^n$ is simply the *equilibrium constant* for the two states ("α" and "coil") of an n-residue-long sequence.

Prior to discussing the ways of experimentally determining the σ and s values, let us see whether under varying conditions (temperature, solvent, etc.) the helix forms gradually or through an "all-or-none" transition.

On the face of it, such a distinct structure as the α-helix should be "frozen out" of the coil by a phase (i.e., "all-or-none") transition, like ice out of water.

However, Landau's theorem states that first-order phase transitions never occur in a system consisting of two *one-dimensional* phases. Let me try to explain this.

First of all, what does one-dimensionality mean? It means that the size (and hence, the free energy) of the phase interface is independent of the phase sizes. In these terms, both helical and coil conformations of a polypeptide are one-dimensional. Figure 9.1 shows that the interface between the helix and the coil regions is independent of their lengths, unlike the interface of the 3D phases (e.g., of a piece of ice with surrounding water). Consequently, the free energy on the helix boundaries does not depend on the helix size, while the free energy of a 3D phase (ice) increases as $n^{2/3}$ with increasing number n of the particles involved in this phase.

Now what does "forming through a first-order phase transition" mean? This means that at a transition temperature either phase can be stable, but their mixture (e.g., the mixture of ice and water) is unstable owing to increasing free energy. You must not be

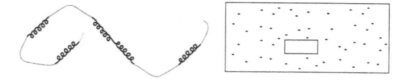

Figure 9.1. Comparison of one-dimensional (coil with helices) and three-dimensional (a piece of ice in water) systems. The size of the interface between the helix and the coil is independent of their lengths, while the interface of the three-dimensional piece of ice with water varies with its size.

misled by the picture of a floating piece of ice: this state is unstable at any temperature (owing to the additional free energy at the interface between ice and water), and with time, at a fixed temperature, ice will either melt or take up the entire water, provided there is no flow of underground heat, streams on other interfering non-equilibrium factors.

Is phase co-existence in a three-dimensional system favorable? *No, it is not.* Why? Let us return to Fig. 9.1 and consider the temperature at which infinite water and infinite ice have equal values of the free energy (that is a condition of the "mid-transition"). If the floating piece of ice consists of n molecules, the interface free energy is proportional to $\xi n^{2/3}$, where $n^{2/3}$ is the characteristic number of interface molecules, and $\xi > 0$ is the interface free energy of each of them. (Note that if $\xi < 0$, the thermodynamically favorable (in this case) "mixing up" occurs on the molecular scale, and the two phases do not emerge at all.) Consequently, the ice surface increases the free energy by $\xi n^{2/3}$. True, the piece of ice also possesses positional entropy, since its position in the vessel can vary. But this entropy never exceeds the value of the order of $k \ln(N)$, if there are N molecules (i.e., N ice initiation points) in this vessel. In total, the free energy of this piece of ice amounts to about $[\xi n^{2/3} - kT \ln(N)]$. However, at large N values, the logarithm grows only slightly. If the piece of ice occupies a considerable part of the vessel (e.g., $n \sim N/10$) and N is very large (e.g., 10 000 000 000), then $\ln(N)$ (here, 23) is very small compared with $(N/10)^{2/3}$ (here, 1 000 000); in other words, the interface term $\xi n^{2/3}$ predominates, and this term is unfavorable for the formation of a piece of ice. Therefore, in a three-dimensional system, macroscopic phases fall apart thereby making possible a first-order transition. (Unstable ice pieces of only a few molecules can be neglected as they are no more than microscopic, local fluctuations in water.)

And is phase co-existence in a one-dimensional system favorable? *The answer appears to be yes.* Let us return to the "mid-transition temperature" at which the helix and the coil have the same free energy, i.e., $f_{EL} = 0$. The free energy of the helix boundaries, f_{INIT}, is independent of both helix and coil lengths. The positional entropy of an n-residue-long helix in an N-residue-long chain is $k \ln(N - n)$. In total, the free energy of the helix floating in the sequence is $f_{INIT} - kT \ln(N - n)$. At large values of N and not too large n, the term containing $\ln(N - n)$ *always* predominates over the constant (f_{INIT}); this logarithmic term reduces the free energy and promotes insertion of the helix into the coil (as well as insertion of the coil into the helix). That is why, in a one-dimensional system, phase division *does not* happen; the phases tend to mix up, and therefore, a first-order transition (i.e., the "all-or-none" type transition) becomes *im*possible, provided the sequence is sufficiently long. Thus, the Landau theorem is proved.

Note. Strictly speaking, unlike α-helix melting, that of the DNA double helix does not come within the Landau theorem, since the double-stranded DNA is not a one-dimensional system: the DNA molten region is a spatial loop closed by double helices at its ends. The loop closing provides an additional contribution to the free energy of the loop boundaries, which grows logarithmically with increasing loop length.

Now we come to the question, "At which characteristic chain lengths do the coil and helical phases begin to mix up?" Or rather, "What characteristic length n_0 of the helical segment corresponds to the midpoint of the helix–coil transition?"

Let us consider an N-residue sequence at mid-transition temperature when the values of the free energies of the helix and coil are equal, i.e., $f_{EL} = 0$. Then the free energy of helix elongation (and coil elongation as well) is zero, that of helix initiation is f_{INIT}, the number of possible positions of a helix in the N-residue chain is about $N^2/2$ (the helix can be started and ended anywhere, the only condition being that it must contain no less than three residues) and the free energy is unaffected either by the helix position or by its length. To obtain a qualitative estimate, minor things (numerals in equations) can be neglected, and only major ones (letters in equations) must be taken into account. Then there are about N^2 helix positions, which gives their entropy as $k \times 2\ln(N)$; and the total free energy of insertion of a portion of the new phase (a helix with fluctuating ends) into the N-residue sequence is approximately $f_{INIT} - 2kT\ln(N)$. If this free energy is greater than zero, the insertion of the new phase will not happen; if it is less than zero, the insertion will happen and may be repetitive. Consequently, the coil and helical phases begin to mix up when the sequence length is $N \sim n_0$, and n_0 is found from $f_{INIT} - 2kT\ln(n_0) = 0$. Thus, at the midpoint of the helix–coil transition the characteristic lengths of their fragments is

$$n_0 = \exp(+f_{INIT}/2kT) = \sigma^{-1/2}. \tag{9.6}$$

Experimentally, the mid-point of this transition is the point (temperature) corresponding to 50% helicity of a very long polypeptide (as mentioned earlier, the helicity of a polypeptide is usually measured using CD spectra; at 50% helicity, the polypeptide CD spectrum represents a half-sum of the spectra of the polypeptide coil conformation and its totally helical conformation). At this point $f_{EL} = 0$, i.e., $s = \exp(-f_{EL}/kT) = 1$.

The n_0 value can be found as the sequence length that provides 12% helicity at $s = 1$. I will not prove this numerical estimation as it is beyond the scope of these lectures. You can try and do it yourself.

I would just like to explain why the helicity of an n_0-residue chain is several times lower than that of a very long sequence (i.e., than 50%). This is because this chain can be either completely in the coil state (in this case, its free energy is zero), or include a helix/coil mixture (with an additional free energy of about $f_{INIT} - kT \ln[(n_0)^2/2] > 0$), where the helix covers only some part of the chain.

Finally, with n_0 known, we can calculate f_{INIT} and σ. For most amino acids, $n_0 \approx 30, f_{INIT} \approx 4\,\text{kcal mol}^{-1}$ and $\sigma \approx 0.001$.

Now we can find the free energy of H-bonding (together with all the interactions accompanying formation of a hydrogen bond in an α-helix): according to eq. (9.2), $f_H = -f_{INIT}/2 \approx -2\,\text{kcal mol}^{-1}$. Also, it is possible to determine the loss of conformational entropy caused by fixation of a residue in the α-helix: according to eq. (9.3), at $f_{EL} = 0$, $TS_\alpha = f_H \approx -2\,\text{kcal mol}^{-1}$.

Both parameters of helix stability, f_{EL} and f_{INIT}, are temperature-dependent, but what greatly influences the stability is the deviation of f_{EL} from 0. The reason is that this deviation is multiplied by the large value n_0 (for an n_0-residue helix), and only in this way does it appear in the free energy of the helix. When the quantity $f_{EL}n_0/kT$ is close to $+1$ (or more precisely: $f_{EL}n_0/kT = +2$), the helicity almost disappears, and with $f_{EL}n_0/kT = -2$ the coil almost ceases to exist. Consequently, in very long ($N \gg n_0$) polypeptide chains the helix–coil transition occurs in a region of $-2/n_0 < f_{EL}/kT < 2/n_0$, i.e., in a region of $-0.07 < f_{EL}/kT < 0.07$ at $n_0 \approx 30$ (Fig. 9.2). This is an example of an abrupt, cooperative, but *not* phase transition (since its width does not tend to zero with increasing length of the chain).

The stability of the α-helix usually decreases with increasing temperature and added polar denaturants; and it increases with added weakly polar solvents (which increase the price of H-bonds).

To measure the effect of amino acid residues on helix stability, short ($\sim n_0$ or less) polypeptides are currently most often used. They can house only one helix, and therefore, the effect of each amino acid replacement on the helicity can be estimated most easily. Now we know that the contribution of an amino acid residue to the helix stability ranges from alanine, the most "helix-forming" residue, with $s \approx 2$, i.e., $f_{EL} \approx -0.4\,\text{kcal mol}^{-1}$, to glycine, the most "helix-breaking" residue, with

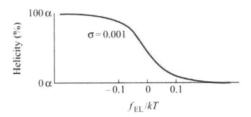

Figure 9.2. A finite width is typical of any helix–coil transitions, even those in infinitely long chains. This is an example of a cooperative transition, which is *not* a phase transition: at a low value of the helix initiation parameter ($\sigma \ll 1$) it is caused by a certain small change (much less than kT) in the value of f_{EL}, which shows that a "transition unit" involves many chain residues but far from all of them.

$f_{EL} \approx +1\,\text{kcal mol}^{-1}$, i.e., $s \approx 0.2$. Proline (an *imino* acid with no NH group to participate in the helix-forming H-bond) has a considerably lower value of s (0.01–0.001; it has not been accurately measured yet).

Earlier, the estimates of this kind were made (in particular, by one of us, O.B.P.) using statistical co-polymers (e.g., chains with a random mixture of 80% Glu and 20% Ala); it is in this way that the first – and hence the most important – estimates were obtained. But with the advent of pre-set sequence synthesis, the use of such random co-polymers became a thing of the past.

Also, potentiometric titration was used to measure the helical state stability (the quantity f_{EL}) in polypeptides containing acidic or basic side chains (e.g., in poly(Glu) or poly(Lys)). The idea of this approach is that by charging a helix, we destroy it (because in the helix the side-chain charges are closer to one another than in the coil, and, hence, their repulsion is stronger). So, the helix stability can be calculated from the dependence of the total charge and helicity of the chain on the medium's pH. Unfortunately, consideration of this interesting method in more detail is beyond the scope of this course.

Using short peptides with pre-set sequences, an estimate can even be made as to how the helicity is affected by each single amino acid replacement at a given position in the peptide, i.e., in fact, as dependent on the residue position about the N- and C-termini of the helix (and on the residues surrounding the residue in question). The side chains, and in particular charged ones, interact with these termini in opposite ways, because, as mentioned, the N-terminus of the helix houses the main-chain NH groups free of hydrogen bonds (and the partial charge of the α-helix N-terminus is equal to $+e/2$), while its other terminus holds free CO groups (with the total partial charge $-e/2$, half the electron charge).

Similar approaches are used to measure the stability of the β-structure in polypeptides. However, they are less developed, since the β-structure aggregates strongly. Currently, β-structure stability is measured right in the protein by estimating the effect on protein stability of replacement of each of its surface β-structural residues. It is shown that contribution of amino acid residues into β-structure stability ranges from ≈ -1 to $\approx +1\,\text{kcal mol}^{-1}$. The ability of various residues to stabilize α- and β-structures will be considered in the next lecture.

Now let us consider *the rate of formation* of the secondary structure in peptides.

α-Helices are formed very rapidly: as shown, within $\sim 0.1\,\mu s$ a peptide of 20–30 residues adopts the helical conformation (such rapid measurements require a pico- or nanosecond laser-induced temperature jump of the solution). Consequently, the rate of helix extension is at least a residue per several nanoseconds.

I said "at least" because the rate of helix formation depends not only on its extension rate but also on how rapidly the first "nuclei" of the helical structure appear. Initiation of the helix requires overcoming the activation barrier; therefore, the formation of the first turn is the slowest step, and subsequent growth of the helix is rapid. Hence, it is possible that nearly the entire observation time might be taken by helix initiation, with its elongation being far more rapid. Let us consider this in more detail.

The typical dependence of the free energy of a helix on its length is illustrated in Fig. 9.3. Even if $f_{EL} < 0$, i.e., when the rather long helix is stable, formation of its first turn requires overcoming an activation barrier as high as f_{INIT}.

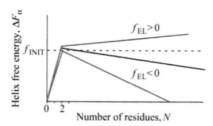

Figure 9.3. Typical dependence of the helix free energy (ΔF_α) on the number (N) of amino acid residues involved in the helix, with various free energies of helix elongation (f_{EL}). When $f_{EL} < 0$, a long helix is stable but its initiation requires overcoming an activation barrier as high as f_{INIT}. When $f_{EL} > 0$, a helix of any length is unstable, and therefore it cannot form.

According to the transition state theory, this first step of helix formation takes the time

$$t_{INIT\,\alpha} = \tau \exp(+f_{INIT}/kT) \tag{9.7}$$

where τ is the time of an elementary step (here, helix elongation by one residue), and the exponent allows for low occupancy of the barrier state possessing (by definition of a barrier) the highest free energy. By definition, $\sigma = \exp(-f_{INIT}/kT)$; thus

$$t_{INIT\,\alpha} = \tau/\sigma \tag{9.8}$$

However, initiation can occur anywhere in the future helix, and its length (even in very long chains) is about $n_0 = \sigma^{-1/2}$. Consequently, the typical time of initiation of the first turn *anywhere* in the future helix is n_0 times less, and $t_{INIT}/n_0 = \tau/(n_0\sigma) = \tau/\sigma^{1/2}$.

Propagation of the helix to all its $\sim n_0$ residues takes as much time as the initiation, $\sim \tau n_0 = \tau/\sigma^{1/2}$. This gives the total helix–coil transition time as approximately $2\tau/\sigma^{1/2}$, and the half-time (i.e., the characteristic time) of this transition is

$$t_\alpha \sim \tau/\sigma^{1/2} = \tau \exp(+f_{INIT}/2kT) \tag{9.9}$$

Half of the time, roughly, is spent on helix initiation anywhere in the sequence, and the rest on elongation, the rate of which has been estimated as about a residue per nanosecond.

The kinetics of α-helix formation is relatively simple: all these are formed rapidly. The kinetics of β-structure formation is far more complex and interesting.

In polypeptides, the β-structure often forms extremely slowly. It may take hours and even weeks, although sometimes the β-structure folds within milliseconds. What is the reason for that? Surprisingly, the folding rate of proteins containing β-structures is not much lower than that of α-helical proteins. How do they manage? And what controls the anomalous rate of β-structure formation in polypeptides: slow initiation or slow elongation?

The "anomalous" (as compared with the helix–coil transition) kinetics of formation of the β-structure is connected with its two-dimensionality (in contrast to a

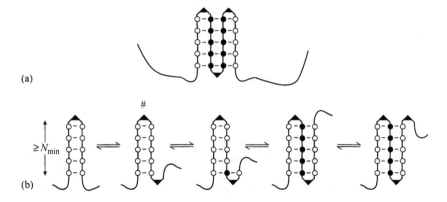

Figure 9.4. (a) Schematic of a β-sheet. The amino acid residues of internal β-strands are indicated by closed circles, and of edge β-strands by open circles; the bends (or loops) connecting the β-strands are indicated by angles. (b) Illustration of the β-sheet growth scenario given in the text for the case when separate β-hairpins are unstable. The most unstable structure on the pathway is marked with #.

one-dimensional helix or coil) (Fig. 9.4), which results in a first-order phase transition. Let us consider this process in more detail.

Edge residues have fewer contacts than internal residues of the sheet. In other words, the edge of a β-sheet (like the boundary of any other phase, like a drop of water, a piece of ice or an α-helix) has a higher free energy. However, like a drop of water and in contrast to the α-helix, the β-sheet is not one-dimensional, i.e., its boundary (and hence, the boundary free energy) grows with increasing number of residues involved in the sheet. Therefore, the transition from a random coil to the β-structure becomes a first-order phase transition like the formation of a water drop or a piece of ice.

Let us show that this provokes the occurrence of a high free-energy barrier (especially in the formation of an only marginally stable β-structure) capable of slowing down the folding initiation zillion-fold.

The edge of a β-sheet consists of (a) edge β-strands, and (b) bends or loops connecting the β-strands (Fig. 9.4a). Let the coil free energy be zero (i.e., the reference point); f_β, the free energy of a residue in the center of the β-sheet; $f_\beta + \Delta f_\beta$, the free energy of an edge β-strand residue (i.e., Δf_β is the edge effect); and U, the free energy of a bend. Since the β-sheet forms, it is stable (i.e., $f_\beta < 0$), and the edge effects prevent it from falling into pieces (i.e., $\Delta f_\beta > 0$ and $U > 0$).

In the kinetics of β-sheet formation we must distinguish between the following two cases:

1. $f_\beta + \Delta f_\beta < 0$, i.e., in itself a long β-hairpin is more stable than a coil. Then only the turn at its top needs to overcome the activation barrier (which is almost identical to the barrier to be overcome in forming an α-helix), and subsequent growth of the β-structure is rapid, like the elongation of an α-helix (see the line with $f_{EL} < 0$ in Fig. 9.3).

2. $f_\beta + \Delta f_\beta > 0$, i.e., in itself the β-hairpin is unstable, and it is only the association of the initiating hairpin with other β-strands into a β-sheet that stabilizes the β-structure. Then the activation barrier is represented by the formation of a "nucleus", that is, such a β-sheet or β-hairpin that provides further growth of the sheet accompanied by an overall decrease of the free energy.

The formation and subsequent growth of the nucleus of a new phase are the most typical feature of first-order phase transitions, β-structure formation among them. However, as we will see, overcoming the nucleus-provoked activation barrier may be an extremely slow process.

Let us consider the following simplest scenario of formation of a stable β-sheet when separate β-hairpins are unstable (Fig. 9.4b): formation of the initiating β-hairpin by a turn and two β-strands N residues long each; formation of the next turn at its end; association of another N-residue β-strand; formation of the next turn; association of another β-strand, and so on.

The formation (in a coil) of a β-hairpin consisting of a turn and two N-residue β-strands contributes as much as $U + 2N(f_\beta + \Delta f_\beta)$ to the free energy of the chain; formation of the next turn makes it still higher by a value of U. Association of the N-residue edge of this hairpin with a new N-residue β-strand decreases the free energy by $N f_\beta$ (since the number of edge residues remains the same, while the number of internal residues increases by N); formation of the next β-turn increases the free energy again by U; association of another N-residue β-strand decreases it by $N f_\beta$, and so on.

The cycle of "association of another β-strand and formation of the next β-turn" changes the net free energy by $N f_\beta + U$. And since this cycle must result in a decrease of the free energy (as a prerequisite of rapid growth), each associating strand should contain not less than

$$N_{\min} = \frac{U}{(-f_\beta)} \tag{9.10}$$

residues.

The "transition" state, i.e., the most unstable state in formation of the β-structure is, according to our scenario, the β-hairpin with a subsequent turn. Since we consider the case where the hairpin stability decreases with increasing length of the hairpin, and since the β-strand of the initiating hairpin must contain at least $N_{\min}$ residues, the minimum free energy of the initiating hairpin and the next turn is

$$F^\# = U + 2N_{\min}(f_\beta + \Delta f_\beta) + U$$
$$= \frac{2(U \Delta f_\beta)}{(-f_\beta)}. \tag{9.11}$$

This is the free energy of the transition state in β-sheet folding *according to our scenario*. It can be very high when f_β is close to zero. Now we have to show that, irrespective of the scenario, there cannot exist transition states of a considerably higher stability.

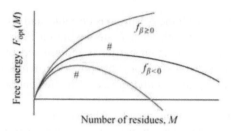

Figure 9.5. β-Sheet minimum free energy F_{opt} as a function of M, the number of residues in the sheet. The curves refer to various free energies (f_β) of internal residues. The maximum of F_{opt} is marked with #. In a growing β-sheet (unlike an α-helix, see Fig. 9.3) this maximum does not correspond to the very beginning of the process.

To do so, let us estimate the changes in the minimum free energy of a growing β-sheet. The general course of the free energy changes is presented in Fig. 9.5. When the sheet is small, edge effects predominate, and the free energy of a growing β-sheet increases. In a large sheet, internal residues predominate, and the free energy of a growing β-sheet decreases.

Now we pass to calculations. First, let us estimate the minimum free energy of an M-residue β-sheet. The free energy of a sheet comprising m β-strands (which are of the same length in order to minimize the edge free energy) and $m - 1$ turns is

$$F(M, m) = Mf_\beta + 2(M/m)\Delta f_\beta + (m - 1)U \qquad (9.12)$$

Varying the number of β-strands m (at a given M) gives the minimum of this free energy from the condition

$$\frac{dF}{dm} = -2(M/m^2)\Delta f_\beta + U = 0. \qquad (9.13)$$

This yields the optimal number of β-strands in this sheet, $m_{opt} = M^{1/2}(2\Delta f_\beta/U)^{1/2}$, and its free energy, $F_{opt}(M) = F(M, m_{opt}) = Mf_\beta - U + 2M^{1/2}(2\Delta f_\beta U)^{1/2}$. Varying the magnitude $F_{opt}(M)$ over M gives its maximum (see Fig. 9.5) from the condition

$$\frac{dF_{opt}}{dM} = f_\beta + M^{-1/2}[2\Delta f_\beta U]^{1/2} = 0. \qquad (9.14)$$

Then the sheet size corresponding to this maximum can be determined as $M^* = 2(\Delta f_\beta U)/(-f_\beta)^2$, and its free energy as

$$F^* = F_{opt}(M^*) = 2\frac{\Delta f_\beta U}{-f_\beta} - U. \qquad (9.15)$$

The quantities F^* and $F^\#$ (see eq. (9.11)) coincide as to their principal term, $2(\Delta f_\beta U)/(-f_\beta)$. It is because of this term that the free energy of the transition state is *always* high when there is a low free energy of stabilization $(-f_\beta)$ of the β-structure, i.e., when the β-structure is only marginally stable. Actually, $F^\#$ tends to infinity when the β-structure approaches thermodynamic equilibrium with the coil, i.e., when $(-f_\beta) \to 0$.

Above, we considered the free energy of initiation of β-sheet folding. According to the theory of the transition state, the time of the process (here, folding initiation) depends on the free energy $F^\#$ of the transition state (i.e., of the most unstable state) as

$$t_{\text{INIT } \beta} \sim \tau_\beta \exp(+F^\#/kT), \qquad (9.16)$$

where τ_β is the time of an elementary step (here, β-sheet elongation by one residue; and there is no reason for a qualitative difference between elongation rates of the β-sheet and α-helix).

The $t_{\text{INIT } \beta}$ is the time necessary to initiate β-folding in one given point of the chain. Therefore, the time of β-folding initiation *somewhere* in a chain of M residues is $t_{\text{INIT } \beta}/M$. The expansion of the β-structure all over the chain from *one* initiation center takes $\sim M\tau_\beta$. The initiation is a time-limiting step for a relatively short chain (when $\tau_\beta \exp(+F^\#/kT)/M > M\tau_\beta$). In this case the time of β-folding is $t_\beta \sim \tau_\beta \exp(+F^\#/kT)/M$.

For long chains (and/or for intermolecular β-structures), the time of β-structure formation is about $\tau_\beta \exp(+F^\#/2kT)$. (This problem can be considered in the same way as helix formation in long chains that we discussed earlier. The main idea is that in this case the number of residues (M_{eff}) included in each independently folding sheet is such that its initiation and expansion times are equal, i.e., $\tau_\beta \exp(+F^\#/kT)/M_{\text{eff}} = M_{\text{eff}}\tau_\beta$.)

Thus, in both cases, the time of β-sheet formation depends on the β-structure stability per residue, $-f_\beta$, as

$$t_\beta \sim \exp[A/(-f_\beta)]. \qquad (9.17)$$

when the β-structure stability is low. No matter what the numerical value of the constant A may be, it clearly follows from eq. (9.17) that the time of β-structure formation is enormous at low $-f_\beta$ (in the limit, it is infinity; see Fig. 9.6).

This formula gives an explanation both for the experimentally observed extremely low rate of β-structure formation in non-aggregating polypeptides (where the stability

Figure 9.6. The dependence of the time needed to form a β-sheet on the stability of the β-structure (general view).

of the β-structure is always low) and for a drastic increase of this rate with increasing stability of the β-structure.

Thus, a β-structure of low stability must form very slowly not because of slow elongation but because of slow initiation, as it is difficult to accumulate the free energy of thermal motions in order to overcome the high initiation barrier.

However, β-sheets and hairpins of high stability (which are observed in proteins) must form almost as rapidly as the α-helix.

A very slow initiation is a common feature of first-order phase transitions when the emerging phase is only marginally stable. Recall the overcooled liquid or the overcooled vapor... All these effects are connected with a large area and, hence, the high free energy of interface between the phases. And the β-structure is formed through a first-order phase transition with all its consequences...

In contrast, the α-helix *avoids* the first-order phase transition (remember, the helix boundary, unlike that of the β-structure (or of a piece of ice) *does not* increase with its increasing size), and therefore the barrier to be overcome in helix folding is always of a finite (and small) value; hence, the initiation here takes a fraction of a millisecond.

A concluding remark: what is a "coil"? I have used this term many times meaning a chain without any distinct structure. Indeed, the term "coil" covers a great many conformations without any long-range order in the chain. However, some weak short-range order (within a few consecutive residues only) cannot be ruled out.

The most interesting features of the coil (experimentally observed by using hydro-dynamic techniques and light- and X-ray scattering) are its extremely low density and large volume, and a most peculiar dependence of its radius and volume on the chain length.

To shed light upon this peculiarity let us consider the simplest model of a coil, the so-called "loose joint chain" (Fig. 9.7). Its "links" are represented as sticks (each link can include a few chain monomers); the main distinctive feature of this model is that each stick can freely turn on the joint about the neighboring sticks. Let us assume that there are M sticks in the chain, and the length of each stick is r.

Such chain can be described as a sequence of vectors $\mathbf{r}_1, \mathbf{r}_2, \ldots, \mathbf{r}_M$ (may I remind you that in math a vector is a directed straight segment). These vectors have identical lengths r, and each of them is directed from the previous joint to the next one. The

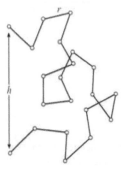

Figure 9.7. The loose joint chain model of a coil.

sum of these vectors,

$$\mathbf{h} = \sum_{i=1}^{M} \mathbf{r}_i,$$ (9.18)

is just a vector running from the beginning of the chain to its end. Its squared length is

$$\mathbf{h}^2 = \left(\sum_{i=1}^{M} \mathbf{r}_i\right)^2$$

$$= \sum_{i=1}^{M} \mathbf{r}_i^2 + \sum_{i \neq j}^{M} \sum^{M} \mathbf{r}_i \mathbf{r}_j.$$ (9.19)

Now let us find $\langle \mathbf{h}^2 \rangle$, the average value of $\mathbf{h}^2$; i.e., let us have $\mathbf{h}^2$ averaged over all possible thermal fluctuations of the chain conformation (this averaging is denoted by $\langle \rangle$). In doing so, we have to keep in mind that $\langle \mathbf{r}_i^2 \rangle$, the average squared vector $\mathbf{r}_i$, is just r^2, and the average scalar product of any vectors $\langle \mathbf{r}_i \mathbf{r}_j \rangle$ is zero when $i \neq j$, since free rotation provides equal probabilities of any direction of these vectors.

Hence, the average squared distance between the ends of a loose joint chain is

$$\langle \mathbf{h}^2 \rangle = \left\langle \left(\sum_{i=1}^{M} \mathbf{r}_i\right)^2 \right\rangle$$

$$= \left\langle \sum_{i=1}^{M} \mathbf{r}_i^2 + \sum_{i \neq j}^{M} \sum^{M} \mathbf{r}_i \mathbf{r}_j \right\rangle$$

$$= \sum_{i=1}^{M} \langle \mathbf{r}_i^2 \rangle + \sum_{i \neq j}^{M} \sum^{M} \langle \mathbf{r}_i \mathbf{r}_j \rangle$$

$$= Mr^2;$$ (9.20)

i.e., the linear dimensions (radius, etc.) of the coil increase with increasing number of chain links M as $M^{1/2}$ (recall diffusion: there, the mean square of the distance depends on time in the same manner). Consequently, the coil volume is proportional to $M^{3/2}$, although the volume of all "normal" (i.e., fixed density) bodies increases only as the number of particles M, that is, much more slowly than $M^{3/2}$. This abnormally strong dependence of the coil volume on the chain length is the most prominent characteristic feature of the coil. Specifically, this feature is responsible for the extremely low density of a coil formed by long chains, and consequently, for nearly zero contacts between distant links in the chain.

In addition, since the coil volume is proportional to $M^{3/2}$ and the probability that the chain ends meet is inversely proportional to the volume occupied by the coil, the probability of chain ends meeting is proportional to $M^{-3/2}$. This means that the free energy of loop closing in a coil-like chain increases with its length as $kT \times (3/2) \ln(M)$.

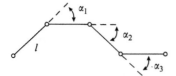

Figure 9.8. Rotational isomeric model of the coil.

The loose joint model presents a far too ideal coil in which any link may rotate by any angle, which is not actually true. However, eq. (9.20) can be generalized as

$$\langle \mathbf{h}^2 \rangle = Lr \qquad (9.21)$$

where $L = Mr$ is the full ("contour") length of the chain proportional to the number of its constituent links, and r is the effective distance between the chain "loose joints", i.e., the characteristic length after which the chain "forgets" its direction (note: in a polypeptide chain this characteristic length, also called "length of the Kuhn segment", is 30–40 Å (when the polypeptide includes neither too flexible Gly nor too rigid Pro), i.e., the Kuhn segment includes about 10 amino acid residues).

The equation for the coil size presented as eq. (9.21) is general enough to be applied in the description of real polymers.

The rotational isomeric model of the coil (Fig. 9.8) gives the following estimate of the r value:

$$r = l(1 + \langle \cos \alpha \rangle)/(1 - \langle \cos \alpha \rangle) \qquad (9.22)$$

where l is the distance between the points where the chain turns (i.e., the covalent bond length) and $\langle \cos \alpha \rangle$ is the average cosine of the turn angle (see Fig. 9.8).

LECTURE 10

Now let us discuss the properties of the side chains of amino acid residues. In particular, I would like to consider the question of what structures stabilize individual residues.

The list of 20 "standard" (i.e., DNA-encoded) amino acid residues is given in Table 10.1, and their structures are presented in Figure 10.1.

As has become known later, sometimes DNA encodes one more (the twenty-first) "non-standard" amino acid, selenocysteine, i.e., cysteine in which a sulfur atom is replaced by a selenium atom. Selenocysteine corresponds to the codon UGA of matrix RNA, although not always (usually UGA serves as a terminating codon) but only in the appropriate nucleotide environment. And selenocysteine-charged tRNA recognizes only such UGA codons.

Now let us consider the structural tendencies of amino acid residues: these have become known after long-term statistical investigations of protein structures. Such investigations answer the question as to what is most likely to happen and what is not.

Table 10.2 may be helpful in putting these answers in order. Along with the abundance in different parts of proteins, I have tabulated such residue properties as the presence of an NH group in the main chain (it is absent only from the imino acid proline), the presence of the C^β-atom in the side group (it is absent only from glycine), the number of non-hydrogen γ-atoms in the side chain, and the presence and type of polar groups in the side chain (dipoles or charges with a sign; the charged state corresponding to a "normal" pH of 7.0 is shown in bold).

Let us try and understand the major features of this table based on what we have already learned. In doing so, we will use the following logical criterion: since the protein as a whole is stable, the majority of its components must be stable, i.e., stable components must be most often observed in its structure, while non-stable ones must be rare.

Why does proline dislike the secondary structure? Because it lacks the main-chain NH group, i.e., its ability of H-bonding is halved, and H-bonds are of primary importance for the secondary structure. Why does it nevertheless like the N-terminus of the helix? Because here, at the N-terminus, NH-groups protrude from the helix,

Table 10.1. The principal properties of amino acid residues.

Amino acid residue	Code		Occurrence in *E. coli* proteins (%)	M.W. at pH 7 (Da)	$\Delta G_{\text{water}\to\text{alcohol}}$ of side chain at $25\,^\circ$C (kcal mol^{-1})
	3-letter	1-letter			
Glycine	Gly	G	8	57	0
Alanine	Ala	A	13	71	−0.4
Proline	Pro	P	5	97	−1.0
Glutamic acid	Glu	E	≈6	128	+0.9
Glutamine	Gln	Q	≈5	128	+0.3
Aspartic acid	Asp	D	≈5	114	+1.1
Asparagine	Asn	N	≈5	114	+0.8
Serine	Ser	S	6	87	+0.1
Histidine	His	H	1	137	−0.2
Lysine	Lys	K	7	129	+1.5
Arginine	Arg	R	5	157	+1.5
Threonine	Thr	T	5	101	−0.3
Valine	Val	V	6	99	−2.4
Isoleucine	Ile	I	4	113	−1.6
Leucine	Leu	L	8	113	−2.3
Metionine	Met	M	4	131	−1.6
Phenylalanine	Phe	F	3	147	−2.4
Tyrosine	Tyr	Y	2	163	−1.3
Cysteine	Cys	C	2	103	−2.1
Tryptophan	Trp	W	1	186	−3.0

All data are from [12] except for those on side-chain hydrophobicity ($\Delta G_{\text{water}\to\text{alcohol}}$), which are from Fauchere I.I., Pliska V., *Eur. J. Med. Chem.-Chim. Ther.* (1983) **18**: 369. The volume (in Å^3) occupied by a residue (in protein or in water) is close to its molecular weight (in daltons) multiplied by 1.3. To be more precise, it is ∼5% higher than M.W. × 1.3 if the residue contains many aliphatic (—CH$_2$—, —CH$_3$) groups and ∼5% lower than M.W. × 1.3 if the residue contains many polar (O, N) atoms.

i.e., they do not participate in hydrogen bonds, and proline loses nothing here . . . On the other hand, in proline, its ring fixes the angle φ at about −60°, i.e., proline is about ready to adopt the helical conformation (Fig. 10.2a).

Why does glycine dislike the secondary structure and prefer irregular segments ("coil")? Because its allowed $\varphi\psi$ region in the Ramachandran map is extremely broad (Fig. 10.2b), and it can easily adopt a variety of conformations other than secondary structure.

In contrast, alanine with its more narrow (but including both α and β conformations) region of allowed conformations (Fig. 10.2b) prefers the α-helix (and partially the β-structure) rather than irregular conformations.

Other hydrophobic residues (i.e., residues without charges and dipoles in their side chains) prefer, as a rule, the β-structure. Why? Because there is more room for their large γ-atoms (Fig. 10.2c). This is of particular importance for side chains with two large γ-atoms, and indeed, these are strongly attached to the β-structure.

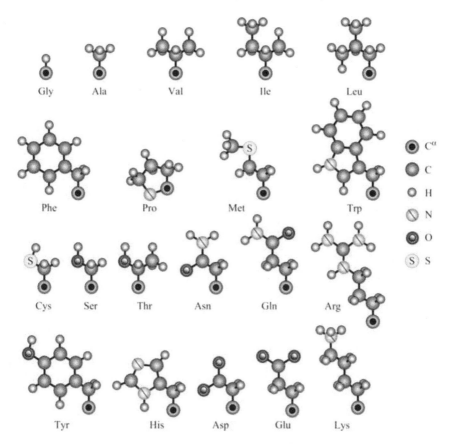

Figure 10.1. The side chains of twenty standard amino acid residues. Atoms involved in the amino acids are shown on the right.

As to amino acids with polar groups in their side chains, they prefer irregular ("coil") surface regions where these polar groups can easily participate in H-bonds with both the irregular polypeptide chain and water. This tendency is most clearly displayed by most polar residues, which are charged at a "normal" pH of 7.0, as well as by the shortest polar side chains whose polar groups are closest to the main chain. By the way, this possibility of additional H-bonding explains the tendency of short polar side chains to be located at both ends of the helix.

Tryptophan and tyrosine are a kind of exception from amino acids having dipoles in their side chains: they have a small dipole and a large hydrophobic part; another exception is cysteine whose SH groups have extremely weak H-bonds. Their behavior is rather similar to that of hydrophobic residues.

As we can see, the negatively charged side chains have a preference for the N-terminus of the helix (or rather, to the N-terminal turn plus one or two preceding residues). At the same time, they avoid the C-terminal turn (plus one or two

Table 10.2. The principal structural properties of amino acid residues.

Residue	Main chain[a] NH	Side chain[a] C^β	Side chain[a] γ Number	Dipole/charge[b]	pK_a^b	Structural occurrence tendency[c] before α_N	in helix α_N	in helix α	in helix α_C	after α_C	after β	in loops	in core
Gly	+	−					−				−	+	
Ala	+	+					+				−	+	
Pro	−	+	1				+	−	−	−	−	+	−
Glu	+	+	1	$COOH \rightarrow CO_2^-$	4.3	+	+				−		−
Asp	+	+	1	$COOH \rightarrow CO_2^-$	3.9	+	+				−	+	−
Gln	+	+	1	$OCNH_2$							−		−
Asn	+	+	1	$OCNH_2$		+		−		+	−	+	−
Ser	+	+	1	OH		+						+	−
His	+	+	1	$NH;\ \text{and}\ N \rightleftharpoons NH^+$	6.5	−	−				−		
Lys	+	+	1	$NH_2 \rightleftharpoons \mathbf{NH_3^+}$	10.5	−	−		+	+		+	−
Arg	+	+	1	$\mathbf{HNC(NH_2)_2^+}$	12.5	−	−		+	+	−		−
Thr	+	+	2	OH		+						+	
Ile	+	+	2								+	−	+
Val	+	+	2								+	−	+
Leu	+	+	1					+			+	−	+
Met	+	+	1					+			+	−	+
Phe	+	+	1								+	−	+
Tyr	+	+	1	$\mathbf{OH} \rightleftharpoons O^-$	10.1			−			+		+
Cys	+	+	1	$\mathbf{SH} \rightleftharpoons S^-$	9.2			−			+		+
Trp	+	+	1	NH							+		+

[a] See text for a definition of these factors.

[b] Bold type in the "Dipole/charge" column shows the state of the ionizable group at the "neutral" pH 7; pK is that pH value where the group can be in the charged and uncharged states with equal probability (see Fig. 10.5 below).

[c] These columns refer respectively to the tendency to be: immediately before the helix N-terminus; in the α-helix (the N-terminal turn, the body and the C-terminal turn); immediately after the C-terminus; in the β-structure; in irregular structures (loops), including β-turns of the chain; and in the hydrophobic core of the globule, rather than on its surface. The "tendency" is measured as a concentration of a residue in a given structure relatively to the average concentration of this residue in studied proteins. The difference in occurrence between "+" and "−" is approximately a factor of two. A particularly strong tendency is shown by a larger, bold symbol.

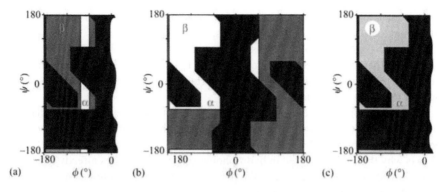

Figure 10.2. Disallowed and allowed conformations of various amino acid residues as a background for α and β-conformations. (a) Allowed proline conformations (□) against allowed alanine conformations (■); ■, conformations disallowed for both of them. (b) Allowed alanine conformations (□) against conformations allowed for glycine only (■); ■, conformations disallowed for all residues. (c) The map of disallowed (■) and allowed (□, □) conformations of larger residues. □, the region where all side-chain χ^1 angles are allowed; (□), the region where some χ^1 angles are disallowed.

Figure 10.3. Favorable positions of charged side chains near the N- and C-ends of the helix.

subsequent residues). The preference of the positively charged groups is just the opposite. What is the reason for that? It is NH groups protruding from the N-terminus and creating a considerable positive charge that attracts "minuses" and repels "pluses" of the side chains (Fig. 10.3). In contrast, the C-terminus is charged negatively, and therefore it is attractive for "pluses" of the side chains, while their "minuses" avoid it.

As to the residue location in the globule, the general tendency is that polar (hydro*philic*) side chains are on the protein surface, in contact with polar water molecules ("like dissolves in like"). Separation from water molecules negatively affects polar groups since then they lose their H-bonds. This negative effect is especially profound when it concerns the charged groups because their transition from the high permittivity medium (water) to that of low permittivity (protein core) is accompanied by a drastic increase in the free energy. Indeed, ionized groups are almost completely absent from the protein interior (nearly all exceptions are connected with either coordinate bonding of metal ions, or with active sites which are, in fact, the protein's focus. . .).

In contrast, the majority of hydrophobic side chains are in the protein interior to form the hydrophobic core (again, "like dissolves in like"). As we have learned, the hydrophobicity of a group grows with its non-polar surface that is to be screened from water. For purely non-polar groups, the hydrophobic effect is directly proportional to their total surface, while in the case of polar admixtures it is proportional to their surface less the surface of these admixtures.

Adhesion of hydrophobic groups is the main, although not the only, driving force of protein globule formation: it is also assisted by H-bonding in the secondary structure (as discussed earlier) and by tight quasi-crystal packing within the interior of the protein molecule (to be discussed later).

To form the protein hydrophobic core, the chain must enter it with hydrogen bonds already formed (or formed in the process), because the rupture of H-bonds between polar peptide groups and water molecules occurring in any other way is expensive. That is why the chain involved in hydrophobic core formation has already formed its secondary structure (or forms it in the process), and thereby saturates the hydrogen bonds of the main-chain peptide groups.

The core must comprise only hydrophobic side chains from secondary structures, while polar side chains of the same secondary structures must remain outside; therefore, both α-helices and β-strands located at the surface have hydrophobic and hydrophilic surfaces created by alternating appropriate groups of the protein chain in a certain order (Fig. 10.4).

All the regularities just discussed are used to build artificial (*de novo*) proteins and to predict the secondary structure of proteins from their amino acid sequences, and also to predict internal and surface portions of protein sequence segments either deeply immersed in the protein or positioned on its surface. We will discuss this later.

In conclusion, a few more words about ionized side groups. Increasing pH (i.e., decreasing H^+ concentration) always shifts the charged state of a group "in the negative direction", i.e., a neutral group acquires a negative charge, and a positive group becomes discharged (see Fig. 10.5). Different groups change from the uncharged to the charged state or vice versa at different pH, but the transition width always remains the same – about 2 units of pH (within this range the charged/uncharged state ratio changes from 10 : 1 to 1 : 10).

Special attention should be paid to groups that change their uncharged state to a charged one at a pH of about 7.0 typical of proteins in a living cell: these easily rechargeable groups (histidine in particular) are often used in protein active sites.

As it has been already mentioned, an ionizable group easier penetrates into a non-polar medium (e.g., protein or membrane interior) in its uncharged form. Indeed,

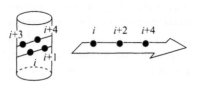

Figure 10.4. The positions at which nonpolar side groups can form continuous hydrophobic surfaces on α-helices and β-structures. The numbers show residue positions in the chain. Similar combinations of polar groups result in the formation of hydrophilic areas on opposite surfaces of α-helices and β-strands.

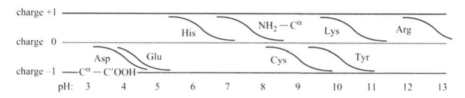

Figure 10.5. The polarization of ionized side groups, as well as the N- and C-termini of the polypeptide chain (NH_2—C^α and C^α—$C'OOH$, respectively) in water at various pH values. The pH value corresponding to the "half-charged" state of a group is the pK value of its ionization (cf. Table 10.2). The ratio of probabilities of charged and uncharged states is $10^{(pK-pH)}$: 1 for a positively charged group, and $10^{-(pK-pH)}$: 1 for a negatively charged group.

the estimated cost of an ion penetration into such a medium is as high as several dozen kcal mol^{-1}. And what does the discharging cost? This can be easily estimated from Fig. 10.5. The probability of uncharged state is $W_0 = 1/[1 + 10^{(pK-pH)}]$ for a positively charged group, and $W_0 = 1/[1 + 10^{-(pK-pH)}]$ for a negatively charged group in water. Thus, the free energy of uncharging is $F_0 = -kT \ln W_0$. That is, for a positively charged group $F_0 \approx 0$ at pH > pK, and $F_0 \approx 2.3kT(pK-pH)$ at pH < pK; for a negatively charged group $F_0 \approx 0$ at pH < pK, and $F_0 \approx -2.3kT(pK-pH)$ at pH > pK. Thus, the free energy of discharging does not exceed several kcal mol^{-1} (at a "normal" pH $\approx$ 7) for all the ionizable groups shown in Fig. 10.5.

Part IV

PROTEIN STRUCTURES

LECTURE 11

Now that we know the features of polypeptide secondary structures and the properties of amino acid residues, we can at last pass to proteins.

The "living conditions", structure-stabilizing interactions and overall architecture of proteins provide the basis for classifying them as (1) fibrous proteins, (2) membrane proteins and (3) water-soluble globular proteins.

In this lecture we will consider fibrous proteins.

The function of fibrous proteins is mostly structural. They form microfilaments and microtubules, as well as fibrils, hair, silk and other shielding textures; they reinforce membranes and maintain the structure of cells and tissues. Fibrous proteins are often very large. Among them, there is the largest known protein, titin, of about 30 000 amino acid residues.

Fibrous proteins often form enormous aggregates; their spatial structure is mostly highly regular, composed of huge secondary-structure blocks, and reinforced by interactions between adjacent polypeptide chains. The primary structure of fibrous proteins is also characterized by high regularity and periodicity, which ensures the formation of vast regular secondary structures.

We shall consider some typical representatives of fibrous proteins.

(a) β-structural proteins like silk fibroin. As we know, periodicity of a β-sheet is manifested by residues pointing alternately above and below the sheet (Fig. 11.1).

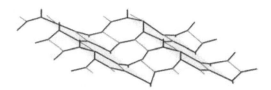

Figure 11.1. A β-sheet with its pleated structure and periodicity emphasized. Hydrogen bonds between the linked β-strands are shown in light blue. Adapted from [12].

In silk fibroin, the major motif of the primary structure is an octad repeat of six-residue blocks, each consisting of alternating smaller (Gly) and larger residues, for example,

$$\sim (/\ \overset{\text{Ser}}{\underset{\text{Gly}}{\diagdown\ \diagup}}\ \overset{\text{Ala}}{\underset{\text{Gly}}{\diagdown\ \diagup}}\ \overset{\text{Ala}}{\underset{\text{Gly}}{\diagdown\ \diagup}}\ /\ )_8\sim$$

and this octad repeat occurs about 50 times, separated by less regular sequences.

Antiparallel (like those in Fig. 11.1) β-sheets of silk fibroin are placed onto one another in the "face-to-face, back-to-back" manner: a double sheet of glycines (the distance between the planes is 3.5 Å) – a double sheet of alanines/serines (as clearly seen by X-rays, the distance between the planes is 5.7 Å) – a double sheet of glycines – and so on.

In a silk fiber, these quasi-crystals consisting of many β-sheets are immersed in a less-ordered matrix formed by irregular parts of fibroin, as well as by sericin, a special disordered matrix protein, that is S—S-bonded into a huge aggregate.

(b) α-structural fibrous proteins formed by long coiled-coil helices (Fig. 11.2). In α-keratin or tropomyosin such helices cover the entire protein chain, and the major part of the myosin chain also forms a fibril of this type. These structures are also observed in some silk (not in the silkworm product considered above but in silk produced by bees and ants).

Associated helical chains form a superhelix known as a "coiled-coil" (Figs 11.2 and 11.3). The coiled-coil is usually formed by parallel α-helices. In different proteins there may be two, three or more α-helices forming the coiled-coils.

As we know, a regular α-helix has 3.6 residues per turn, while the residue repeat of coiled-coil helices is 7.0, i.e., 3.5 residues per turn (Figs 11.3 and 11.4). The typical primary structure of a supercoil-forming chain has the same 7.0 residue repeat (Fig. 11.4; here, lettering in bold corresponds to hydrophobic amino acids forming the main interhelical contacts, while other letters refer to hydrophilic amino acids).

Interestingly, a slight increase in the hydrophobicity of "intermediate" residues e and g turns the double supercoiled helix (Fig. 11.3a) into a triple one (Fig. 11.3b), a greater increase turns it into a quadruple helix, and so on.

The next higher structural level.is the association of supercoiled helices (shown in Fig. 11.2) into fibrils; this happens often, though not always, e.g., it happens in myosin but not in tropomyosin.

It is also of interest that mechanical tension of a wet fiber formed by α-helices may result in its transformation into the β-structure, and when the tension is released and moisture decreased, the α-helical structure restores.

Let us consider in more detail how the helices associate. The α-helix has several "ridges" formed by side groups coming close together (Fig. 11.5, left). The periodicity of some of these ridges is of the 1–4–7–$\cdots$ type (ridges "$i, i + 3$", or simply "+3"),

Figure 11.2. Right-handed coiled-coil α-helices. In the complex they are parallel and slightly wound around each other to form a left-handed supercoil with a repeat of 140 Å. The interhelical contacts are formed by amino acid residues at repeating chain positions **a** and **d** (see Figs 11.3 and 11.4).

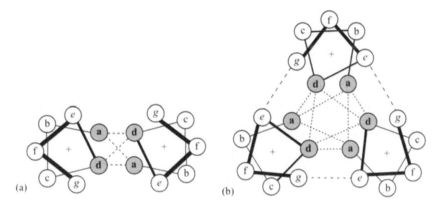

Figure 11.3. Interactions of α-helices in double (a) and triple (b) superhelices (as viewed along the helix axis). In the double helix, only residues **a** and **d** are in immediate contact with another helix, while in the triple helix e and g residues are also involved in contacts (although to a lesser extent). Adapted from [4].

$$\cdots-a-b-c-d-e-f-g-a-b-c-d-e-f-g-a-b-c-d-e-f-g-\cdots$$
$$1\ \ 2\ \ 3\ \ 4\ \ 5\ \ 6\ \ 7\ \ 8\ \ 9\ \ 10\ 11\ \ 12\ 13\ 14\ 15\ \ 16\ 17\ 18\ 19\ 20\ 21$$

Figure 11.4. Typical 7-residue repeats in primary structures of α-supercoil-forming chains.

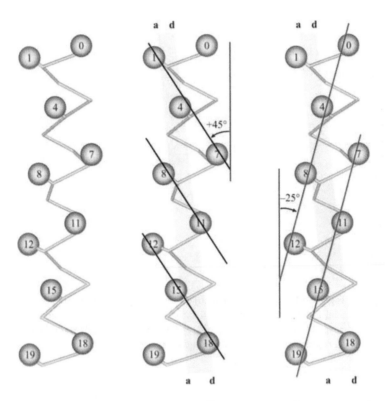

Figure 11.5. (a) The α-helix (main chain and C^{β}-atoms). (b), (c) Two types of ridges (thin lines) formed by side groups on its surface that are close in space. (b) Ridges of "$i, i+3$" type; (c) ridges of "$i, i+4$" type. The bands show the "contact surface" acting in the coiled-coil; its edges are formed by the lines of residues **a** and **d**. Typical angles between the helix axis and ridges $i, i+3$ and $i, i+4$ are given (the angles in the picture look smaller than the typical angles given, since the typical ridges run not through centers of C^{β}-atoms but through massive side groups). Adapted from [1a].

and the contact-zone-involved parts of these ridges consist of residue pairs a_1–d_4, a_8–d_{11}, $\ldots$ (Fig. 11.5, center). Other ridges have periodicity of the 0–4–8–12–$\cdots$ type (ridges "$i, i+4$", or simply "$+4$"), and the contact-zone-forming parts of these ridges consist of the residue pairs d_4–a_8, d_{11}–a_{15}, $\ldots$ (Fig. 11.5, right).

The angle between the helix axis and the "$i, i+4$" ridges is about $-25°$, that between the axis and the "$i, i+3$" ridges is about $+45°$ (Fig. 11.5). If one helical surface is turned about the vertical axis and superimposed on the other surface (Fig. 11.6)

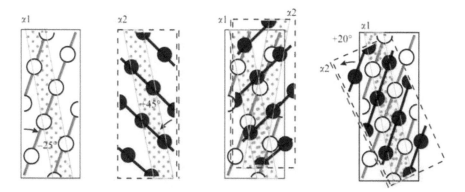

Figure 11.6. To ensure close packing of helix side-chain ridges, a +20° turn of one helix about the other is required. The contact area is viewed through the $\alpha2$ helix turned around its axis. Residues of the "lower" ($\alpha1$) helix are light circles, of the "upper" ($\alpha2$) helix, dark circles. The lines of contact for the residues **a** and **d** are shown in light blue. Adapted from Chothia C., et al., Proc. Natl. Acad. Sci. USA (1977) **74**: 4130–4134.

with a subsequent turn by +20° around the vertical axis, the 1–4–7 ridges of one helix will fit between the 0–4–8 ridges of the other helix (and *vice versa*) and ensure a close contact between the helices (Fig. 11.6, right). Then **a**-groups of one helix fit between **d**-groups of the other, and a slightly twisted contact line arises on the surface of each helix. And when these helices become intertwined (Fig. 11.2) and coiled around the common axis, the interaction area becomes straight and appears in the middle of a slightly twisted helical bundle.

This is not the only way to ensure a close contact of helices; others will be considered when discussing globular proteins. But this is the only good way for the very long helices typical of fibrous proteins. It was predicted by Crick in 1953, the same year that he and Watson predicted the DNA double helix.

(c) Collagen ("glue forming", in Greek). This is the major structural protein amounting to a quarter of the total mass of proteins of vertebrates. It forms strong non-soluble fibrils. A collagen molecule is formed by a special superhelix consisting of three polypeptides (Fig. 11.7) that are free of intra-chain H-bonding and supported by inter-chain hydrogen bonds only.

The conformation of all residues of each collagen chain is close to that of a polyproline (or rather, a poly(Pro)II helix), but collagen chains form additional inter-chain H-bonds. This helix is a left-handed helix with a 3.0 residue repeat. Accordingly, in collagen, the main motif of the primary structure is the repeated triad of residues (Gly-Pro-Pro)$_n$ or rather (Gly-something-Pro)$_n$. Gly is essential for hydrogen bonding in collagen since it (unlike Pro) has an NH group and no side chain; and any side chain would be unwanted in the middle of a tight collagen superhelix where the H-bonding glycine is positioned.

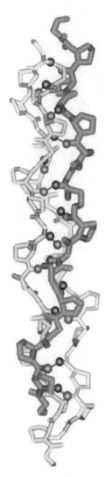

Figure 11.7. A model of the triple collagen helix with the (Gly-Pro-Pro)$_n$ repeat. Each chain is colored differently. H-atoms of NH-groups of glycines (blue) and O-atoms of the first proline in the Gly-Pro-Pro repeat (red) are shown to participate in hydrogen bonds. Gly of chain "1" binds to chain "2", while its Pro binds to chain "3", and so on. Each chain is wound around two others forming a right-handed superhelix. It is called a *super*helix because at a lower structural level, the level of conformation of separate residues, each individual collagen chain is already helical (it forms a *left-handed* "microhelix" of the poly(Pro)II type, with three residues per turn, which is easily seen from the alignment of the proline rings). A collagen molecule is about 15 Å wide and about 3000 Å long.

Interestingly, the collagen chain-coding exons always start with glycines, and the number of their codons is always a multiple of three. As you may remember, eukaryotic genes contain protein-coding exons separated by introns that are cleaved from mRNA (and therefore do not encode proteins).

At the next higher structural level collagen superhelices associate into collagen fibrils.

The biosynthesis of collagen, its subsequent modification and the formation of the mature structure of a collagen fibril have been well studied (Fig. 11.8).

Correct collagen folding needs to be initiated by *pro*collagen that contains, apart from collagen chains, globular heads and tails. Without these heads and tails, the collagen chains fold into "incorrect" triple helices that lack the heterogeneity typical of native collagen (which contains one α_2 and two α_1 chains), its inherent register (i.e., the correct shift of chains about one another), etc. Thus, a separately taken

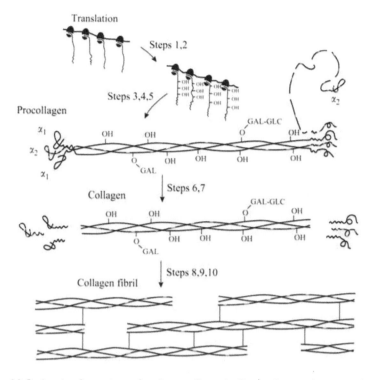

Figure 11.8. *In vivo* formation of collagen. Step 1. Biosynthesis of pro-α_1-chains and pro-α_2-chains (1300 residues long each) in the ratio of 2:2. Step 2. Enzymatic co-translational hydroxylation of some Pro and Lys residues positioned prior to Gly. Step 3. Association of sugars (GLC-GAL) with some hydroxylated residues. Step 4. Formation of the tetramer from C-terminal globules of two pro-α_1 and two pro-α_2 chains; subsequent degradation of one pro-α_2 chain and formation of the procollagen (pro-α_1-α_2-α_1 trimer) with S—S-bonds between globular ends. Step 5. Formation of a triple helix in the middle of procollagen. Step 6. Procollagen secretion into extracellular space. Step 7. Cleavage of globular parts. Steps 8–10. Spontaneous formation of fibrils from the triple superhelices, final modification of amino acid residues and crosslinking (caused by a special enzyme) between modified residues of collagen chains. Adapted from [12].

collagen triple helix is incapable of spontaneously self-organizing its correct spatial structure *in vitro*. In this respect, it is similar to silk fibroin and distinct from some of the previously described α-helical coiled-coils and especially from globular proteins that we will discuss later.

As temperature is increased, the collagen helix melts (this is how gelatin is formed). The collagen melting temperature is strongly dependent on the proportion of proline and oxyproline (the higher the concentration of these rigid residues, the higher the melting temperature, naturally). This melting temperature is usually only a few degrees higher than the body temperature of the host animal. Please note this fact: we will return to it in a future lecture.

Collagen folding is particularly interesting because a number of hereditary diseases are known to be associated with mutations in collagens, the best characterized of which is *osteogenesis imperfecta* or "brittle bone" disease. The most common cause of this syndrome is a single base substitution that results in the replacement of glycine by another amino acid. This breaks the $(Gly-X-Y)_n$ repeating sequence that gives rise to the characteristic triple-helical structure of collagen. It appears that at least one effect of this is to slow folding and allow abnormal post-translational modifications to occur. In order to overcome the difficulty of studying collagen itself, the effect of mutations relevant to disease is explored through the study of highly simplified repeating peptides. This allows detailed biophysical investigations to be carried out, including "real-time" NMR experiments, in order to probe the nature of the folding steps at the level of individual residues. Interestingly, this example illustrates both the increasingly well-established link between protein misfolding and human disease, and the power of properly applied structural methods in probing its molecular origins.

In conclusion, I would like to emphasize that fibrous proteins are often structurally simple owing to the periodicity of their primary structure and, hence, of their secondary structure as well.

However, one more addition would not be out of place here.

Proteins forming huge aggregates without any distinct inner structure are also often classified as fibrous proteins. They form a chemically linked elastic matrix in which other more structural proteins are immersed.

Elastin is a typical matrix protein. It plays an important role in building artery and lung walls, etc. Its long chain is rather hydrophobic and consists of short residue repeats of several types. The resultant product resembles rubber: each elastin chain forms a disordered coil, and together these chains form a net linked by enzyme-modified lysines, four per knot. I cannot but mention that the disturbed function of lysine-modifying enzymes causes the loss of elasticity of vessel walls, and sometimes even a rupture of the aorta.

And one final remark.

Sometimes fibrils formed by globules (e.g., actin or amyloid fibrils) are regarded as fibrous proteins. We will not consider them here – globular proteins are to be discussed later on.

LECTURE 12

Let us now focus on membrane proteins. As concerns their transmembrane parts, these proteins are almost as simple as fibrous proteins.

Membranes are films of lipids (fat) and protein molecules (Fig. 12.1). They envelop cells and closed volumes within them (the so-called compartments). The peculiar role of membrane proteins (they amount to half of the membrane weight) is to provide transmembrane transport of various molecules and signals. The membrane is a kind of "insulator" while its proteins (or rather, as we will see later, protein channels) act as "conductors". These conductors are specific, each ensuring transmembrane transport of molecules of a particular kind or signals from particular molecules (probably by a slight change in the protein's conformation).

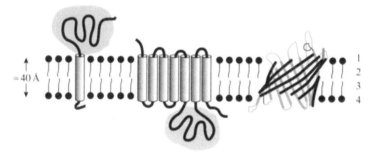

1. Polar "heads" of lipids of one membrane layer
2. Non-polar "tails" of lipids of one membrane layer
3. Non-polar "tails" of lipids of second membrane layer
4. Polar "heads" of lipids of second membrane layer

Figure 12.1. Membrane-embedded proteins. Extramembrane domains are shown in gray. Protein sequences within the membrane are virtually free of irregular segments. In eukaryotes, chain portions projecting from the membrane out of the cell are strongly glycosylated and therefore more hydrophilic.

True membrane proteins reside within the membrane where there is virtually no water. The intramembrane parts have a regular secondary structure, and their size is determined by the membrane thickness.

Let us consider the structure of membrane proteins using several examples.

As a matter of fact, only a handful (a couple of dozen) of such proteins have been solved so far. This is due to their poor solubility in water (detergents have to be used, etc.) and difficulties of crystallization caused by their tendency to disordered association.

Figure 12.2 gives the structure of bacteriorhodopsin which pumps protons across the membrane. This structure was originally determined from many high-resolution electron microphotographs, because it was too difficult to obtain good 3D crystals of bacteriorhodopsin.

As we see, the transmembrane portion of bacteriorhodopsin comprises seven regular α-helices that form a membrane-spanning bundle slightly tilted with respect to the plane of the membrane, while the single β-hairpin and all irregular segments (connecting loops) protrude from the membrane.

The highly regular arrangement of the transmembrane chain backbone is only natural. Each H-bond is expensive in the fatty, waterless environment. Therefore, the protein chain has to adopt a structure with fully accomplished hydrogen bonding, i.e., either the α-helix or the β-cylinder (see below).

The hydrophobic groups positioned on the bacteriorhodopsin α-helices are turned rather "outwards", towards lipids that are also hydrophobic, while the few polar groups face the interior and form a very narrow proton-conducting channel. The proton flow is assisted by a cofactor, which is the retinal molecule bound inside the bundle of helices. It blocks the central channel of bacteriorhodopsin.

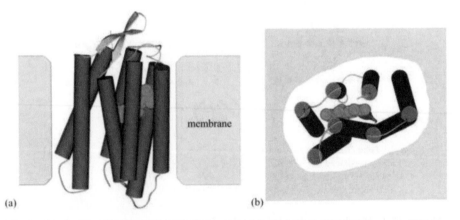

Figure 12.2. Membrane-embedded bacteriorhodopsin: (a), as viewed along the membrane; (b), as viewed from above. Its seven helices are shown as cylinders. The connecting loops are also shown together with the bound retinal molecule (light blue). The lipid layer is schematic.

Having accepted a photon, retinal changes its form from *trans* to *cis*, bends (with a simultaneous slight change of the protein body conformation) and moves a proton from one end of the seven-helix bundle to the other. Then it regains its previous shape, but this time without a proton.

Similar (but usually wider) pores arranged like a hollow bundle of helices can be formed in other cases from separate α-helical transmembrane peptides.

By the way, helical bundles are also abundant in quite different membrane proteins. These do not transport molecules across the membrane but conduct signals.

I am now speaking about receptors, and specifically about hormone receptors (Fig. 12.3). Let us briefly consider one of them.

Having bound a hormone, this receptor somehow changes the conformation of its transmembrane helical bundle (the mechanism is as yet unknown) thereby "announcing" the hormone's arrival. This signal causes the α-subunit of the receptor-associated *G-protein* (Guanine-binding protein) to release its own guanosine diphosphate (GDP) molecule and take up a guanosine triphosphate (GTP) molecule from the surrounding cytosol. Then this α-subunit leaves both its fellow subunits and the receptor, thus providing an opportunity for another GDP-loaded G-protein α-subunit to contact the receptor and then, in its turn, to leave it having exchanged its GDP for GTP.

The α-subunit of G-protein can cleave GTP but, importantly, the process is slow. Meanwhile, together with GTP, the α-subunit (with its "tail" buried inside the membrane) drifts along the membrane, reaches adenylate cyclase and binds to it; as a result, adenylate cyclase starts functioning and converts many molecules of adenosine triphosphate (ATP) into cyclic adenosine monophosphate (cAMP). This initiates a physiological reaction of the cell in response to hormone binding. But the α-subunit's impact upon adenylate cyclase eventually ends when α-subunit-bound GTP turns into GDP. Then, with this GDP, the α-subunit drifts along the membrane and eventually

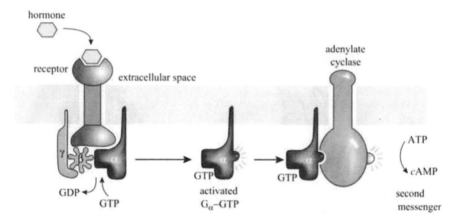

Figure 12.3. Activation of adenylate cyclase by G-protein, which, in turn, is activated by the hormone-binding transmembrane receptor. Reproduced from [1b], with permission.

comes back to the host receptor. If the latter is still bound to hormone, the cycle repeats; if not, it is over.

Hence, the signal of a molecule of hormone is enhanced manyfold, but its duration is finite. This outlines a peculiarity of all G-proteins (not only those activating adenylate cyclase): they function until GTP is cleaved, and a slow GTP cleavage serves as a peculiar timer.

Now we will consider porin (Fig. 12.4), another transmembrane protein. Its structure is highly regular too and looks like a wide cylinder built up from β-structures. Note that here the β-sheet forms a *closed* β-cylinder, thus avoiding the "free" H-bond donors and acceptors typical of edges of a planar β-sheet. The cylinder comprises 16 very long β-segments, and the diameter of a pore in its center is about 15 Å. The side groups of polar residues pertaining to the β-strands face the pore, while nonpolar residues alternating with them in the strand face the membrane.

Porin is responsible for the transport of polar molecules, but it is not very selective.

The transport selectivity, i.e., specificity of function, of membrane proteins is strongly determined by the fact that separate polar groups, not to mention charged ones, can hardly penetrate inside the membrane by themselves.

As you remember, the free energy of a charge q in the medium with permittivity ε is equal to $+q^2/2\varepsilon r$, where r is the van der Waals radius of the charge. It can be easily estimated that, with q equal to the electron charge and $r = 1.5$ Å (the typical radius of a singly charged ion), the value of $+q^2/2\varepsilon r$ is close to 1.5 kcal mol^{-1} at $\varepsilon = 80$ (i.e., in water), while at $\varepsilon_{membr} = 3$ (i.e., inside a "purely lipid" membrane) this value will amount to nearly 37 kcal mol^{-1}. In total, an increase in the free energy ΔF of $+35$ kcal mol^{-1} results. According to Boltzmann, the probability of accumulation of such free energy is $\exp(-\Delta F/kT) = \exp(-35/0.6) = 10^{-25}$. This means that only one in 10^{25} ion attacks on the membrane will be successful. Given the attack time is no less than 10^{-13} s (as you know, this is the thermal fluctuation time), for

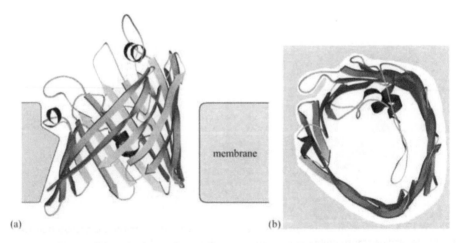

(a)   (b)

Figure 12.4. Porin, as viewed (a) along and (b) across the membrane plane.

an ion passing through a purely lipid membrane would take at least $\sim 10^{12}$ s, that is about ten thousand years... Thus, a purely lipid membrane appears to be virtually impermeable for ions.

It's quite another matter if the membrane includes protein with a more or less broad water-filled channel where ions at least partly enjoy the high permittivity of water, although to some extent restricted by the surrounding membrane. Roughly, the membrane-caused increase in the ion's free energy amounts to about $+q^2/[(\varepsilon_{membr}\,\varepsilon_{water})^{1/2}R]$, where R is the channel radius, $\varepsilon_{membr} = 3$ and $\varepsilon_{water} = 80$. It is easy to calculate that with $R \approx 1.5$ Å, it takes an ion a fraction of a second to pass through the channel, and with $R \approx 3$ Å the time is a tiny fraction of a millisecond.

The channel sites that can attract the ion and thereby reduce the barrier to be overcome regulate the selectivity of ion transfer across the membrane. For example, the presence of a positive charge near the channel accelerates transport of negatively charged ions and strongly hampers transport of positively charged ions (Fig. 12.5). In the case of a negatively charged channel, the transport of positive ions is accelerated while that of negative ions is hampered. This effect (and, of course, the pore size) underlies the selectivity of membrane proteins.

Now let us focus on the photosynthetic reaction center (Fig. 12.6). Its function is to ensure the transport of light-released electrons from one side of the membrane to the other, thereby creating the transmembrane potential that underlies photosynthesis.

The photosynthetic reaction center comprises cytochrome with four hemes (actually, this protein is not a membrane protein: it is outside the membrane in the periplasmic space) and three membrane subunits, L, M, and H (though the transmembrane part of the last is represented by one α-helix only). Subunits L and M are very much alike.

All transmembrane parts are α-helical. As usual, they are long (equal to the membrane thickness) and regular. There are no irregular loops inside the membrane. The outer chain portions are considerably less regular and contain many loops; in

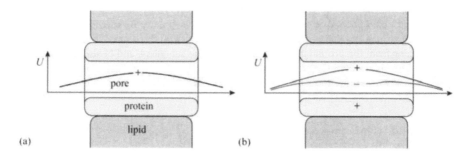

Figure 12.5. Schematic diagram of a transmembrane pore (here the membrane is vertical and the pore is horizontal) and the electrostatic free energy U for positively ($- + -$) and negatively ($- - -$) charged ions: (a) with no charge on the inner pore surface; (b) with a positive charge close to the pore.

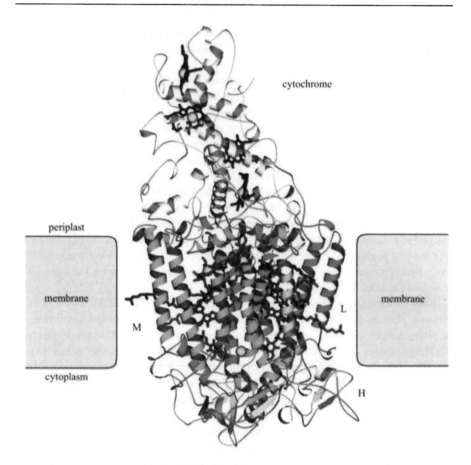

Figure 12.6. A photosynthetic reaction center. The membrane is shown schematically. The transmembrane subunit M appears in light blue, L in green, and H (with its single transmembrane helix) in yellow, while cytochrome is gray. Notice how distinctly more regular the transmembrane chain portions are compared with those outside the membrane. The subunits L and M bind photosynthetic pigments, chlorophylls (shown in red with a yellow spot for the magnesium ion) and pheophytins (shown in dark blue). Each has a long hydrophobic tail projecting from the protein into the membrane. The subunits L and M also bind the two quinones, shown in violet. Cytochrome positioned outside the membrane binds four hemes (grayish-black with yellow spots for iron ions). All co-factors are shown as wire models; see also Fig. 12.7.

fact, their fold is the same as that of "ordinary" water-soluble proteins, which we will discuss later.

Notice the many rather small cyclic molecules, pigments, embedded in this protein: these form the "conductors", i.e., the pathways for the electron flow (the flow of electrons can be followed by the changing electron spectra of pigments during electron transport). The polypeptide only serves as a "shaping insulator".

First a light quantum displaces an electron from the "special pair" of chloro-phylls (see the schematic diagram, Fig. 12.7). Seemingly, having passed through the "accessory" chlorophyll B_A, this electron instantly, in $\sim 10^{-12}$ s, joins to pheophytin P_A (note: P_A and not P_B); $\sim 10^{-10}$ s later, it arrives at quinone Q_A. Then it spends about 10^{-4} s to reach Q_B. We still do not know why the electron prefers this round-about way to Q_B, and why its association with B_A is not observed from electron spectra.

An electron coming from the cytochrome heme replaces the electron released from the special pair of chlorophylls. This completes the first half-cycle of the reaction.

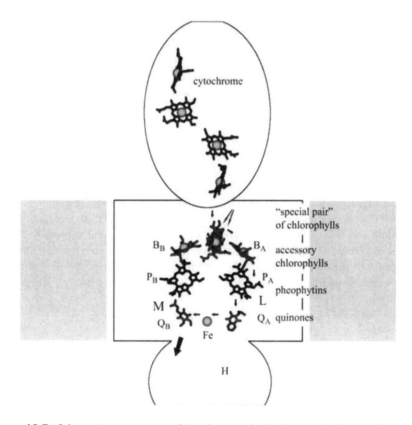

Figure 12.7. Schematic arrangement of the photosynthetic pigments in the reaction center. The reaction center orientation is the same as in Fig. 12.6. The pseudo-twofold symmetry axis of the L and M subunits passes through the "special pair" of chlorophylls and the Fe ion. The pigments' long "tails" are omitted as unnecessary details. Electron transfer proceeds preferentially along the branch associated with the L subunit (on the right of the figure). The electron pathway is shown as small arrows. The large arrow indicates release of the quinone with two consecutively accepted electrons. The left chain is not used. Presumably, it was used in the past, but in present-day reaction centers it is an appendix-like relic.

The other similar half-cycle brings another electron to Q_B making it doubly charged and therefore capable of leaving the membrane more easily to participate in further photosynthetic events.

Thus, the photosynthetic reaction center performs electron transport from the upper (in Figs 12.6 and 12.7) to the lower compartment, i.e., against the apparent difference in electric potential between the compartments. The efficiency is about 50% (in other words, 50% of the captured light is converted into the energy of separated charges, which is not bad at all).

The following two important physical aspects must be emphasized:

1. All chlorophylls (Fig. 12.8), pheophytins, quinones, and other pigments contain partial double (p-electron) bonds of the $\cdots-C=C-C=\cdots$ type. In other words, owing to Pauling resonance ($\cdots C=C-\cdots \Leftrightarrow \cdots C-C=\cdots$), each of these molecules is covered by a common electron cloud, and on such a molecule electrons move as on a piece of metal. This provides a potential well where electrons are "delocalized", i.e., where they can shift by distances greater than the atom diameter. (In passing, note that it is electron delocalization that underlies the typical pigment colors: an electron localized in a separate covalent bond is excited by short-wave UV light, whereas a delocalized electron is excited by "ordinary" visible light of a longer wavelength.)

2. Electron transfer from one "piece of metal" (pigment) to another requires no direct contact of these pigments. It is performed by quantum *tunneling* (see Fig. 12.9).

   The main point about tunneling (also called the "sub-barrier" transition because it is as if the electron passes *under* the energy barrier) is that in accordance with quantum mechanics, an electron (like any other particle, and especially a light particle) "projects" slightly beyond the potential well in which it resides. The host molecule of the electron (chlorophyll, pheophytin, etc.) serves as the "potential well", i.e., the area of low potential energy U. When outside the "well" (Fig. 12.9)

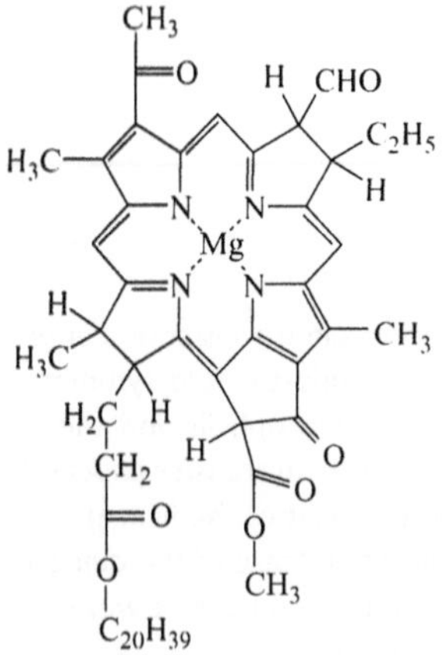

Figure 12.8. A molecule of bacteriochlorophyll.

Figure 12.9. Schematic representation of tunneling. The profile of the electron potential energy U is shown as a solid line. The dashed line shows the level of total (potential + kinetic) energy of the electron. The bell-shaped line (with hatching below) represents the density of the electron cloud ρ. Initially, the electron stays in the left well (as shown in the figure) with the edge of its cloud (although at an extremely low density) reaching the right well, where the electron can move with time if the energy proves to be lower there. Given a well depth of a few electron volts or about 100 kcal mol^{-1} (the typical energy required for molecular ionization), the typical distance at which the density of the electron cloud is reduced by an order of magnitude is about 1 Å.

the electron's potential energy is higher than its total (potential + kinetic) energy when in the well. But for the quantum effect, this deficient energy would not let the electron leave the well. However, owing to the quantum effect the electron wave function (or, simply, the density of the electron cloud) extends beyond the potential well, although the value of this density decreases exponentially with growing distance from the well.

The latter point is another manifestation of the same quantum effect that prevents an electron from falling onto the nucleus: although this would decrease the potential energy of the electron, its kinetic energy would increase still more. The thing is that if the distance between the electron and the nucleus (Δx) tends to zero, the potential energy of the electrostatic interaction of the electron with the nucleus tends to minus infinity as $1/(\Delta x)$, while according to the Heisenberg *uncertainty principle*, the electron's kinetic energy tends (at $\Delta x \to 0$) to plus infinity far more rapidly, as $1/(\Delta x)^2$.

Indeed, according to the Heisenberg principle, uncertainty in speed (Δv) and uncertainty in coordinate (Δx) of a particle are related as $m\Delta v\Delta x \sim \hbar$, where $\hbar$ is Planck's constant divided by 2π, and m is the mass of the particle. In other words, the absolute value of the particle's speed (v) in the well of a Δx width is about $\hbar/(m\Delta x)$ (with complete uncertainty of the direction in which the particle moves at the given moment). Hence, the kinetic energy of the particle, $E = mv^2/2$, is a value of about $m[\hbar/(m\Delta x)]^2 = (\hbar^2/m)/\Delta x^2$.

The same is true for an electron staying in the potential well: provided it does not project a straw beyond the well, its total energy would be higher.

That is why the electron slightly "comes out" of the potential well and its density decreases exponentially, like the electron cloud of an atom. The typical distance at which the cloud density becomes an order of magnitude (10-fold) lower is about 1 Å (which, as we know, is the typical radius of an atom).

Hence, the electron cloud density becomes 1000 times lower at a distance of 3 Å from the "home well". This means that the probability of electron's moving as far as 3 Å off its "home well" during one vibration is about 10^{-3} (for 5 Å it is about 10^{-5}, for 10 Å about 10^{-10}, and so on). When in the "well" (in a pigment molecule), an electron performs $\sim 10^{15}$ vibrations per second (it vibrates at visible light frequencies: this is seen from light absorption spectra of such molecules). Hence, the typical time of its transition into another "well" (another pigment molecule) 3 Å away is about 10^{-15} s$/10^{-3} = 10^{-12}$ s; for 5 Å it is about 10^{-10} s, for 10 Å about 10^{-5} s and for 15 Å about 1 s. This relationship between transition rates and distances is in agreement with what we observe in the photosynthetic reaction center.

The following points deserve your special attention.

First. The total distance of the transition is about 40 Å. This distance cannot be covered by one tunnel jump (it would take $\sim 10^{-15}$ s$/10^{-40} \sim 10^{25}$ s or $\sim 10^{17}$ years, which is beyond the lifetime of the Universe). However, because of the protein arrangement this great jump is divided into four small jumps from one electron-attracting pigment to another; as a result, an electron covers the entire 40 Å distance during a fraction of a millisecond.

Second. To prevent an electron's prompt return to the first pigment from the second one, and to promote its further movement to the third pigment and so on, its total (potential + kinetic) energy must decrease along the pathway; in other words, every step of the electron path must be a descent from a high-energy orbital to a low-energy one. The arrangement of the photosynthetic reaction center is believed to provide such a decrease in electron energy from pigment to pigment.

Third. An electron spends no energy on tunneling (since there is no "friction" here). The energy is decreased by the electron-conformational interaction. Specifically, when arriving at the next pigment, an electron faces the pigment's conformation corresponding to the energy minimum *without* the newcomer. In its presence the energy minimum corresponds to another, slightly deformed, conformation of the pigment (i.e., another location of the nuclei of its atoms). When adopting this new conformation, the pigment atoms rub against the surroundings, and the excess energy dissipates. As a result, at each step the electron changes a high-energy orbital for a low-energy orbital, its energy decreases and appears to be spent on making the tunnel transition "efficient", i.e., irreversible.

And finally: A tunneling ("sub-barrier") transition can be distinguished from an ordinary activation mechanism that requires the overcoming of an energy barrier. The rate of tunneling is virtually temperature-independent (and therefore tunneling does not disappear at low temperatures), whereas the rate of an activation transition (proportional to $\exp(-\Delta E^{\#}/k_B T)$, where $\Delta E^{\#}$ is the energy of the activation barrier, and T is temperature) decreases dramatically with decreasing temperature.

LECTURE 13

Now we will focus on globular proteins or, rather, on water-soluble globular proteins. They are the best-studied group: for hundreds of them the spontaneous self-organization is known; for thousands, their atomic 3D structure. Therefore, it is this type of protein which is usually meant when "the typical protein structure", "the regularities observed in protein structure and folding", etc. are discussed. After this necessary remark, let us consider the structures of globular proteins.

First X-ray, and later two- and many-dimensional NMR studies by hundreds of laboratories have yielded, after half a century of intensive work, the atomic structure of about 3000 proteins (and taking mutants and various functional states into account, the number is increased about five-fold).

Inner voice: Does the structure seen by X-rays in the crystal coincide with the protein structure in solution?

Lecturer: It virtually does, as a rule. This is supported by three groups of data. First, it can often be shown that a protein preserves its activity in the crystal form. Second, sometimes one protein can form different crystals, with its structure virtually unchanged. Finally, the NMR-resolved structure of a protein in solution and its X-ray-resolved structure in a crystal are virtually the same. However, a reservation should be made that some flexible portions of proteins (some side chains, loops, as well as inter-domain hinges in large proteins) may have a changed structure after or due to crystallization. But this only concerns either minor details of protein structures or connections between domains and sub-globules in large proteins, rather than small single-domain protein globules, which are virtually solid.

I should add that X-rays see not only the "static", averaged structure of a protein (which is the subject of this lecture and the next) but also thermal vibrations of protein atoms, which will be discussed briefly later.

So, what is the bird's-eye view of water-soluble globular proteins?

We see that short chains (of 50–150 or, less frequently, 200–250 residues) pack into a compact 25–40 Å globule (Fig. 13.1), and that larger proteins consist of a few such subglobules, or domains (Fig. 13.2). The protein chain is packed into a globule as

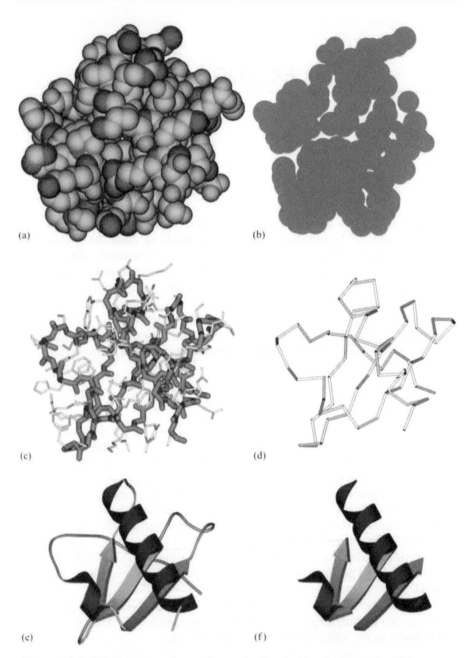

Figure 13.1. The structure of a small protein, interleukine 8, shown in different ways. (a) The atomic model (only "heavy", non-hydrogen atoms are shown: nitrogens in blue, oxygens in red, carbons in gray); because of the close packing of the chain, we see only the protein surface. (b) The cross-section of the atomic model emphasizes the close packing.

tightly as organic molecules into a crystal. This is clear when you look at both the protein surface (Fig. 13.1a) and the cross-section of a protein globule shown in Fig. 13.1b. However, when examining a protein we will not focus on the closely packed electron clouds (or van der Waals surfaces) of atoms: then nothing can be seen inside the protein; instead, we will inspect the atom-"flesh"-free (Fig. 13.1c) and even side-chain-free (Fig. 13.1d) skeletons (wire models) of protein molecules. But do not submit to the impression (often created by drawings) that a protein molecule is a loose structure.

The spatial structures of nearly all globules (domains) are composed of the regular secondary structures already familiar to us: α-helices and β-sheets (Fig. 13.1e) stabilized by regular H-bonds in the regular main chain. In globular proteins the total proportion of α- and β-structures amounts to 50–70% of the number of residues. By the way, Pauling, Corey and Branson theoretically predicted these secondary structures prior to the resolution of atomic structures of protein molecules. The "stack" composed of these structures (Fig. 13.1f) determines the main features of protein architecture.

The hydrophobic core (or cores) of the protein is surrounded by α- and β-structures, while irregular loops are moved towards the edge of the globule. The loops almost never enter the interior of the protein. This can be easily explained by the necessity for their peptide groups, uninvolved in secondary structures, to preserve

Figure 13.2. Globular domains in γ-crystallin. The pathway of the chain is traced in color (from blue at the N-terminus via green in the middle to yellow and red at the C-terminus).

Figure 13.1. (c) Wire model of the main chain (dark line) and side chains (the lighter projections). (d) The pathway of the main chain. (e) Diagram of the protein fold showing the secondary structures involved (two α-helices and one β-sheet consisting of three β-strands). (f) Structural framework ("stack") of the protein globule built up from secondary structures. The projection and scale are the same for all the drawings.

their H-bonds to water, otherwise the globule's stability would be compromised. Note, in passing, that X-rays often find H-bonds between water molecules and loops, α-helix ends and β-sheet edges.

The structural features of the main chain are the basis for subdivision of globular proteins into "pure" β-proteins, "pure" α-proteins, and "mixed" α/β and $\alpha + \beta$ proteins. Strictly speaking, this classification refers to small proteins, as well as to separate domains (i.e., to compact subglobules forming large proteins); large proteins can contain, say, both β- and α-domains.

Of particular interest to us are (1) the architecture of packing of α- and β-structural segments into a compact globule (Fig. 13.1f), and (2) the pathway taken by the chain through the globule (Fig. 13.1e) or, as it is often called, "the topology of the protein globule".

We will frequently use simplified models of protein structures (Fig. 13.3). The simplification implies not only focusing on secondary structures (with details of loop structures neglected) but also paying no attention to the difference in size of these structures or to details of their relative orientation (in this way we pass from "folds" (Fig. 13.3a) to "folding patterns" (Fig. 13.3b) of protein chains).

The simplification is justified by the change in the details of loop structures and precise sizes and orientations of structural segments (and even of some small structural segments themselves) that occurs when the protein is compared with another one of similar sequence (i.e., with its close relative of the same origin) – for example, when hemoglobin α is compared with hemoglobin β (Fig. 13.4).

The next, higher level of simplification necessary for classification of protein structures is restricted to the packing of structural segments in a globule, i.e., the stacks are composed of secondary structures, with no attention paid to loops connecting these secondary structures in a molecule (Fig. 13.3c).

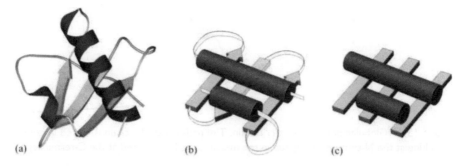

(a)   (b)   (c)

Figure 13.3. Simplified models of protein structures. (a) A detailed *fold* describing the positions of secondary structures in the protein chain and in space (see also Fig. 13.1e). (b) The *folding pattern* of the protein chain with details of loop pathways, the size and exact orientation of α-helices (shown as cylinders) and β-strands (shown as arrows) omitted. (c) *Packing*: a stack of structural segments with no loop shown and omitting details of the size, orientation and direction of α-helices and β-strands (which are therefore presented as ribbons rather than as arrows).

Figure 13.4. Two close relatives: horse hemoglobin α and horse hemoglobin β (both possessing a heme shown as a wire model with iron in the center). Find similarities and differences. (Tips: they are highly similar although they have some differences in details of loop conformations, in details of the orientation of some helices, and in one additional helical turn available in the β globin, on the right).

I will purposely use such simplified models of chain folds and packing along with computer-produced "true" protein structures. It might seem pointless to use the simplified models when a computer can describe the structure "as it is". However, this "as it is" picture has a lot of unnecessary details, while models embody the main features that are the same in similar proteins. Therefore, the models are useful both in classifying protein structures and in outlining their major typical features. When scrutinizing a picture of a protein, we cannot but outline its typical features in our minds – exactly what is done by models, which simply help viewers to systematize their intuitive perception. Besides, models allow us to compose the "verbal portrait" of a protein. Because these models and "verbal portraits" embody the main features and omit details, they will be of practical use as soon as you would like to find out how a protein under consideration resembles others. Of course the omitted details can be basic in protein functioning (as we will see later), but this only emphasizes that the function of a protein is relatively independent of the folding pattern of its chain.

Chain "packings" and "folding patterns" do not bring into focus all possible (loose, open-work, etc.) complexes of structural segments but only those closely packed. Thus they outline the configuration areas corresponding to close (although free of steric overlapping) packing of the protein chain into a globule, i.e., the vicinity of sufficiently deep energy minima of non-bonded interactions. These areas allow us not only to classify known protein structures but also to predict new ones yet to be detected.

It is not out of place to mention that when speaking about classification of protein structures, about similarities displayed, and so on, I will not mean the commonplace that all globins are alike irrespective of whether their host is a man or a lamprey. This is certainly true, and proteins can be divided into phylogenetic classes within which their functions and, importantly, their amino acid sequences do not vary much. However, similar structural features are often intrinsic to proteins that, as revealed by numerous tests, evolutionarily have nothing in common. And I will concentrate on this purely structural (not genetic) similarity.

We begin with the architecture of β-proteins. Structurally, β-structural domains turn out to be simpler than others: two (or sometimes several) β-sheets composed of extended chain segments are stacked one onto another. In other words, the "stacks" of secondary structures look quite simple in β-proteins. The *anti*parallel β-structure predominates in β-proteins.

Since proteins are composed of asymmetric (L) amino acids, the extended β-strands are slightly twisted individually: as you may remember, the energy minimum of an extended conformation is positioned above the diagonal in the Ramachandran plot. The twisted β-strands are H-bonded into β-sheets that are arranged in a propeller-like assembly (Fig. 13.5). The angle between adjacent extended strands of the β-sheet is about $-25°$. This propeller-like assembly looks *left*-handed when viewed across the β-strands (Fig. 13.5a) and *right*-handed when viewed along the β-strands (Fig. 13.5b). The latter is a common viewpoint (*along* the β-strands), and hence the β-sheet is said to have a *right*-handed twist.

There are two basic packing types for two β-sheets, namely, orthogonal packing and aligned packing (Figs 13.6a and 13.6b). In both cases the sheets pack "face-to-face" around a hydrophobic core of the domain, though their relative arrangement is different: in the second case the angle between the sheets is about $-30°$ only ($\pm10-15°$), while in the first case it amounts to $90°$ ($\pm10-15°$); angles beyond these two ranges (specifically, angles of about $+30°$) are rare.

In *orthogonal* packing (Fig. 13.6a), the β-strands are twisted and usually slightly bent, so that the overall architecture of the "stack" resembles a cylinder with a significant angle between its axis and the β-strands. This type of β-sheet packing is often called the β-cylinder or the β-barrel, although in β-cylinders composed of *anti*parallel

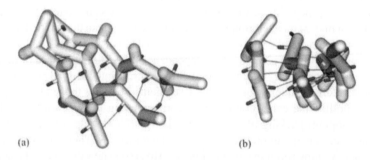

(a)   (b)

Figure 13.5. The β-sheet as viewed (a) across and (b) along its β-strands. The sheet is pleated (which is emphasized by projecting the C^β-atoms shown in green) and usually has a right-handed (viewed along the strands) propeller-like twist. H-bonds between β-strands are shown by thin lines.

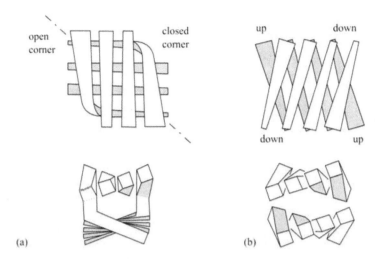

Figure 13.6. The orthogonal (a) and aligned (b) packing of β-sheets viewed face on (top) and from their lower end (bottom). In the face view, the β-strands are wider as they approach the viewer. The dashed line shows the axis of the orthogonal β-barrel to which both "open" corners belong. Here the two β-sheets are most splayed. At the two "closed" corners the sheets are extremely close together; here the chain bends and passes from one layer to the next. In the orthogonal packing the hydrophobic core is almost cylindrical. In contrast, in the aligned packing, the core is flat, the distance between the twisted sheets remains virtually unchanged, and the relative arrangement of the sheets allows the hydrophobic faces of twisted β-strands to make contact over a great length. Adapted from Chothia C., Finkelstein A.V. *Annu. Rev. Biochem.* (1990) **59**: 1007–1039.

β-structures (unlike in those of parallel β-structures to be discussed later) the two β-sheets are usually clearly distinguished, because on opposite sides of the barrel the H-bond net is often fully or partially broken. In the "closed" corner of this packing, β-segments of both sheets come close together, which enables the chain to pass from one sheet to the other at the expense of a 90° bend; it can be said that a single sheet is bent with one part imposed on the other. At the opposite ("open") corners, the β-sheets splay apart, the open corner being filled with either an α-helix or irregular loops, or even with an active site, as in the case of retinol-binding protein (Fig. 13.7).

The *aligned* packing (Fig. 13.6b) is typical of non-bent sheets with a propeller-like twist. This packing type is usually called a β-sandwich. Its ends are covered with irregular loops (as you can see in Fig. 13.8). In some β-sandwiches the edge strands of the β-sheets are so close (and sometimes even H-bonded) that the packing acquires the shape of a cylinder with a small angle between its axis and the β-strands.

It should be stressed that the folding patterns are much less numerous than the known protein globules, and the types of stacks are in turn much fewer than the folding patterns of the protein chain. One and the same stack, i.e., one and the same packing of structural segments, can conform to various patterns of chain folding in a globule; in other words, these segments can be arranged in the polypeptide chain in

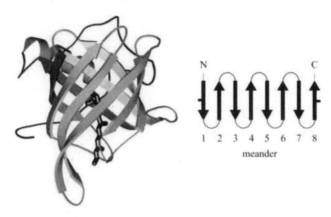

Figure 13.7. Retinol-binding protein exemplifies β-sheet orthogonal packing. The chain pathway resembles the "meander" pattern (see the topological diagram, the planar presentation of the β-sheet, on the right). In this diagram β-strands are shown as arrows. The "meander" results from the fact that β-strands adjacent in the chain are also adjacent at the surface of the cylinder; there are H-bonds between them (the H-bonding between the edge (in the planar diagram) β-strands is shown by small lines). The retinol-binding site is at the cylinder axis. Retinol is shown in violet. The numbers in the topological diagram reflect the order of structural segments in the chain.

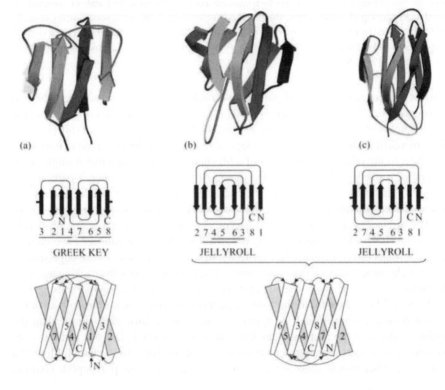

different ways. Figure 13.8 shows how one and the same β-sandwich (with a variety of topologies, i.e., with different pathways taken by the chain through this sandwich) serves as the structural basis of three distinct protein domains: (a) a domain of γ crystallin, (b) the β-domain of catabolite activating protein, and (c) the virus coat protein. And the folding patterns of the last two proteins are the same, i.e., in these proteins the chain takes the same pathway through identical β-structural stacks (this is emphasized by the brace in Fig. 13.8).

Examples of such structural similarity in the absence of any other apparent relationship between the proteins are plentiful.

Here I cannot resist the temptation of showing you another β-sandwich-based protein. It is immunoglobulin; each domain of this large protein is arranged in (or close to) the way shown in Fig. 13.9.

Similar folding patterns are intrinsic to the chains of about fifty other superfamilies bearing no sequence similarity to immunoglobulin (although some are responsible, like immunoglobulin, for specific binding to certain agents, for example in the course of cell recognition).

Some members of these superfamilies are somewhat different from the "standard" structure presented in Fig. 13.9. So, in some proteins β-strand 1 forms a parallel β-structure with strand 7 (and then it can even lose contact with strand 2); in others β-strand 4 moves to strand 3 from strand 5; sometimes an additional β-hairpin forms

---

Figure 13.8. Examples of β-sheet aligned packings. The chain folding patterns for (a) γ crystallin (see also Fig. 13.2), (b) the β-domain of catabolite activating protein and (c) the coat protein of satellite tobacco necrosis virus. The topological diagrams for these proteins are shown below. For all these proteins, the typical topology is a "Greek key" (a "long hairpin bent in two") where four β-strands are adjacent in the chain and antiparallel, and there are H-bonds between the first and the fourth strand. In the topological diagrams structures of the two-layer β-sandwich are drawn in one plane (as for a cylinder cut along its side and unrolled flat). The place of cutting is chosen to stress the symmetry of the chain fold. For example, in γ crystallin, the slit between β-strands 3 and 8 stresses the similarity of the first (strands 1–4) and the last (strands 5–8) halves of the domain. A shorter distance between the β-strands implies their H-bonding; H-bonds (if any) between the edge β-strands are shown as small projecting lines. A gap between the β-strands separates two β-sheets of the sandwich (if there are no H-bonds between them). The domain of γ crystallin contains a repeated Greek key: one formed by strands 1–4, and the other by strands 5–8; still another Greek key is composed of the β-strands 4–7. The proteins shown in drawings (b) and (c) have Greek keys formed by strands 3–6 and 4–7. Moreover, their topology can be described as a repeatedly bent hairpin (usually called a "jellyroll") where β-strand 1 is H-bonded to 8, and 2 to 7, in addition to the H-bonds between strands 3 and 6, 4 and 5, typical of Greek keys. Note that for the purpose of enveloping the globule core by the chain a Greek key proves to be better than "meander" topology (Fig. 13.7) because apart from the β-structure on the sides, it provides loops from below and above. Usually, β-sheet aligned packings are β-sandwiches (a, b), but some of them (e.g., the coat protein of satellite tobacco necrosis virus and coat proteins of some other viruses) can also be seen as β-cylinders with co-linear β-strands (c).

Figure 13.9. The aligned packing of β-sheets in the constant domain of the light chain of immunoglobulin κ. On the left, a detailed diagram of the protein is shown; the chain pathway is traced in color (from blue to red) from the N- to the C-terminus. The topological diagram (in the center) accentuates the "Greek keys". On the right, the protein is shown as viewed from below (i.e., from the butt-ends of structural segments; the butt-ends shown as rectangles). The cross corresponds to the strand's N-end (i.e., "the chain runs from the viewer"), and the dot to the C-end (i.e., "the chain runs towards the viewer"). The segment-connecting loops close to the viewer are shown by black lines, and those distant (on the opposite side of the fold) by light lines. Note that such a diagram allows the presentation of the co-linear packing of these segments (β-strands) in the simplest possible way. It also visualizes the spatial arrangements of "Greek keys" and makes evident that two of them (formed by strands 2–5 and 3–6, respectively) differ in their spatial arrangements. The structure formed by two Greek keys that overlap as shown in this figure is sometimes called a "complete Greek key".

in the connection between β-strands 3 and 4. But the core of the fold, which involves β-strands 2, 3, 5, 6, 7, remains unchanged.

Apart from the pleasure of showing you this very popular folding pattern, I also aim to show that the easiest way to illustrate the folding pattern of a protein with more or less co-linear packing of its structural segments is to use the diagram giving a view from the butt-ends of β-strands.

Now let us come back to the orthogonal packing of β-sheets and see some other common architectures. Figure. 13.10 shows how one and the same β-cylinder (with two different chain topologies, i.e., with two different pathways taken by the chain through the orthogonal packing of the β-sheets) serves as the basis for both a serine protease like chymotrypsin (a) and an acid protease like pepsin (b).

So far, we have considered the most significant "basic" arrangements of β-proteins. However, there are other "basic arrangements", for example the "multiple-blade propeller".

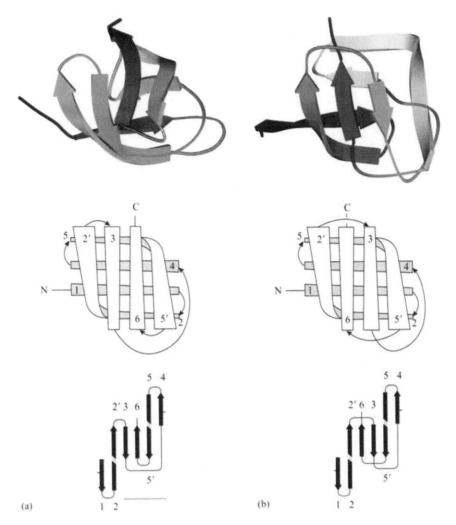

Figure 13.10. The chain folding patterns in a serine protease such as chymotrypsin (a) and in an acid protease such as pepsin (b). For the latter, the loops are shortened and rather schematic. The orthogonal packings of β-sheets in these proteins are shown along with the β-sheet topology diagrams. In the trypsin diagram, the "Greek key" is underlined. In both folding patterns the β-sheets bend, such that their edges move away from the reader and stick together by H-bonding (marked as short lines); the places where the β-sheet bends are colored lighter in the topology diagrams (see β-strands 2 and 5).

In the neuraminidase "propeller" (Fig. 13.11), six inclined β-sheets form a rosette (in other proteins of this kind there may be as many as eight sheets). If considered in pairs, the sheets form β-sandwiches; therefore the "propeller" can be described as a supercylinder built up from β-sandwiches.

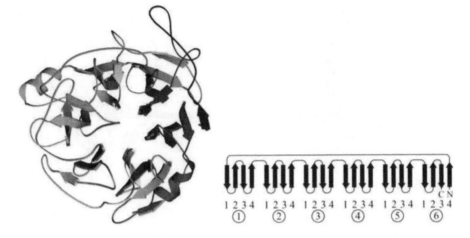

Figure 13.11. The β-structure in the form of a "six-blade propeller" in neuraminidase, and a topological diagram of this protein, which is composed of six antiparallel β-sheets. Adapted from [1a] with minor modifications.

As you can see, the axis is not covered with loops and the "indent" at the super-cylinder axis contains the active site. We have already seen one similar position of the active site: in the retinol-binding protein (Fig. 13.7) it is also located in an indent in the middle of the cylinder; and we will see it again later.

The structural arrangement of the "β-prism" type (also called the "β-helix") (Fig. 13.12) is of interest mostly due to its regularity. The three facets of this prism are formed by three β-sheets (note: parallel β-sheets) such that the chain takes its pathway through them continuously passing from one sheet to the next. The chain appears to coil around the axis of this prism and forms either a right-handed helix, which is typical for joining parallel β-strands, or (in the other prism) a left-handed helix, which is extremely rare for the case of joining β-strands in other proteins.

Now it would not be out of place to discuss the *topology* of β-proteins. We have observed that β-proteins are built up from mostly antiparallel β-structures. The structure of the majority of β-proteins that have been discussed so far is purely antiparallel. Although sometimes a minor admixture of the parallel structure was observed (see Fig. 13.10b), proteins built up from a purely parallel β-structure are extremely rare, although they do exist (see Fig. 13.12).

The fact that an admixture of parallel and antiparallel structures rarely occurs is not surprising since parallel and antiparallel β-structures have somewhat different conformations, and therefore their connection is likely to be energetically unfavorable. The extent of correlation between the unfavorability and the uncommon occurrence of various structures will be discussed in a later lecture, but in principle it is clear that a stable system (like protein) must be composed of mostly stable elements and avoid those internally unstable.

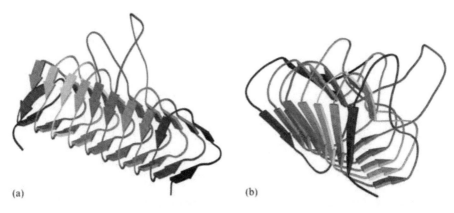

Figure 13.12. The β-prism in acyl transferase (a) and in pectate lyase (b). Notice the hand-edness of the chain's coiling around the axis of the prism: it is unusual, *left*, in (a) and common, *right*, in (b). Also note that when the chain's coiling is left-handed, as in (a), the common twist of the β-sheet is absent. This common twist, i.e., the right-handed (viewed along the β-strands) propeller twist, is seen in (b) and in Figs 13.5–13.11.

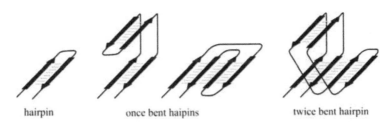

hairpin   once bent haipins   twice bent hairpin

Figure 13.13. Antiparallel β-hairpins.

The mostly *anti*parallel character of β-sheets in β-proteins is closely connected with the fact that their architecture is usually based on β-hairpins (Fig. 13.13). These hairpins are often bent and sometimes may even have two or three such bends (see Fig. 13.13 and also Fig.13.8b, c).

The pathways of loops connecting β-segments usually start and finish on the same edge of the fold (i.e., the loops do not cross the "stack" but cover its butt-end). This is well seen from almost all drawings. The loops, even long ones, tend to connect ends of β-segments that are close in space. That is why, as a rule, β-segments adjacent in the chain are not parallel and tend to form antiparallel β-hairpins.

Also note that "overlapping" loops (or "crossed loops") occur rarely (an exception of this kind is shown in Fig. 13.10b), probably because one of the crossed loops must have an energetically unfavorable additional bend (to avoid a collision or dehydration). The avoidance of loop overlapping is a general structural rule for proteins.

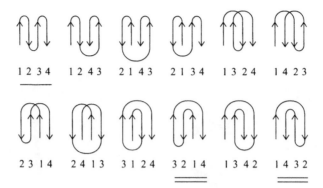

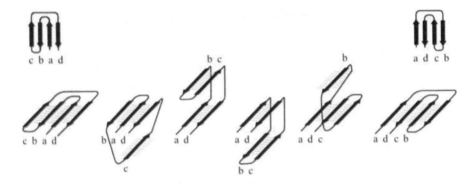

Figure 13.14. Possible topologies of sheets composed of four β-strands. The scheme includes only the sheets where two β-strands adjacent in the chain are oppositely directed. Among these, the common topologies are "meander" (underlined once) and two "Greek keys" (underlined twice), the latter two being different only in the direction of the chain turn from the hairpin consisting of strands 1 and 4 to the hairpin consisting of strands 3 and 2. The "meander"-containing protein is exemplified by retinol-binding protein (see Fig. 13.7); the examples of "Greek key"-containing proteins are γ crystallin and other proteins shown in Fig. 13.8, or trypsin (Fig. 13.10).

Figure 13.15. Various spatial Efimov's "abcd" structures with Greek key topology. Notice the *right-handedness of the superhelix* that consists of two parallel β-strands from one β-sheet and one β-strand (between them) from another β-sheet. It is seen, for example, in the superhelices b-c-d (the second drawing from the left in the lower line) and a-b-c (the second drawing from the right in the lower line). The "right-handed" connection of parallel strands of the same β-sheet is typical for proteins; the reverse ("left-handed") is extremely rare.

Among a variety of configurations of a β-sheet formed by a continuous chain (Fig. 13.14), those really abundant are only three: the "meander" and two "Greek keys" (underlined in the drawing): these folding patterns have none of the disadvantages mentioned above.

By the way, "Meander" is the name of a very winding river in Greece; in the meander pattern, β-strands adjacent in the chain are also adjacent in space (Figs 13.14 and 13.8) and usually linked with H-bonds.

It is characteristic of the "Greek key" pattern (which can be seen on ancient vases and garden railings), that four β-strands adjacent in the chain are antiparallel, and that there are H-bonds between the first and the fourth strands. Actually, the second and/or the third β-strand of the "Greek key" usually belong to another β-sheet rather than to the same one (as it may appear from Fig. 13.14). This gives rise to various spatial structures (the so-called Efimov's *"abcd" structures*) with the same Greek key topology but with different shapes in space (Fig. 13.15). Look for them in Figs 13.8 and 13.10.

These typical protein structures (hairpins, meanders, Greek keys, abcd structures, etc.), which are built up from elements of the β- (and/or α-) structure that are adjacent in the chain, are often called "supersecondary" structures.

LECTURE 14

Now we pass to α-proteins, which are proteins built up from α-helices. They are more difficult to classify than β-proteins. The reason is that the arrangement of β-strands in the sheets is stabilized by hydrogen bonds of the main chain (which is everywhere the same), while the arrangement of α-helices in a globule is maintained by close packing of their side chains, which vary greatly in size. This is why, unlike β-strands, α-helices do not pack into more or less standard sheets.

The α-proteins composed of long α-helices have the simplest structure. This structure is a bundle formed by almost parallel or antiparallel (in a word, co-linear) long α-helices. We have already met such bundles when considering fibrous and membrane proteins.

Figure 14.1 presents three four-helix proteins. These proteins, structurally very close, have different functions: cytochrome binds an electron, hemerythrin binds oxygen, and tobacco mosaic virus coat protein binds molecules that are much greater in size – other coat proteins and RNAs. The first two proteins may have something in common in their function because they both act as carriers in the respiratory chain. This resemblance is far from close, though: in cytochrome, the polypeptide binds the heme that binds an iron that binds an electron, while in hemerythrin the polypeptide binds iron ions without any mediating heme, and two iron ions bind an oxygen. Thus, functionally, hemerythrin and cytochrome bear some resemblance, although only a minor one, and they have no common function with the RNA-binding virus coat protein – in spite of the fact that all three of them are very similar structurally. The similarity is not restricted to the overall architecture (a four-helix bundle) but also involves the pathway taken by the chain through this bundle, i.e., the folding pattern of the chain. The latter is well illustrated by a common topological diagram (Fig. 14.1, bottom) that shows a view along the bundle axis.

Thus, proteins with the same folding pattern may have utterly different functions. In contrast, hemerythrin and the classical oxygen-binding protein myoglobin (Fig. 14.2) have identical functions (the former in worms and the latter in vertebrates, including ourselves), while their architectures are utterly different except that they are both α-proteins. However, in hemerythrin all the α-helices are parallel, while in myoglobin they are assembled into two perpendicular layers. This is another example

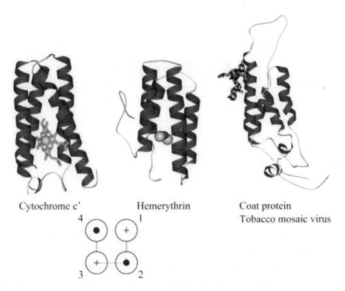

Cytochrome c'   Hemerythrin   Coat protein
Tobacco mosaic virus

Figure 14.1. Three α-proteins that are similar in architecture ("four-helix bundle") but different in function: cytochrome c′, hemerythrin and tobacco mosaic virus coat protein. Both the protein chain and co-factors are shown: wire models represent the heme (in cytochrome) and an RNA fragment (in virus coat protein), orange balls are for iron ions (in the cytochrome heme and in hemerythrin), and the red ball is for iron-bound oxygen (in hemerythrin). The overall architecture of such "bundles" resembles the co-linear packing of β-sheets. The topological diagram (below) shows all these proteins as viewed (in the same orientation) from their lower butt-ends. The circles represent the ends of α-helices. The cross corresponds to the N-end of the segment (i.e., the segment goes away from the viewer); the dot corresponds to its C-end (i.e., the segment comes towards the viewer). The loops connecting the structural segments are shown by the black line (if the loop is close to the viewer) and by the light line (if it is on the opposite side of the fold). The numerals indicate the order of structural elements in the chain (from the N- to the C-terminus).

showing that proteins with different architectures can carry out similar functions, while similarly arranged proteins may have different duties.

Again, I am drawing your attention to non-trivial cases of the lack of relationship between protein structure and function, because undoubtedly you know that the kindred proteins (e.g., myoglobin and other globins) are similar in both architecture and function.

Besides, by comparing myoglobin with hemerythrin, I want to draw your attention to the fact that in both cases the active site (for the former, it is the heme with an iron ion inside; for the latter, two iron ions) is localized in the "architectural defect" of the protein structure, namely, in a crevice between the helices.

The "bundles" described above are typical of quite long α-helices. They are observed in water-soluble globular proteins, as well as in fibrous and membrane proteins. The protein core enveloped by the helices has an elongated, quasi-cylindrical

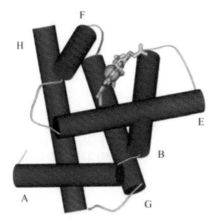

Figure 14.2. The structure of globin: crossed layers of three α-helices each. The helices A, E and F (lettered in accordance with their sequence positions) belong to the upper layer, while H, G and B to the lower layer. The short helices (of 1–2 turns each) C and D are not shown since they are not conserved in globins. A crevice in the upper layer houses the heme. Such "crossed layers" resemble the orthogonal packing of β-sheets. (The orthogonal contact of helices B and E is especially close, since both helices have glycine-formed dents at the contact point.)

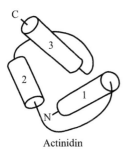

Actinidin

Figure 14.3. Typical packing of helices in a globular protein exemplified by the N-terminal domain of actinidin (the loops are traced very schematically). Note that the architecture of this domain cannot be described in terms of co-linear and orthogonal packings of α-helices. Adapted from Chothia C. *Nature* (1989) **337**: 204–205.

shape. It is hydrophobic in water-soluble globular and fibrous proteins and hydrophilic in membrane ones. The "crossed layers" are also formed by rather long helices; they have a flat hydrophobic core.

However, relatively short helices are more typical of globular proteins. For these helices (with a length of about 20 Å), a quasi-spherical packing around a ball-like core is more typical.

Figure 14.3 illustrates a typical packing of helices in a globular protein. This packing cannot be described in terms of a parallel or perpendicular helical arrangement because the angles between helices are usually 40–60°.

However, even such intricate packings can be described and classified accurately enough using the "quasi-spherical polyhedron model" proposed by A.G. Murzin

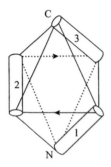

Figure 14.4. The α-helix positions on the ribs of a quasi-spherical polyhedron that models the N-terminal domain of actinidin. The pathways of helix-connecting loops are shown by arrows. Adapted from Chothia C. *Nature* (1989) **337**: 204–205.

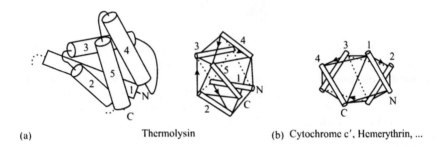

(a)　　　　　Thermolysin　　　　(b) Cytochrome c', Hemerythrin, ...

Figure 14.5. More examples showing how the geometry of helix packings in globular proteins can be described by the quasi-spherical polyhedron model. (a) The C-terminal domain of thermolysin and its model showing the helix positions on the polyhedron ribs (adapted from Chothia C. *Nature* (1989) **337**: 204–205); (b) the model of the four-helix globule presented in Fig. 14.1.

and A.V.F. For example, let us see (Fig. 14.4) how this model describes the α-helical globule of Fig. 14.3. Incidentally, these two figures are from a review of this work by A.G.M. and A.V.F., published in *Nature*.

I will not deny myself the pleasure of showing you another pair of figures from the same review (Fig. 14.5a).

By the way, the quasi-spherical polyhedron model is also sufficiently good for describing rather long helical bundles (like those we saw in Fig. 14.1). This is illustrated by Fig. 14.5b.

Essentially, the quasi-spherical polyhedron model focuses on the positioning of α-helices packed around the ball-like core of the globule (Fig. 14.6).

The model only takes into account that α-helices, solid extended particles, surround the core closely, and that the polar helix ends are located on the globule surface. The geometry of any helix packing can be described by a polyhedron (Fig. 14.6) where each vertex corresponds to half of the helix. The most compact, "quasi-spherical" polyhedra (Fig.14.7) describe compact globules. The helix packings actually observed in globular α-proteins are close to these ideal packings. For a given number of helices, there is only one most compact polyhedron; it allows for a number (from two to ten)

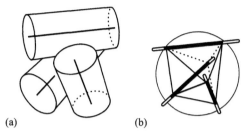

(a)   (b)

Figure 14.6. This figure illustrates how the geometry of helix packings can be described by a polyhedron. (a) Three packed helices are shown as cylinders of diameter 10 Å (their axes are also shown). (b) To construct the polyhedron, a sphere of radius 10 Å is drawn from the center of the packing; the polyhedron vertices occur at its intersection with the helix axes. The sections of the ribs enclosed by the sphere are shown as dark lines. Each vertex corresponds to one half of one helix. The helix axes form one set of the ribs of the polyhedron; it is completed by another set of ribs formed by connections linking the helix ends.

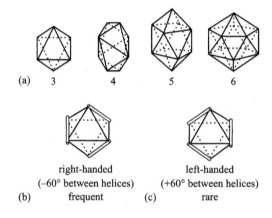

(a)   3   4   5   6

right-handed   left-handed
(−60° between helices)   (+60° between helices)
(b)   frequent   (c)   rare

Figure 14.7. Quasi-spherical polyhedra describe the compact packing of three, four, five and six helices. Larger assemblies of helices cannot be placed around a spherical core. Each polyhedron describes several packing arrangements, i.e., several types of "stacks" of helices; the stacks differ in helix positioning on the polyhedron ribs. For example, three helices form two different arrangements: (b) a right-handed bundle (like that in Figs 14.3 and 14.4); (c) a left-handed bundle. Four helices form ten arrangements, five helices form ten arrangements, and six helices form eight arrangements ("stacks" for four-, five- and six-helix globules are not shown, but you can easily construct them by placing the helices on the polyhedral ribs in all possible ways such that each vertex corresponds to one end of a helix). The packings with inter-helical angles favorable for close helix contacts (see Fig. 14.9) are observed in proteins more often than others.

variants of helix positioning on the ribs of this polyhedron. The previously considered "helix bundles" and "crossed layers" are among these arrangements.

Interestingly, in the observed architectures of α-proteins it is not only helices that fit on the ribs of quasi-spherical polyhedra but also, as a rule, the helix-connecting irregular loops (see Figs 14.4 and 14.5). In other words, a typical protein chain envelops its hydrophobic core as if taking a continuous path along the ribs of a quasi-spherical polyhedron.

Now let us see how close packing forms in a protein globule. The existence of such packing follows from experiments that show that protein is as compact and solid as an organic crystal. However, it is still to be explained how it comes about that such packing emerges regardless of the vast variety of most intricate shapes of side groups of a protein chain.

Actually, the outline of close-packing formation is more or less clear only for α-helices, and that is why it would not be out of place to consider it here.

The first model of the close packing of α-helices, that of the "knob (side chain) into hole (between side chains)" type, was proposed by Crick in 1953, earlier than the solution of the first 3D protein structure. Later, this model was further developed by Efimov, and independently, by the Chothia–Levitt–Richardson team, and by now it has acquired the "ridge (of side chains) into groove (between them)" description.

According to this model, side chains in the surface of an α-helix tend to form ridges separated by grooves. The "ridges and grooves" prove to be a bit better in describing the reality than "knobs and holes" because a turn of one knob (one side chain) towards another (another side chain) can make this or that "ridge composed of knobs" more definite. There are two types of ridges (and their parallel grooves): those of the "+4" type formed by side chains of residues at sequence positions "i", "$i + 4$", "$i + 8$", etc. (in other words, separated in sequence by four chain residues), and ridges of the "+3" type formed by side chains of residues at sequence positions "i", "$i + 3$", "$i + 6$", etc. (i.e., separated by three residues). Figure 14.8 shows that ridges of these two types form angles of opposite signs with the helix axis.

The close packing brings the ridges from one helix into the grooves from another. This gives two kinds of possible packing (Fig. 14.9).

In the first type, "+4" ridges of one helix fit into grooves between similar "+4" ridges of the other (Fig. 14.9a; as seen, the close packing results from superimposing the overturned helix $\alpha 2$ onto helix $\alpha 1$ and turning it further until the "+4" ridges of both helices become parallel). In such packing the angle between the helix axes is close to $-50°$. This is the most typical angle formed by helices in α-helical globules. Also, it is typical for α/β and $\alpha + \beta$ proteins to be discussed a little bit later. The thing is, such an angle provides for a twist of the α-helix layer (in which the twist angle is close to $-50°/10\,\text{Å}$, where $-50°$ is the angle between axes of adjacent helices, and $10\,\text{Å}$ is the width of an α-helix), which is in good agreement with the typical twist of a β-sheet (characterized by the same $-25°/5\,\text{Å}$ value, where $-25°$ is the angle between axes of adjacent β-strands, and $5\,\text{Å}$ is the width of a β-strand).

In the second type, "+3" ridges of one helix fit into grooves between "+4" ridges of the other (Fig. 14.9b). In such packing the angle between the helix axes is close to $+20°$. This is the most typical angle for helix contacts in long bundles that occur in α-helical globules, as well as in fibrous and membrane proteins.

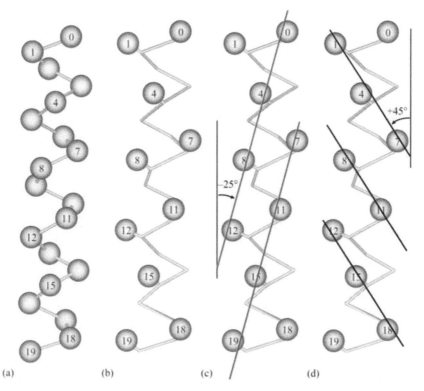

Figure 14.8. Ridges at the surface of the α-helix. The C^α-atoms (a) and C^β-atoms (b–d) are shown. Numbered residues face the viewer. Two kinds of ridges (thin lines in the helix face) from close side groups are shown (c, d). The ridges "+4" from side groups "i"–"$i+4$"–"$i+8$"··· are inclined at $-25°$ to the helix axis (c), the ridges "+3" from groups "i"–"$i+3$"–"$i+6$"··· are inclined at $+45°$ (d); in the drawing these angles look smaller because typical ridges pass through massive side groups, while in (c) and (d) they run through the centers of the C^β-atoms. Adapted from [1a].

In addition, "+3" ridges of one helix can fit into grooves between similar "+3" ridges of the other, thereby forming an extremely short contact of almost perpendicular helices. The contact is so short that I did not show it in Fig. 14.9, although it is quite typical of α-helical globules.

Concluding the consideration of close packing, I would like you to note that the actual inter-helical angles in proteins may differ from the above given "ideal" values because side chains vary considerably in size. For the same reason, the picture of the ridge-into-groove fitting is slurred over in β-structures (where the side chains project much less: unlike the cylindrical α-helix, the β-sheet has a rather flat surface) and is clearly observed only in rare cases.

Finally in this section, let us see how the close packing of helices conforms to the positioning of helices on the ribs of quasi-spherical polyhedra. As a matter of fact, it

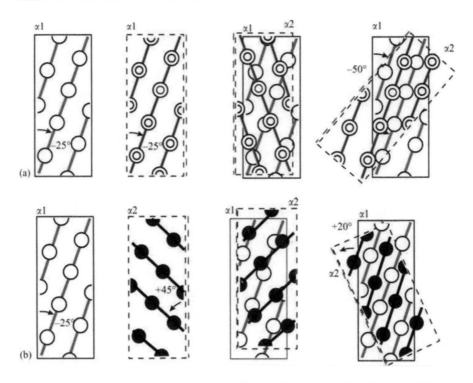

Figure 14.9. Two basic variants of close packing of side chains: with helix axes inclined at −50° (a) or +20° (b). We look at the contact area through one helix (through α2 turned over through 180° around its axis). The residues of the "lower" (α1) helix are shown as light circles and those of the upper helix (α2) by dark circles. Adapted from [1a].

does so quite curiously. The "polyhedral" packings with angles between helices close to −50° and/or +20° (which are favorable for close packing of helices) are observed frequently; others are rare, although occasionally these can be observed too. For example, the first of the two three-helix packings shown in Fig. 14.7, a bundle with a right-handed twist, causes inter-helical angles of −60° (close to the −50° angle optimal for close packing, see Fig. 14.9a). This three-helix bundle is observed very frequently. The other packing, with the left-handed twist, causes inter-helical angles of +60° (which differs greatly from all angles optimal for close contacts, i.e., −50°, +20° and 90°), and this bundle can be observed an order of magnitude less frequently.

Now let us consider "mixed" proteins built up from β-sheets and α-helices. Typically, they consist of separate α- and β-layers, and never have "mixed" ones, which would cause energetically unfavorable dehydration of H-bonds at the β-sheet edge (Fig. 14.10).

There are α/β ("α slash β") and $\alpha + \beta$ ("α plus β") proteins (or rather, domains), and sometimes they are combined into the common class of $\alpha\&\beta$ (i.e., "α and β") or $\alpha-\beta$ proteins.

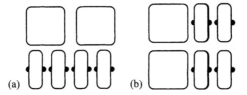

Figure 14.10. (a) The layer structure of mixed (α/β and $\alpha + \beta$) proteins viewed along the α-helices and β-strands to stress their close packing (helix ends are shown as squares and strand ends as rectangles). (b) α-Helices and β-strands cannot belong to the same layer because this would cause dehydration of H-bonds at the β-sheet edge (H-bond donors and acceptors in the β-sheet are shown as dots).

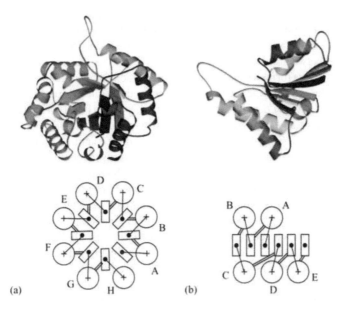

Figure 14.11. Typical folding patterns of α/β proteins and their simplified models as viewed from the β-layer butt-end: (a) the "α/β-cylinder" in triose phosphate isomerase; (b) the "Rossmann fold" in the NAD-binding domain of malate dehydrogenase. The detailed drawing of the former shows a viewer-facing funnel formed by rosette-like loops and directed towards the center of the β-cylinder. The latter has a crevice at its upper side; the crevice is formed by loops going upwards and downwards from the β-sheet.

In α/β domains the β-structure is parallel, the α-helices are also parallel to one another (and antiparallel to the β-strands), and along the chain they alternate: $-\alpha-\beta-\alpha-\beta-\alpha-\cdots$

Two folding patterns are most typical for α/β proteins: the α/β-cylinder where the β-cylinder lies inside a cylinder formed by α-helices (Fig. 14.11a) and the "Rossmann fold" where a more or less flat (except for the ordinary propeller twist) β-layer is

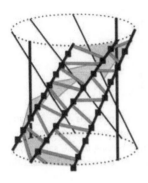

Figure 14.12. The closed β-cylinder. H-bonds (the blue lines) are shown for one strand only. One line of H-bonding is shown as a shaded band. The shear number is equal to 8 in the given case. Adapted from [8].

sandwiched between two α-helix layers whose twist is complementary to that of the β-layer (Fig. 14.11b). Unlike previously considered domains, α/β domains usually have two hydrophobic cores: in the Rossmann fold they are between the β-sheet and each α-layer; in the α/β-cylinder the smaller core is inside the β-cylinder, while the larger one is between the β- and α-cylinders.

β-cylinders are formed by straight β-strands. Each pair of neighboring (H-bonded) strands has the usual propeller twist. Therefore, the strands form an angle with the cylinder's axis, and the β-cylinder has a hyperbolic shape (Fig. 14.12). The β-cylinder is rigid, being stitched up with a closed hydrogen bond network. The H-bonds are perpendicular to the strands. Going from one residue to another along the line of H-bonds (along the shaded band in Fig. 14.12), one returns to the initial strand but not to the initial residue (because the strands are tilted with respect to the cylinder's axes). The shear between the two ends of the hydrogen bond line is expressed as a number of residues. This number is even, since the H-bond directions alternate along the strand. Two digits, the number of β-strands and the "shear number", allow a precise discrete classification of the closed β-cylinders given by Lesk and Chothia.

By the way, there also exist α/β "almost" cylinders that do not complete the circle, and hence, have no closed hydrophobic core inside the β-cylinder. They are known as "α/β-horseshoes" and contain up to a dozen and a half α/β repeats.

Usually, an α/β-cylinder contains eight α- and eight β-segments, and the topologies of all α/β-cylinders (and of "α/β-horseshoes", too) are alike: β- and α-segments form a right-handed superhelix, where β- and α-segments adjacent in the chain or separated by one other segment are in contact with each other. Presumably this overall structure provides particular stability to the protein globule, since numerous protein globules with such architecture (10% of all proteins) display close similarity in their shape, although most of them have nothing in common as concerns their origin or functions.

No common functions, no similarity in the structure of their active sites; but when it comes to *location* of the active site, α/β-cylinders have much in common: each architecture has a special place (a dent) as if specially designed for the active site, no matter what function it performs.

I would like to draw your attention to the "funnel" on the axis of the β-cylinder (Fig. 14.13a). As you see, it is a dent in the overall protein architecture; this dent

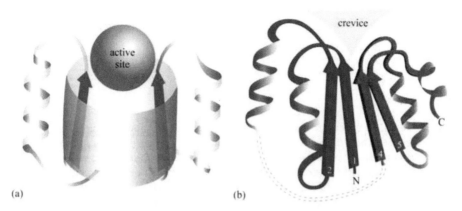

Figure 14.13. Typical locations of the active site in α/β proteins: (a) in the "funnel" on the α/β-cylinder axis; (b) in the crevice formed by separated loops in the "Rossmann fold". Reproduced from [1a] with permission.

is determined by the folding pattern, and is not covered with loops. This is where the active site is located. Or rather, one of two such "funnels" (to be found at both ends of the β-cylinder) is used to house the active site – it is the one where C-termini of β-strands and N-termini of α-helices are directed. These termini (connected with relatively short loops and having numerous open NH groups at N-ends of the helices) are believed to be most useful in binding various substrates. This is still to be studied, though.

In the Rossmann fold the active site is located similarly: it is in a dent, in the crevice, and again in the crevice to which the C-termini of β-strands and N-termini of α-helices are directed. The only difference is that this crevice is formed not by loops drawn outwards from the cylinder center but by loops some of which are drawn to the upper and some to the lower α-layer (Fig. 14.13b).

Inner voice: You should not believe that the active site always occupies an obvious dent. It happens often but far from always.

Lecturer: Right. As a rule (in about 80% of instances) the active site occupies the largest dent on the globule; in turn, this dent is usually determined by the globule architecture built up by the secondary structures. Nevertheless, in many cases a search for the active site took much time – even with the spatial structure of the protein known – and was not always successful . . .

Now let us consider $\alpha + \beta$ proteins. They are based on the antiparallel β-structure (in contrast to α/β proteins based on the parallel β-structure).

The $\alpha + \beta$ proteins can be divided into two subclasses. Those of the first subclass (also known as "$\alpha\beta$-plaits") resemble α/β proteins in that the α-layer is packed against the β-sheet. Like α/β proteins, they are characterized by a regular alternation (though distinct from that of α/β proteins) of α- and β-regions both in the chain and in

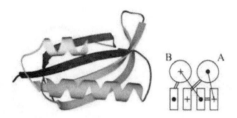

Figure 14.14. A typical structural motif for $\alpha + \beta$ proteins: the $\alpha\beta$-plait in the ribosomal protein S6. The $\alpha\beta$-plait is distinct from other $\alpha + \beta$ proteins because it has a more
regular alternation of secondary structures in the chain (in this case, the alternation is
$\beta\alpha\beta\beta\alpha\beta$). S6 represents an example of the so-called "ferredoxin fold". The rainbow coloring
(blue-green-yellow-orange-red) traces the pathway of the chain from the N- to the C-terminus.
On the right, a schematic diagram of this protein as viewed along its almost co-linear structural
elements. The helices are lettered. An α- or β-region going away from the viewer (i.e., viewed
from its N-terminus) is marked with "+", and that approaching the viewer with a dot.

space. Proteins of the other subclass ("usual" $\alpha + \beta$ proteins) have no such alternation;
their α-structures are more or less separated from β-structures in the chain.

The typical alternation of α- and β-regions in the $\alpha\beta$-plait is either
$\cdots \alpha-\beta-\beta-\alpha-\beta \cdots$ or $\cdots \alpha-\beta-\beta-\beta-\beta-\alpha-\beta-\beta \cdots$ (Fig. 14.14). Here separate
α-helices are placed between β-hairpins or β-sheets composed of an even number
of β-strands. The β-strands adjacent in the sequence form *anti*parallel β-sheets; and
because of the even number of β-strands between α-helices (in contrast to the *odd*
number of these observed in α/β proteins), as well as the general co-linearity of the
strands and helices, α-helices form *anti*parallel hairpins too. The "pleated" protein
structure is observed as one of the most abundant protein architectures; it is observed,
in particular, among ferredoxins and . . . RNA-binding proteins.

In "normal" $\alpha + \beta$ domains (Fig. 14.15) α- and β-regions alternate irregularly
and tend to form something like blocks. They commonly look like a β-sheet (which
is often bent on itself thus forming a subdomain) covered by separate α-helices or
by an α-helical subdomain. The β-structure is mostly antiparallel in $\alpha + \beta$ proteins
(as in "pure" β proteins).

A very typical feature of α/β and $\alpha + \beta$ proteins (as well as β-proteins) is
the *right*-handed (i.e., counterclockwise, when approaching the viewer) topology of
connections between parallel β-strands of the β-sheet (see Figs 14.11 and 14.13 to
14.15). In α/β and $\alpha + \beta$ proteins such a connection usually contains an α-helix
(Fig. 14.16). In β-proteins (and sometimes in $\alpha + \beta$ too) such a connection contains,
as you may remember from the previous lecture (on "abcd" structures and so on), a
β-strand from another sheet, and sometimes even a separate β-sheet. It also happens,
though rarely, that the connection between parallel β-strands contains neither α- nor
β-structures; but in this, as well as in all other cases, it usually appears to be a *right*-
not left-handed connection.

It will become clear from the next two lectures that such handedness of the
connection usually contributes to protein stabilization thus allowing a greater variety

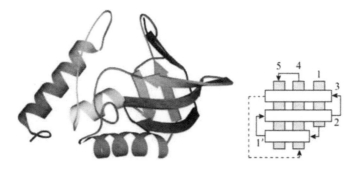

Figure 14.15. A typical structural motif for $\alpha + \beta$ proteins: staphylococcus nuclease. This "normal" $\alpha + \beta$ protein is characterized by a less regular (compared with α/β proteins or $\alpha\beta$-plaits) alternation of secondary structures in the chain (in this case, $\beta\beta\beta\alpha\beta\beta\alpha\alpha$), and these α and β structures are more separated in space. The folding pattern observed in the β-sub-domain of the nuclease is called the "OB-fold" (i.e., "Oligonucleotide-Binding fold"). On the right, a schematic diagram of the OB-fold (the orthogonal packing of β-strands is viewed from above), which is abundant in various multi- and mono-domain proteins. The β-strands are marked with numerals. The first strand is bent (actually, it is broken); its two halves are marked as 1 and 1'. Notice the "Russian doll effect": one characteristic fold (the OB-fold) is a part of another characteristic fold (the nuclease fold).

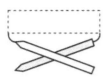

Figure 14.16. Typical *right*-handed topology of connections between parallel β-strands of the same sheet. The connection usually contains an extra secondary structure.

of coding, i.e., of the structure-stabilizing sequences; that's why the *right*-handed connection is quite frequently observed in different proteins, while the left-handed connection is extremely rare.

In conclusion of this brief outline of globular protein structures, I would like to stress again that the same or very close architectures are often observed in proteins quite different both functionally and phylogenetically. This finding underlies the physical (also known in literature as "rational") classification of proteins. This will also be discussed in the next lecture.

LECTURE 15

This lecture is an attempt to explain why the majority of proteins fit a small set of common folding patterns, which should be your impression from the previous lectures.

Actually here we come across the "80% : 20%" law. In its initial form, this law suggests that 80% of the total amount of beer is consumed by only 20% of the population.

As for the proteins, 80% of protein families are covered by only 20% of observed folds. In the previous lectures I took the liberty of focusing mainly on these typical structures.

So, why do most proteins fit a limited set of common folds? And why not all of them (like DNA chains)? And what is behind this limited number of common folds: common ancestry? common functions? or the necessity to meet some general principles of folding of stable protein structures? Also, at what structural level is the similarity of proteins of distinct ancestry and function displayed?

For now we will consider these questions only qualitatively, passing to more strict answers in the next lecture, and when we know more about protein folding I'll add a couple of words on the matter.

When only a few protein structures were known (approximately up to the middle of the 1970s) each tertiary structure was believed to be absolutely unique, i.e., proteins of evolutionarily different families were thought to share no similarity at all. However, with increasing information on the spatial structure of protein molecules it became more and more clear that there are "standard designs" for protein architectures. The architectures of newly solved proteins (or at least of their domains) more and more often appeared to resemble those of known proteins, although their functions and amino acid sequences were utterly different. This generated the idea discussed more than once in our previous lectures, namely, that similarity of protein tertiary structures is caused *not only* by evolutionary divergence and *not* (or not only) by functional convergence of proteins, but simply by restrictions imposed on protein folds by some physical regularities.

By the end of the 1970s it became absolutely clear that there is an intermediate structural level sandwiched between two "traditional" ones, i.e., between the secondary structure of a protein and its detailed 3D atomic structure. This intermediate

level is already known to us as the "folding pattern" determined by the positions of
α- and/or β-regions in the globule, and it is at this level that we observe similarities
in proteins having no common ancestry or function. Unlike the detailed 3D atomic
protein structure, folding patterns are surprisingly simple and even elegant (Fig. 15.1).

The finding that the same or very similar architectures are often observed in
proteins utterly different functionally or phylogenetically sets the basis of physical
(or "rational") classification of proteins.

The most complete computer classifications of protein folds are "Dali/FSSP"
developed by Holm and Sander, "CATH" (Class–Architecture–Topology–Homology)
by Thornton's team, and, perhaps the most popular among them, "SCOP" (Structural
Classification of Proteins) developed by Murzin after he left Pushchino for Cambridge.

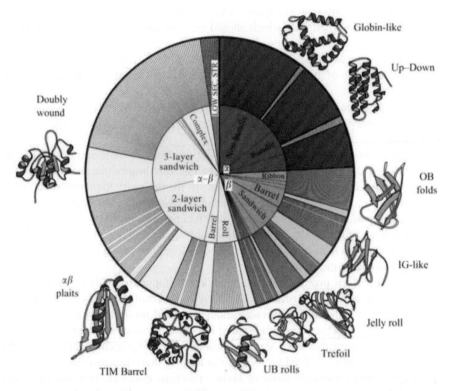

Figure 15.1. Structural classes of proteins ("α", "β", "α–β" and "Low Secondary
Structure"), typical architectures ("Non-bundle", "Bundle", etc.), and typical folding patterns
(topologies) according to physical classifications of proteins (CATH). The sector width shows
the abundance of structures of the given type in non-homologous proteins. Note that α- and
β-structures are layered, and that each layer comprises exclusively α-helices or β-strands
and never houses both structures. Reproduced with a minor modification (kindly permitted
by C.A. Orengo and J.M. Thornton) from Orengo C.A., Michie A.D., Jones S., Jones D.T.,
Swindells M.B., Thornton J.M., *Structure* (1997) **5**, 1093.

Actually, the classified folds refer to protein domains that are compact globules existing either separately or as a part of a multi-domain protein. Classification begins (Fig. 15.1) with structural *classes* (α, β, etc.). The classes are subdivided into *architectures* of protein frameworks built up from α- and/or β-regions. In turn, the architectures are subdivided into *topologies*, i.e., pathways taken by the chain through the frameworks; in other words, they are subdivided into folding patterns.

Further on, the folding patterns are subdivided into superfamilies displaying at least some sequence homology (a trace of common ancestry); those, in turn, into families with clearly displayed homology, and so on, down to the separate proteins of concrete organisms.

The physical classification of protein structures (class–architecture–topology) allows not only the systematizing of studied structures but also the prediction of protein structures yet to be found. For example, for a long time only β-proteins composed of antiparallel β-structures were known, while a lacuna gaped where β-proteins composed of parallel β-structures should be. However, later it was filled in (remember β-prisms?).

Figure 15.2 exemplifies such a classification of the structures of protein globules, which explicitly leaves "vacancies" for possible but not (yet?) found chain folds.

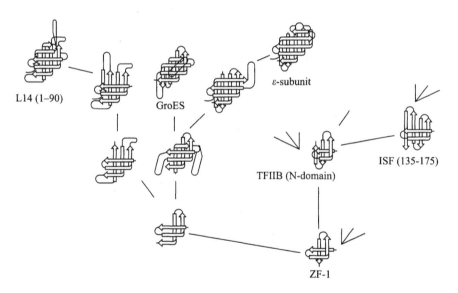

Figure 15.2. A fragment of the "tree of structures" (Efimov A.V. *FEBS Letters* (1997) **407**: 37–46) based on the initial small "nuclei" of various types gradually growing and becoming more and more complicated. Notice the "Russian doll effect", i.e., that a simpler structure is contained within a more complex one. The branch fragment shown relates to β-proteins "growing" from the "β-angles" (bent β-hairpins). In this fragment the known native structures are named, while the remaining unnamed structures have not so far been identified in native proteins.

Interestingly, this classification ("Efimov's tree") is based on the initial small "nuclei" of various types that gradually "grow" and become more and more complicated. This scheme can be interpreted as an imitation of protein folding; here it should be stressed that the modern idea of this process (supported by experimental and theoretical data to be discussed in a later lecture) is based on the concept of folding started by the formation of a small folded part of the native globule. However, these "trees" may also be interpreted as an imitation of the evolutionary history of proteins. The connection between these two phenomena, folding process and evolutionary history, is yet to be understood.

Now we will discuss the question that always excites a biologist: do we see the evolution of protein structures?

Actually, this question contains two questions: (1) whether we see a microscopic evolution of proteins, i.e., whether we see (apart from a simple "drift", i.e., change from organism to organism) some connection between a change in the protein structure and a change of the entire organism; (2) whether there is a macroscopic evolution of proteins, i.e., whether their structure becomes more complex with increasing complexity of the organism.

The first question definitely has a positive answer. Although far from all the changes occurring in a protein play a clear functional role (which is stressed by Kimura's "neutral evolution theory"), in some cases the functional role of changes in a protein is understood and well studied. For example, hemoglobin from a llama (mountain animal) binds oxygen more strongly than hemoglobin from its animal relatives living on the plain. Such adaptation to living conditions is still more clearly illustrated by comparison of hemoglobins from adult animals with fetal hemoglobins: the latter has to derive oxygen from the mother organism, so oxygen binding to fetal hemoglobin must be stronger. And M. Perutz showed which microscopic changes in the hemoglobin structure are responsible for this strengthening.

It is believed that evolution often occurs through amplification of a gene with subsequent mutations of its copies, so that one copy of this gene keeps maintaining the "previous" function (and the organism's life as well), while another copy, or other copies, become free to mutate in a (random) search for a change that could adapt the protein's function to a biological need. For example, α-lactalbumin of milk undoubtedly originated from lysozyme at the advent of the vertebrates. It is known that there is usually only one gene copy of each major protein (or rather, two identical copies, with diploidy taken into account); however, the living conditions are capable of changing the situation. An "almost fatal" dose of poison can provoke multiplication of copies of the gene responsible for its elimination. And then random mutations of these copies go into operation, then the selection . . .

Evolution of proteins is supported by their domain structure. It is known that the domain-encoding genes can migrate, as a whole, from one protein to another, sometimes in various combinations and sometimes individually. Closely related domains are often observed both as parts of different proteins and as separate proteins (examples: the calcium-binding domain of calmodulin, parvalbumin, etc.; various kringle domains, and so on). Presumably, such exchange is facilitated by the intron–exon structure of genes (specifically, this is well seen in immunoglobulins); however, the hypothesis that the role of a "module" in the exchange is generally played by an exon rather than by the whole domain seems to lack confirmation.

The other question (whether there is macroscopic evolution of proteins, i.e., whether their structures become more complex with increasing complexity of the organism) must most probably be answered negatively. A review of protein structures shows that the same folding patterns (specifically, those shown in Fig. 15.1) are observed both in eukaryotes and prokaryotes, although the distribution of the most "popular" folds in eukaryotes is somewhat different from that in prokaryotes. In other words, we do not see that proteins become more complicated with increasing organism complexity, as happens, for example, at the cellular level, as well as at the level of chromatin and organelles, down to ribosomes. (On the contrary: sometimes, we see that fibrous proteins having the most simple structure are typical for higher organisms rather than for prokaryotes. The simplest proteins are not typical for the simplest organisms. This is strange, is it not?)

Although there are some vague data that eukaryotic proteins and their domains are larger than prokaryotic proteins sharing the same folding pattern, this still remains to be elucidated.

However, there exists one more important "macroscopic" structural difference, though it is not connected with the chain folds. It is as follows: the proteins of eukaryotes, of multicellular ones in particular, are much more liable to co- and post-translational chemical modifications (such as glycosylation, iodination, etc.). Modification sites are marked by the primary structure, while the modification is carried out by special enzymes, and often only partially, which causes a variety of forms of the same proteins, although their biochemical activity usually remains unchanged. An alternative splicing, the privilege of eukaryotes, also contributes to the diversity of their proteins.

Inner voice: Nevertheless, there is evidence that eukaryotic proteins are not only larger in size but also contain a greater number of domains than prokaryotic proteins (a typical eukaryotic protein contains four or five domains, while a prokaryotic protein only two).

Lecturer: True. However, probably the general idea that eukaryotic proteins are larger in size is connected with the fact that higher organisms have many large multidomain "outer" proteins like immunoglobulins rather than with changes of the "housekeeping" cellular proteins.

Also, it should be mentioned that the investigation of the "macroevolution" of protein chain folds is strongly hampered by the possibility of horizontal gene transfer, which may result in penetration of "new" proteins into "old" organisms.

Inner voice: I should like to come back to the "drift" of protein structures (that you had left aside) and to a problem of the origins of protein folds. There is a hypothesis that the "jumping elements" of protein evolution are short sequences of about

10 residues rather than whole domains. After all, a difference between the folds of distantly similar sequences is often caused by addition or deletion or displacement of such a short chain region. And since it often contains α- or β-structure, kindred protein domains are sometimes attributed to different folding patterns. Thus, a "drift" of protein structures includes transitions from one folding pattern to another due to additions or deletions or displacements of structural modules. If so, is it possible that protein folds originated from the association of small structural modules?

Lecturer: These questions are intensively discussed from time to time, but no definite general answer has been achieved so far. The structural modules considered are so small, and their sequences are so diverse, that it is impossible to prove their relationship. Only short fragments with similar functions (for example, some heme binding fragments) have sufficiently similar sequences. However, the same similarity is sometimes observed for "discontinuous" active sites (e.g., for sites of hydrolysis) formed by remote chain residues in proteins having completely different folds. In this case one can hardly say that these sites are "transferred" from one protein to another. Therefore, coming back to fragments that possibly serve as functional and structural modules: maybe, they are transferred from one protein gene to another; maybe, they arise anew in each protein family. This is not yet known.

However, it is more correct to say that this question is still open for globular proteins. Fibrous proteins, as I have already mentioned, definitely look like multiple repeats of short fragments. Thus, their origin from "modules" looks highly probable, the more so as the repeated structural modules of fibrous proteins are often coded by separate exons. There is every reason to believe that current genomic, and especially structural genomic, projects will answer all these exciting questions.

With a huge number of filed and systematized protein structures available, there arise philosophical questions such as (1) what is the physical reason for the simplicity and regularity of typical folding patterns? and (2) why are the same folding patterns shared by utterly different proteins, and what are the distinctive features of these patterns?

In scientific terms, we would like to elucidate what folding patterns are most probable in the light of the protein physics laws we have studied, how numerous these are and to what extent they coincide with folding patterns observed in native proteins. To answer these questions, we will first of all study the stability of various structures. This approach – to study stability prior to folding – is justified by the fact that the same spatial structures of proteins can be yielded by kinetically quite different processes: *in vivo* (in the course of protein biosynthesis on ribosomes or during secretion of more or less unfolded proteins through membranes) and *in vitro* when the entire protein refolds from a completely unfolded state. This means that a detailed sequence of actions does not play a crucial role in protein folding.

Let us start with a simple question: why do globular proteins have the layered structure that we discussed in the previous lecture? In other words, let us try and see why the stability of a dense globule requires that the protein framework should look like a close packing of α- and β-layers, why it requires that α- and β-regions should extend from one edge of the globule to the other, and why it requires that the irregular regions should be outside the globule.

In principle, we have already discussed that. Hydrogen bonds are energetically expensive and therefore must be saturated in any stable protein structure. Hydrogen-bond donors and acceptors are present in the peptide group of any amino acid residue. They can be saturated when participating either in H-bonds to water molecules or in the formation of secondary structures. That is why only the secondary structures of a stable (if it wants to be so) globule but not the irregular loops have the right to be out of contact with water and belong to the molecule's interior, and why the elements containing free (from intramolecular H-bonds) NH- and CO-groups, that is, irregular loops, bends, edges of β-sheets, and ends of α-helices, should emerge at the surface.

For the sake of globule stability, extended α- and β-structures must closely surround the hydrophobic core created by their side chains, thereby screening it from water. At the same time, α-helices and β-sheets cannot share the same layer because in this case edge H-bonds of the β-sheet edges would be lost. This means that globule stability demands the formation of purely α-layers and, separately, purely β-layers (Fig. 15.1). In other words, separate α- and β-layers are stable elements of the globular protein structure, while α- and β-structures mixed in the same sheet would be a *structural defect*, or, more accurately, an *energy defect* of the protein globule. It is also evident that a *stable* globule must contain a majority of stable elements and avoid structural defects. Since we can observe only stable globules (unstable ones fall apart and therefore cannot be observed), the observed protein structural elements must be mostly stable, and defects must be only occasionally observed.

In particular, this is true for α- and β-layers. They are stable if not mixed. And as we have seen, such layers (usually they are not flat but twisted, cylindrical, and even quasi-spherical, as in α-helical globules) are indeed typical of protein globules. The layered structure simplifies protein construction, and the large majority of domains can be represented by two-, three- or four- (rarely) layer packings.

Some proteins (especially those containing metalorganic complexes or numerous S—S-bonds) are sometimes observed to deviate from the "layered packing" scheme (Fig. 15.3), but such deviating proteins are very rare.

Domains with more than four layers are extremely rare, and in principle, it is clear why. They would contain too many residue positions screened from water, which means (for the 1 : 1 ratio of polar and non-polar side chains typical of globular water-soluble proteins) that many polar residues would be brought into the interior of the globule. This is energetically most unfavorable, and such a protein would be unstable. That is why very large (and hence, many-layered) compact globules of a "normal" amino acid composition must be unstable, and therefore, large proteins have to be divided into the sub-globules that we know as domains.

Actually, a chain consisting mostly of hydrophobic amino acids could pack into a very large stable globule, but such sequences are many times less numerous that those

Figure 15.3. An unusual globule with no α- and almost no β-structure (the protein huristasin, a representative of the "Low Secondary Structure" class). This protein has a very special sequence with many Cys residues that form S—S bonds (their side chains are shown as yellow rods).

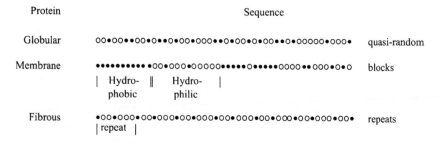

Protein   Sequence

Globular   oo•oo••oo•o••o•oo•ooo••o•oo•o•oo••o•ooooo•ooo•   quasi-random

Membrane   ••••••••••oo•ooo•ooooo••••o••••oooo••ooo•o•o   blocks
            | Hydro- ‖ Hydro- |
              phobic   philic

Fibrous   •oo•ooo•oo•ooo•oo•ooo•oo•ooo•oo•ooo•oo•ooo•oo•   repeats
            | repeat |

Figure 15.4. Typical patterns of alternation of hydrophobic (•) and polar (o) amino acid residues in the primary structures of water-soluble globular proteins, membrane proteins and fibrous proteins.

of the mixed "hydrophobic/hydrophilic" type, and, moreover, such a chain would be brought into the membrane instead of acting as a "water-soluble globular protein".

In principle, a sequence can be suggested in which some specially positioned polar side chains would provide the "cure" for all "defects", e.g., for all broken hydrogen bonds between the main chain and water molecules that result from immersion of a loop or the β-sheet edge in the interior of the globule. Or a sequence can possibly be proposed that would compensate for the broken bonds with some powerful interactions, e.g., with covalent (Cys—Cys) or coordinate (through the metal ion) bonds. *In principle*, this seems to be possible. But these sequences would be *very special*, and hence, *very rare*...

Perhaps this is the heart of the matter: maybe, "common" globular proteins are formed by "normal" (not too strictly selected) sequences rather than by those "strictly selected", which, therefore, are simply very rare.

Let us consider the primary structures of proteins (Fig. 15.4). Statistical analysis shows that the sequences of water-soluble globular proteins appear to be "random". That is, in these sequences various residues are as mixed as would be expected for the result of random co-polymerization. Certainly, each sequence does not result from random biosynthesis but is gene-encoded. Still, the sequences of water-soluble globular proteins look like "random" ones: they *lack the blocks* typical of membrane proteins (where clearly hydrophobic regions that must stay within the membrane alternate with more hydrophilic ones that have to form loops and even domains projecting from the membrane), and also they *lack the periodicity* characteristic of fibrous proteins (with their huge regular secondary structures).

Inner voice: I cannot but note that a coded message may also look like a random sequence of letters, although this is not at all the case . . .

Lecturer: Of course, the amino acid sequences of globular proteins are not truly random (that would imply that any sequence can fold into a globular protein). Protein sequences are certainly selected to create stable protein globules. But the shape of these globules may vary greatly. Therefore, the set of observed primary structures includes the entire spectrum of regularities inherent to all these shapes, i.e., a vast set of various "codes". And when calling the primary structure "random" (or rather, "quasi-random") we mean only that in the totality of primary structures of globular proteins, the traces of selection of protein-forming sequences are not seen as clearly (and therefore are not as restrictive) as traces of selection for periodicity in fibrous proteins or traces of selection for blocking in membrane proteins. This is what is meant when I say that amino acid sequences of water-soluble globular proteins look like random sequences.

And what is it like "to look like a *random* sequence"? This means to look like the *majority* of all possible sequences. Then in considering water-soluble globular proteins it would certainly not be pointless to try to find out which spatial structures are usually stabilized by the most common, random sequences or by those similar to them (by "quasi-random" sequences).

Still more. If a protein globule has a "structural defect" (e.g., immersion of an irregular loop or the edge of a β-sheet in the hydrophobic core) then its stability can be ensured only by an extremely thorough selection of the amino acid sequence (to collect as many structure-supporting interactions as possible). The greater the "defect", the more rigorous the selection. And if there is no defect, then less rigorous selection is required. In other words, a "defect-free" structure can be stabilized by many sequences, a structure with a minor defect by a few, and a structure with a great defect can be stabilized only by a vanishingly small number of sequences.

And (if only physics is taken into account) the structures coded by many sequences must be observed quite often, while those coded by a small number of sequences only rarely. This is how the "physical selection" of protein structures can occur.

Since typical packings, "stacks", of secondary structures of globular proteins (Fig. 15.1) look like stable packings of random or almost random sequences should look, it is evident that, at least at the packing level, the observed result of natural (biological) selection of packings does not conflict with physical selection.

Let us proceed by considering the *folding patterns* of protein chains from the same viewpoint, from the viewpoint of structural defects and the physical selection of structures that are "defect-free" (and hence, "eligible" for many sequences).

As we have seen, protein folding patterns are often most elegant. The pathway of protein chains often resembles the patterns on pottery ornaments (Fig. 15.5). And

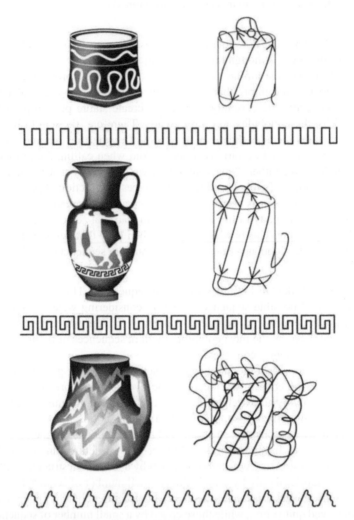

Figure 15.5. Folding patterns of protein chains and ornaments on American Indian and Greek pottery: two solutions to the problem of enveloping a volume with a non-self-intersecting line. On top, the meander motif; in the center, the Greek key motif; at the bottom, the zigzag "lightning" motif. Reprinted with permission from the cover of *Nature*, (1977) **268**(5620) (© 1977, Macmillan Magazines Limited), where a paper by J. Richardson on folding patterns of protein chains is published.

according to the neat idea of Jane Richardson who discovered this resemblance, this is not a coincidence: both the ornament line and the protein chain aim to solve *the same* problem, that is, how to envelop a volume (in protein, its hydrophobic core) by a line avoiding self-intersection.

In proteins, this effect is achieved by surrounding the core (or two cores, as typical for α/β proteins) with secondary structures and with loops sliding over the core surface. It is also important that the loops connect *anti*parallel (and *not* parallel) secondary structures (Fig. 15.6a), and that they *do not intercross* (Fig. 15.6b). Note that the latter virtually rules out knots in the chain.

Why are parallel connections worse than antiparallel? Is it perhaps because then a too long irregular loop (unsupported by H-bonds) is required? Or is it because the rather rigid polypeptide chain has then to be bent, which is energetically expensive?

And what is wrong with loop crossing? After all, we do not mean that one loop runs into another, we only mean that one loop passes over another. Perhaps the problem is that the "lower" loop is pressed to the core and loses some of its hydrogen bonds to water molecules. And to compensate for this loss (the "energy defect"), again a "rare" sequence is needed...

It should be noted that structures with relatively small defects like loop crossing are still observed in proteins (unlike structures with the large packing defects discussed above). But "faulty" protein structures are rare, which is especially significant because a "structure with a defect" can be formed in many more ways than a "defect-free" one. Thus, we see that "structural defects" have a great effect on the occurrence of various protein structures in nature.

Here, however, we might be confused by the fact that only one or two H-bonds are lost by loop crossing, i.e., the loss is energetically inexpensive and amounts to only 3–5 kcal mol^{-1}. This is very much less than the total energy of interactions within the globule, which amounts to hundreds of kilocalories (as follows from protein melting

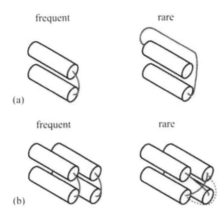

Figure 15.6. As a rule, loops connect *anti*parallel (and *not* parallel) adjacent regions of a secondary structure (a), and any loop crossing is rarely observed in proteins, no matter if one loop covers another or by-passes it (b).

data that we will consider in later lectures). Moreover, it is considerably less than the usual "margin of stability" of the native globule, i.e., the free energy difference between the folded and unfolded protein, which amounts to about 10 kcal mol^{-1} under native conditions (according to the same data). Then, why does an "energy defect" of only 5 kcal mol^{-1} virtually rule out loop crossing in native protein globules?

And one more question: why cannot the upper loop make an additional bend (dashed line in Fig. 15.6) to avoid the crossing (i.e., to have it replaced by bypassing)? Perhaps the matter is, again, that the polypeptide chain is rigid and an additional bend would cost, as estimated, a few (again a few!) kilocalories?

Let us postpone answering these questions till the next lecture, and consider now another characteristic feature of protein architectures, namely, that connections between parallel β-strands are almost always *right*-handed and not *left*-handed (Fig. 15.7).

In this case the stability criterion allows us to point out the "better" of these two asymmetrical connections. Their difference is based on asymmetry of native amino acids. It causes, as you remember, a predominantly *right*-handed twist (shown in Fig. 15.7) of β-layers composed of L-amino acids. The angle between adjacent β-strands is close to 30°, such that the total rotation angle is close to 330° for a right-handed connection, while it is close to 390° for a left-handed one. As a result of polypeptide chain rigidity, the right-handed connection is more favorable: its elastic free energy is lower, although – again – by a couple of kilocalories only.

Here any person knowing polymer physics must stop me and say: "Polymer elasticity is *not* an energy but rather an entropy effect. That is, a significantly bent chain is not as free in its fluctuation as an extended or slightly bent chain. In other words, a smaller number of conformations are possible for a bent chain than for a straight chain. Thus, you must be speaking about a fluctuating loop, whereas in the protein globule loops are *not* fluctuating, they are fixed, and no matter what pathway they take, they have a single conformation. Then, what do the entropy losses have to do with the native protein structure where the entropy of fixed chains is equal to zero anyway!?"

It is not very difficult to answer this question concerning "entropic", to all appearances, defects of too sharp turns of the chain. Here, let us answer it qualitatively.

The fact is that a chain with a limited choice of conformations cannot adjust to its constituent amino acid residues. A certain conformation has low energy only for a

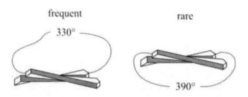

Figure 15.7. The left-handed twist of connections between parallel β-strands is very rarely observed in proteins, while the right-handed twist is quite common. (The connections shown here as simple lines usually incorporate α- or β-regions.)

certain sequence (or a small number of sequences) and high energy for others, i.e., for most sequences. But if the chain can choose among a large set of conformations, then many more sequences can be adequately fitted. Thus, we see that the "entropy defect" can be translated into "energy defect" (or into "decreased number of sequences") language. Entropy effects will be considered more rigorously later on.

We will also consider the postponed question as to why a "defect" of only a few kilocalories per mole plays a significant role in the occurrence of protein structures. We will remember these two problems and consider them in detail in our next lecture.

And right now, not to find the final answer to the latter question but just to drop a hint, let us consider relationships between the energy and other statistical rules known for protein structures. For example, let us consider the immersion of hydrophobic and hydrophilic side groups in the protein globule, different angles of rotation in the chain, and others. Similarly to the previously discussed folding patterns, here "defects" are rare and "good elements" are usual. However, here more detailed statistics allow not only qualitative but also quantitative estimates to be made.

And the estimates obtained show the following. Both the rare occurrence of "defects" (no matter whether it is a "bad" rotation angle or the deep immersion of a polar group in the globule) and the frequent occurrence of "good elements" (e.g., salt bridges formed by oppositely charged groups) are described by the common phenomenological formula

$$\text{OCCURRENCE} \sim \exp(-\text{ENERGY OF ELEMENT}/kT_\text{C}), \qquad (15.1)$$

where T_C is a temperature close to either room temperature or the characteristic protein melting temperature (protein statistics do not allow us to distinguish between 300 K and 370 K).

Expression (15.1) describing the occurrence of protein structural elements is surprisingly similar to Boltzmann statistics *in its exponential shape*, while its physical sense is *absolutely different*. It should be remembered that Boltzmann statistics originate from the particles' wandering from one position to another and staying for a longer time in places where their energy is lower. In contrast, all structural elements of the observed (native) protein globules are fixed, they do *not* appear and disappear, and they do *not* wander from one place to another.

That is, the usual Boltzmann statistics cannot be applied to the occurrence of elements of the native protein structure. And if so, why does the statistics of occurrence of these elements have such a familiar "quasi-Boltzmann" shape?

Let us postpone a detailed answer to this question till the next lecture too. Meanwhile, let us accept, as a phenomenological fact, the estimate that the defect of about kT_C, i.e., of about 1 kcal mol^{-1}, decreases the occurrence of the defect-containing structures by a few times. And, as is clear to us now, this defect must decrease by a few times the number of amino acid sequences maintaining the stability of the defect-bearing protein.

Thus, our conclusions are as follows:

1. "Popular" folding patterns look so "standard", so simple and regular, because the framework of a protein structure is a compact layered packing of extended standard solid bodies (α-helices and β-strands), and their irregular connections slide over the surface of the globule, avoiding intercrossing and crossing the ends of structural segments. Physically, this arrangement is most favorable for the globule's stability because it ensures the screening of non-polar groups from water and H-bonding of all main-chain peptide groups immersed in the compact globule.
2. The number of such "standard" stable folding patterns is not large (numbered in hundreds, while proteins are numbered in tens of thousands); therefore, it is not surprising that some of these "common" structures are shared by proteins different in all other respects.
3. At the same time, other ("defective") folding patterns are not prohibited either. They are simply rare, since only a small number of sequences can ensure their stability. The greater the defect, the lower the occurrence of such folds.
4. Presumably, a "multitude principle" can be proposed to describe structures of domains of water-soluble globular proteins with their typical "quasi-random" amino acid sequences. It would read as follows: *the more sequences fit the given architecture without disturbing its stability, the higher the occurrence of this architecture in native proteins.*

LECTURE 16

Now we will discuss in detail how general structural regularities are connected with protein stability and with the number of protein structure-coding amino acid sequences.

We learned that the framework of a typical protein globule is a compact packing of layers built up from extended solid bodies (α-helices and β-structures), and that irregular connections slide over the surface of the globule virtually never intercrossing or crossing the ends of structural segments (Fig. 16.1).

We concluded that the physical reason for such an arrangement is its contribution to the stability of the globule, since it provides screening of non-polar side chains from water simultaneously with H-bonding of the main-chain peptide groups when they are immersed in the compact globule. In turn, this increased stability allows a greater number of amino acid sequences to fit the given architecture without its destruction (which causes a more frequent occurrence of this architecture in native proteins – we called this "the multitude principle").

As we have mentioned, the peculiarity of typical water-soluble globular proteins consists in the absence of any striking features common for their primary structures (and that is why these sequences are most diverse and numerous, so the "multitude principle" can be applied to them). Indeed, their primary structures are free of any obvious correlation, such as the periodicity characteristic of fibrous proteins or block alternation typical of membrane proteins; their polar groups are rather evenly mixed with non-polar ones. Moreover, in water-soluble globular proteins the amounts of polar and non-polar residues are almost equal. As a result, their primary structures look very much like "random co-polymers" synthesized from hydrophobic and hydrophilic amino acids.

Are these "random" sequences compatible with the compact chain fold in the globule? In particular, are they compatible with the observed secondary structures (whose share in the protein chain is somewhat above a half)? To answer these questions, let us consider hypothetical "protein chains" that result from occasional co-polymerization of equal amounts of polar and non-polar groups.

To be able to fit a compact globule, an α- or β-structural segment should have a continuous hydrophobic surface. An α-helical surface is formed by non-polar residues positioned as $i-(i+4)-\cdots$ (sometimes, as $i-(i+3)$) in the chain, while the alternation

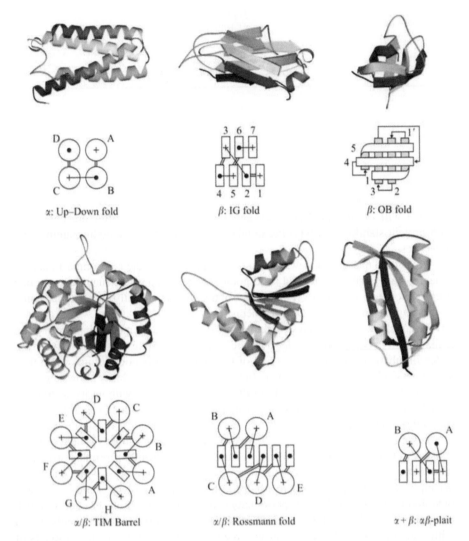

Figure 16.1. Typical folding patterns of the protein chain in α, β, α/β and $\alpha + \beta$ proteins. Simplified diagrams are shown below each. The packings of α- and β-structures are layered, and each layer is composed of either α-helices or β-strands but never contains both structures.

i–$(i + 2)$–$\cdots$ is suitable for the hydrophobic surfaces of β-strands (Fig. 16.2). It can be easily shown that even a random co-polymer contains enough non-polar periodic aggregates for hydrophobic surfaces of α- and β-segments of a medium-sized protein domain.

Let "p" be the portion of non-polar groups in a co-polymer, and "$1 - p$" be the portion of polar ones. Then a periodic sequence of exactly r non-polar groups

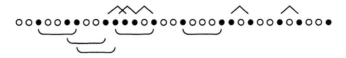

Figure 16.2. The typical pattern of alternation of hydrophobic (•) and polar (○) amino acids in "quasi-random" primary structures of water-soluble globular proteins. Arcs (below) and angles (above) indicate the positions of potential hydrophobic surfaces suitable for α-helical and β-structural segments, i.e., pairs of hydrophobic residues in positions $i, i+4$ and $i, i+2$, respectively.

restricted at its ends by two polar residues, can start at a given point with the probability

$$W(r) = (1 - p)p^r(1 - p) \tag{16.1}$$

The "hydrophobic surface" of the α- and β-segment forms if $r \geq 2$, and the average number of groups involved is

$$\langle r \rangle = \frac{\sum_{r\geq 2}[W(r)r]}{\sum_{r\geq 2} W(r)} = \frac{\sum_{r\geq 2}[rp^r]}{\sum_{r\geq 2} p^r} = 2 + \frac{p}{(1-p)} \tag{16.2}$$

(I have taken the liberty of omitting the summation of series $\sum_{r\geq 2}[rp^r]$ and $\sum_{r\geq 2} p^r$ because this can be found in any mathematical handbook). So, both an "average" α-helix and an "average" β-segment include (at $p = \frac{1}{2}$) $\langle r \rangle = 3$ of regularly positioned hydrophobic groups, i.e., 3 ± 0.5 of the full periods of the α- or β-structure. The expected average numbers of residues in α- and β-segments (their periods are 3.6 and 2) are $\langle n_\alpha \rangle = 11 \pm 2$ and $\langle n_\beta \rangle = 6 \pm 1$, respectively, which practically coincide with the average lengths of α- and β-segments in globular proteins. Interestingly, in random sequences, as well as in primary structures of real proteins, the residue clusters good for α-surfaces often overlap those good for β-surfaces (Fig. 16.2).

Similar estimates show that the average length of loops between secondary structures in a random co-polymer amounts to about $3 + 0.5p^{-2}$, i.e., at $p \sim \frac{1}{2}$, the loops should be somewhat shorter, on the average, than the secondary structure segments – which is indeed observed.

Thus, a random co-polymer provides continuous hydrophobic surfaces that can stick α- and β-segments to the hydrophobic core at least with their one side, while the loops are relatively short. Therefore, "mediocre" random sequences are quite capable of folding into at least a two-layer arrangement of secondary structures.

Inner voice: However, one should not forget that in some, though not many, proteins (for example, in hemagglutinin or leucine zipper) there are extremely long helices that do not fit the above principles. And in some other proteins (e.g., in superoxide dismutase) there are extremely long disordered loops . . .

Lecturer: True. These exceptions look either like blocks borrowed from fibrous helical proteins or (as concerns long loops) like anomalous hydrophilic blocks. But *on the average, in general,* α-, β-, and irregular segments are not too long, and their length is close to that expected for a "random" sequence containing equal proportions of hydrophobic and polar groups.

The harmony between random sequences and compact, potentially stable shapes of globules exists as long as the chain comprises fewer than ~150 residues. However, as the globule increases, as the number of its secondary structure layers grows, its "eligibility" for a random amino acid sequence decreases. This is explained by the fact that the segments belonging to the protein interior must be almost exclusively composed of hydrophobic residues, because otherwise the globule would not survive the presence of numerous water-screened hydrophilic groups and would explode, and the length of such a sequence must be proportional to the diameter of the globule. A small number of such long and almost exclusively hydrophobic segments may also be built up in a "random" co-polymer from ~50% of hydrophobic and ~50% of hydrophilic residues, but only a small number indeed. Therefore, for a random sequence, we can expect not more than two or three, or sometimes four, layers of secondary structures, and this is what we really observe in single-domain water-soluble globular proteins and in the domains of such proteins (Fig. 16.1). And, for the reasons given above, large proteins must be composed of sub-globules, which we know as domains, and this is indeed observed.

Now we have to return to the two questions left unanswered since the previous lecture, namely:

1. Why an "energetic defect" of a few kilocalories per mole, so very minor as compared with the total energy of protein, can virtually prohibit many protein architectures?
2. What do "entropic effects" have to do with the native protein structure where the chain is known to be fixed?

We start with the first question concerning the manifestation of the "defect" energy in protein architecture statistics. First we will aggravate the matter: as you may remember, the stated (at the qualitative level) low occurrence of "structural defects" was supported by the observed "quasi-Boltzmann" statistics of "small elements" of protein structures, which we have to understand as well. And the statistics of "small elements" are exactly what we start with.

For example, let us consider the statistics of the distribution of amino acid residues between the interior and the surface of a protein globule and consider the interrelation between these statistics and the hydrophobicity of amino acid residues.

The hydrophobicity of amino acid residues is usually measured as the free energy of their transfer from octanol (simulating the hydrophobic core of a protein) to water. In Fig. 16.3 this free energy of transfer (divided by RT, at $T \approx 300$ K) is the vertical coordinate, while the horizontal coordinate is the logarithm of the ratio between the frequencies of the outer and inner residues in proteins. As seen, the points are arranged more or less linearly with a slope around 1–1.5.

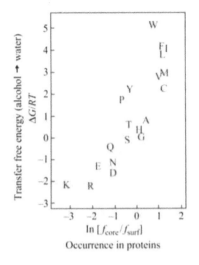

Occurrence in proteins

Figure 16.3. Experimentally found free energy of transfer of residue side groups from a non-polar solvent to water (ΔG, expressed in RT units), and "apparent free energy of transfer of a residue from the protein core to its surface", derived from observed frequencies of residue occurrence in the interior (f_{core}) and on the surface (f_{surf}) of the protein using the formula $\Delta G_1/RT = -\ln[f_{surf}/f_{core}]$. Adapted from Miller S., Janin J., Lesk A.M., Chothia C., *J. Mol. Biol.* (1987) **196**: 641–656.

Thus, the observed statistics of residue distribution between the interior and the surface are quite adequately described by the expression

$$\text{OCCURRENCE} \sim \exp(-\text{FREE ENERGY IN GIVEN MEDIUM}/kT_C) \qquad (16.3)$$

where the "conformational temperature" T_C is close to 300–400 K.

Inner voice: I would be more cautious as concerns the data presented in Fig. 16.3 because in experiment hydrophobicity was derived from residue transfer from water to high-molecular-weight alcohol. But why is the protein hydrophobic core believed to be like an alcohol? As well as the fact that alcohol is liquid while the core is solid, purely hydrophobic cyclohexane could be proposed as probably a better model of the hydrophobic core. And since the solubility of polar groups in cyclohexane is extremely low, the transfer free energy coordinate in Fig. 16.3 would inevitably be much more *extended* ... True, the correlation between hydrophobicity and occurrence will also be preserved in this case, but the slope of the "cyclohexane line" drawn through the experimental points will appear to be much greater, around 3–4, or so. Incidentally, this is close to the slope, shown in the upper part of Fig. 16.3, that refers to hydrophobic amino acids which, by the way, would be least affected by replacing the core-simulating agent ...

Lecturer: What is to be used to model the hydrophobic core really deserves consideration. I would say, an alcohol is still better than cyclohexane because of the presence of polar (NH and especially CO) groups in the protein core (although these usually participate in H-bonds within the core-surrounding secondary structure, the CO group is still capable of forming a "fork-like" H-bond, one branch of which remains unsaturated). So the hydrophobicity coordinate is hardly to be adjusted to purely non-polar cyclohexane. On the other hand, speaking of quantitative estimates, we have to bear in mind that Fig. 16.3 reflects a rough division of all side groups into two classes (those immersed in the protein and those located at its surface), and therefore their exposures differ by only about a half of the group surface. Accordingly, the experimentally derived hydrophobicities are to be *decreased* approximately two-fold, which (in contrast to alcohol replacement by cyclohexane) would decrease the tilt of the interpolation line... In principle, I do agree that the presented *numerical* data are to be treated cautiously, but the qualitative relationship between the energies of various elements and their occurrence in proteins should receive full attention.

Thus, the statistics of the occurrence of amino acid residues in the interior of a protein and at its surface exhibit a surprising *outward* similarity to Boltzmann statistics. This was first noted by Pohl in 1971 for rotamers. Later, the same was shown for the statistics of many other structural elements: for occurrence of ion pairs, for occurrence of residues in secondary structures, for occurrence of cavities in proteins, and so on, and so forth. To date, this analogy has become so common that protein structure statistics are often used to estimate the free energy of a variety of interactions between amino acid residues.

However, it should be stressed that the protein statistics resemble Boltzmann statistics only in the *exponential form* but *not* in the physical sense. As you may remember, the basis of Boltzmann statistics is that particles move from one position to another and spend more time at the position where their energy is lower, whereas in native proteins *no* residue wanders from place to place. For·example, Leu72 of sperm whale myoglobin is *always* inside the native globule and *never* on its surface. And although, according to the statistics, 80–85% of the total amount of Leu residues belong to the protein interior and 15–20% to its surface, this does not mean that each Leu spends 80–85% of the time inside the native globule and 20–15% on its surface. Rather, it means that natural selection has fixed most Leu residues at positions that belong to the interior of the globule.

That is, the usual Boltzmann statistics, the statistics of fluctuations in the usual 3D space, have nothing to do with the distribution of residues between the core and the surface of a protein. A globule has no fluctuations that could take each Leu (in accordance with its hydrophobicity) to the surface for 15–20% of the time and then take it back to the interior of the globule and keep it there for 80–85% of the time. That is, for *each separate* Leu there is *no* Boltzmann distribution determined by its particular hydrophobicity. Then how can we explain that the occurrence of the *total amount* of leucines inside and outside proteins agrees with the Boltzmann distribution determined from leucine's particular hydrophobicity?

Let us change the viewpoint.

Why is the predominant internal location of leucines favorable? Because it contributes to globule stability. Then why were not all leucines fixed inside the protein by natural selection? Presumably, because 80–85% of internal leucines are already *enough* to ensure protein stability, and dealing with the rest would be too expensive for selection.

Let us give up the psychology of natural selection as a pointless and non-scientific topic and pursue the matter on how the internal free energy of a protein structural element affects *the number of amino acid sequences* capable of stabilizing the protein that contains the structural element in question.

For example, let us see how the Leu→ Ser mutation in the protein interior can change the number of fold-stabilizing sequences.

The native (observed) structure is stable if its free energy is lower than that of the unfolded chain. Let us assume, for simplicity, (1) that the observed fold competes only with the unfolded state rather than with other compact folds; (2) that the residue's contribution to the native state stability is determined only by the residue's hydrophobicity; (3) that the internal residues are completely screened from water, and the external residues are completely exposed and (4) that the residues in the unfolded protein are completely exposed to water. All these statements are only approximately correct. Therefore, the following theory is rather rough – but it has the important advantage of simplicity.

The transfer free energy of a Ser side group from the hydrophobic surroundings into water is about 0, while that of leucine is about $+2\,\mathrm{kcal\,mol^{-1}}$. Let us put aside the difference in the Leu and Ser volumes and shapes and consider only their hydrophobicity. Leu is more hydrophobic than Ser. When Leu is inside, the protein fold is more stable against unfolding than the same fold with Ser inside. This means that the sequences which can stabilize a fold with Ser inside can *also* stabilize the fold with Leu inside – but the fold with Leu inside will, *in addition*, be stabilized by some sequences which cannot stabilize the fold with Ser inside.

How does the number of fold-stabilizing sequences change when a more stable structural element ("Leu inside") is replaced by a less stable one ("Ser inside")? Consideration of this problem will help us to understand why occurrence of various elements depends exponentially on their free energy and what the sense of the temperature T_C in eq. (16.3) is.

I apologize in advance for giving calculations in the most simplified form, which implies their incomplete accuracy. I am aiming to describe the essence of the matter without losing you in the math labyrinth through which one of us (A.V.F.) used to ramble a lot together with A.M. Gutin and A.Ya. Badretdinov.

Let $\Delta\varepsilon + \Delta F$ be the free energy difference between the given chain fold and the unfolded state of the chain. Here $\Delta\varepsilon$ is the free energy difference for the concerned element in the fold (including all the element's interactions with the surroundings, e.g., the Leu's hydrophobic free energy in the core), and ΔF is the free energy difference for the remaining chain. The fold is stable against unfolding when $\Delta F + \Delta\varepsilon < 0$, that is, when

$$\Delta F < -\Delta\varepsilon \qquad (16.4)$$

The values ΔF and $\Delta\varepsilon$ depend on the amino acid sequence. Let us consider all those sequences which preserve the $\Delta\varepsilon$ value (e.g., all sequences with Leu (or Ser) at the given point of the chain; for the core positions, $\Delta\varepsilon \approx 2\,\mathrm{kcal\,mol^{-1}}$ for Leu (and ≈ 0 for Ser), while for the surface positions, $\Delta\varepsilon \approx 0$ for all residues). The value ΔF will change with the sequence. The probability P^* that $\Delta F < -\Delta\varepsilon$ is

$$P^*(\Delta F < -\Delta\varepsilon) = \int_{-\infty}^{-\Delta\varepsilon} P(\Delta F)\,\mathrm{d}(\Delta F) \qquad (16.5)$$

where $P(\Delta F)$ is the probability of the given ΔF value for a randomly taken sequence.

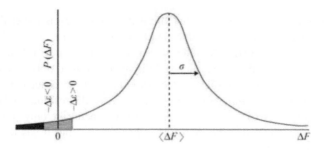

Figure 16.4. The typical Gaussian curve for ΔF distribution among random sequences. ΔF contains the entire free energy difference between the given fold and the unfolded state of the chain, except for the fixed $\Delta \varepsilon$ value of the structural element in question. The values of $\Delta F < -\Delta \varepsilon$ (i.e., those satisfying the condition that $\Delta F + \Delta \varepsilon < 0$) meet the requirements of a stable fold. The area shown in black corresponds to $\Delta F < -\Delta \varepsilon$ values at $\Delta \varepsilon > 0$, while the "red + black" area is for $\Delta \varepsilon < 0$. The latter is larger, which means that a greater number of random sequences stabilize the fold when the free energy of the element in question is below zero ($\Delta \varepsilon < 0$) as compared with its being above zero ($\Delta \varepsilon > 0$).

The ΔF value is composed of energies and entropies of many residues that independently mutate in random sequences. Therefore (according to the Central Limit Theorem of mathematical statistics), $P(\Delta F)$ has a simple, so-called Gaussian (see Fig. 16.4) form:

$$P(\Delta F) = (2\pi\sigma^2)^{-1/2} \times \exp[-(\langle \Delta F \rangle - \Delta F)^2 / 2\sigma^2] \qquad (16.6)$$

Here $\langle \Delta F \rangle$ is the mean (averaged over all the sequences) value of ΔF, and σ is the root-mean-square deviation of ΔF from the mean $\langle \Delta F \rangle$.

I would like to remind you that the Central Limit Theorem describes the expected distribution of the sum of *many* random terms and answers the question as to the probability of this or that value of the sum. Here, "the sum of many random terms" is ΔF (composed of many interactions in a randomly taken, i.e., "random", sequence). As mathematics state, for the majority ($\approx 70\%$) of sequences, the ΔF value must lie between $\langle \Delta F \rangle - \sigma$ and $\langle \Delta F \rangle + \sigma$, and the probability $P(\Delta F)$ decreases dramatically (exponentially) with increasing difference between ΔF and $\langle \Delta F \rangle$ (see Fig. 16.4). Since the great majority of random sequences are obviously incapable of stabilizing the fold, the values of P and P^* must be far less than 1 near the "stability margin" ($\Delta F \approx 0$). Thus, the $\langle \Delta F \rangle$ value is not only positive but also *large*, much larger than σ.

The value of $(\langle \Delta F \rangle - \Delta F)^2$ is equal to $\langle \Delta F \rangle^2 - 2\langle \Delta F \rangle \Delta F + \Delta F^2$, i.e., at small (as compared with large $\langle \Delta F \rangle$) ΔF values, $(\langle \Delta F \rangle - \Delta F)^2 \approx \langle \Delta F \rangle^2 - 2\langle \Delta F \rangle \Delta F$, and consequently,

$$P(\Delta F) \approx [(2\pi\sigma^2)^{-1/2} \times \exp(-\langle \Delta F \rangle^2 / 2\sigma^2)] \times \exp[\Delta F \times (\langle \Delta F \rangle / \sigma^2)] \quad (16.7)$$

I would like you to believe (or better – to take the integral and check up) that in this case the probability of $\Delta F < -\Delta\varepsilon$ is

$$P^*(\Delta F < -\Delta\varepsilon) \equiv \int_{-\infty}^{-\Delta\varepsilon} P(\Delta F)\, d(\Delta F) \approx \text{const} \times \exp[-\Delta\varepsilon/(\sigma^2/\langle\Delta F\rangle)], \quad (16.8)$$

where the constant, equal to $[(2\pi\sigma^2)^{-1/2} \times \exp(-\langle\Delta F\rangle^2/2\sigma^2)] \times (\sigma^2/\langle\Delta F\rangle)$, is of no interest to us, while the really important term **$\exp[-\Delta\varepsilon/(\sigma^2/\langle\Delta F\rangle)]$** demonstrates that an element's free energy ($\Delta\varepsilon$) has an *exponential* impact upon the probability that a random sequence is capable of stabilizing the fold with the given element. Thus, increasing $\Delta\varepsilon$ decreases the number of fold-stabilizing sequences *exponentially*.

Note in addition: Our previous belief was that "eligible" sequences are all those providing $\Delta F + \Delta\varepsilon < 0$, i.e., at least a minimal stability of the protein structure. After a few more lectures you will know that the protein structure should possess a certain "stability reserve". Otherwise, it would melt in our hands and be incapable of rapid and unique folding. Therefore, to be more accurate, we have to consider as "eligible" the sequences that provide $\Delta F + \Delta\varepsilon < -|F_{\min}| < 0$, i.e., $-\Delta F < -\Delta\varepsilon - |F_{\min}|$. To ensure stability of the native globule, it is enough to have $F_{\min}$ of only a few kilocalories per mole, i.e., it can be believed that $|F_{\min}| \ll \langle\Delta F\rangle$. Then eq. (16.8) has the form

$$P^*(\Delta F < -\Delta\varepsilon - |F_{\min}|) \approx \text{const} \times \exp[-(\Delta\varepsilon + |F_{\min}|)/(\sigma^2/\langle\Delta F\rangle)]$$

$$= \text{const}^* \times \exp[-\Delta\varepsilon/(\sigma^2/\langle\Delta F\rangle)] \qquad (16.9)$$

In other words (since a change of the pre-exponential constant is of no interest to us), we come again to the same idea that an element's energy has an exponential impact on the probability that the element will be fixed in the protein structure as a result of the "selection for stability". Note that a requirement of enhanced stability of the native fold (i.e., an increase in the $|F_{\min}|$ value) decreases the number of appropriate sequences also exponentially, in proportion to $\exp[-|F_{\min}|/(\sigma^2/\langle\Delta F\rangle)]$.

The dependence obtained is as exponential as the Boltzmann formula but it has the term $\Delta\varepsilon$ divided not by the temperature of the environment (kT) but by the as yet unknown value of $\sigma^2/\langle\Delta F\rangle$.

What is this value? First of all, note that $\sigma^2/\langle\Delta F\rangle$ is independent of the protein size. Indeed, according to the mathematical statistics laws, the mean value ($\langle\Delta F\rangle$) is proportional to the number of terms summarized in ΔF (which is, in our case, approximately proportional to the protein size), while the mean square deviation from σ is proportional to the square root of the number of these terms, i.e., σ^2 is also approximately proportional to the protein size.

The fact that $\sigma^2/\langle\Delta F\rangle$ *does not increase with increasing protein size* is most important: this means that the "defect's" free energy $\Delta\varepsilon$ must be compared (using eq. (16.8) *not* with the total protein energy but rather with some characteristic energy $\sigma^2/\langle\Delta F\rangle$, which is something like the average energy of non-covalent interactions per residue in the chain (which is also independent of the protein size).

Taken together with the exponential form of eq. (16.8), this provides an immediate answer to the question as to why $\Delta \varepsilon$ of only a few kilocalories per mole can produce a considerable effect upon the occurrence of protein structure elements (e.g., why inside a protein globule Leu residues are an order of magnitude more numerous than Ser ones). This happens because the number of fold-stabilizing sequences decreases e-fold (approximately 3-fold) when $\Delta \varepsilon$ increases by a value of $\sigma^2 / \langle \Delta F \rangle$.

Thus, $\sigma^2 / \langle \Delta F \rangle$ is something like chain energy per residue. One can present $\sigma^2 / \langle \Delta F \rangle$ as kT_C, where T_C is a certain temperature (as you may remember, the "heat quantum" kT also has the sense of characteristic heat energy per particle). Exactly what temperature is T_C? For a "random" globule, there is only one characteristic temperature – that of "freezing out" its most stable fold. A protein chain has one characteristic temperature as well, that is, its denaturation temperature ≈ 350 K (which is close to that of protein's life, ≈ 300K). Eventually, it can be shown that as long as only the minimum protein structure stability is required (i.e., if selection of primary structure is determined only by this minimal restriction), the protein melting temperature is close to the freezing temperature of a random chain. Therefore, it can be suggested that the value of "conformational temperature" T_C is determined by the temperature T_M of protein melting. In other words, $\sigma^2 / \langle \Delta F \rangle$ can be considered as amounting to about 0.5–1 kcal mol^{-1}.

This can be not only believed but shown as well. The words "can be shown" imply a quite complex theoretical physical proof, the Shakhnovich–Gutin theorem, which I shall spare you. [Nevertheless, I would like to outline the main ideas involved. (i) Random replacement of one sequence by another changes its free energy essentially in the same way as a random replacement of one fold by another. (ii) $kT_C = \sigma^2 / \langle \Delta F \rangle$ is determined by the growth in the number of random sequences close to the edge separating the fold-stabilizing sequences from the others. (iii) T_M is essentially determined by the growth in the number of folds close to the low-energy edge of energy spectrum of a random sequence. These two edges are close since (vi) the native and denatures states of a chain are close in stability, and (v) the energy gap between the most stable chain fold and its close competitors is not wide (this will be discussed in Lectures 17, 18); therefore, T_C and T_M are also close.]

Thus, we have cleared up why a defect of only a few kilocalories per mole (against the background of the much higher total energy of the protein) can virtually prohibit many motifs of protein architecture. This happens because the defect's free energy is to be compared with $kT_C \sim 0.5$–1 kcal mol^{-1} rather than with the total energy of the protein; and then we see that any defect of 1 kcal mol^{-1} decreases roughly 5-fold the number of sequences "eligible" for this protein, a defect of 2 kcal mol^{-1} decreases them roughly 20-fold, and so on.

Specifically, that is why the crossing of irregular connections (this defect usually costs 2–3 but never more than 5 kcal mol^{-1}) is rarely observed (Fig. 16.5).

Inner voice: I cannot but cut in. The entire logic of your narration is based on the assumption that sequences unable to fold into a stable structure are rejected. This is quite probable. But then, are those yielding *too stable* structures rejected too?

common   rare

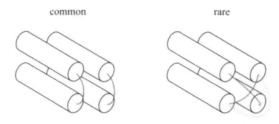

Figure 16.5. Loop crossing is rarely observed in proteins.

Actually, the observed stability of native structures is *never* very high. Moreover, a correlation between the denaturation temperature of a protein and the life temperature of the host organism has been mentioned. Then we have to conclude that *too* stable protein structures (and, according to your logic, *too* stable structural elements as well) are to be rejected. Seemingly, you do not take this into account at all.

Lecturer: It is difficult to make a stable protein. This has been demonstrated by the entire experience of designing *de novo* proteins. That is, sequences capable of folding into something stable are rare. It is still more difficult to create a "superstable" protein because its eligible sequences occur still more rarely. Figure 16.4 also serves as an illustration of how small the fraction of sequences eligible for creating "superstable" protein structures (i.e., those with the extremely low free energy ΔF) is; see also the exponential effect of $|F_{min}|$ on P^* in eq. (16.9). The sequences good for "superstable" proteins constitute only a minor fraction of the sequences that are good for stable proteins... If the selection does not insist on "superstability" and has nothing against it, the fraction of such proteins *automatically* appears to be negligible. This explains quite satisfactorily all the experimental facts and correlations you have pointed out...

Now we pass to the other question as to the role played by entropy effects in the stability of the native protein structure, although the chain there is fixed.

Let us again compare right-handed and left-handed connections between parallel β-strands (see Fig. 16.6). As we know, a *right*-handed connection has to turn by $360° - 30° = 330°$, while a *left*-handed connection would have to turn by $360° + 30° = 390°$. As stated by polymer physics, the more bent the chain, the less numerous its possible conformations. This means that the right-handed (less bent) connection is compatible with a greater number of chain conformations than the left-handed one. This is clear. But what is the impact of the number of possible conformations on the number of sequences stabilizing this or that connection pathway?

A randomly taken sequence can provide the highest stability (among all other chain structures) of either one of the numerous conformations corresponding to the less-bent connection or one of a few conformations corresponding to the more bent

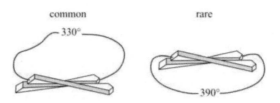

Figure 16.6. In proteins, left-handed (more bent) connections between parallel β-strands are very rare, while right-handed ones are more common.

connection; alternatively, it may be unable to ensure stability of any of them. The latter is certainly the most probable, since randomly composed amino acid sequences possess no stable three-dimensional structure. However, if we have to choose only between the left- and right-handed connections, which of them has a better chance to become the more stable?

Each separate conformation containing a less bent connection is neither better not worse than each separate conformation with a more bent connection, provided that these conformations are equal in compactness, in secondary structure content, etc. Still, the less bent conformations are much more numerous ... Actually, here, we have a kind of lottery with a small chance of winning the main prize (that is creating a stable 3D structure); in this lottery the right-handed (less bent) connection has many tickets (possible conformations), while the left-handed connection has only a few. Which of them, if either, will win the prize? Almost certainly the right-handed connection that has many more tickets and hence a better chance of success, since the probability of winning is directly proportional to the available tickets (conformations).

In other words, the broader the set of *possible* conformations, the more frequently (in direct proportion to the set range) this set contains the most stable structure of a random sequence. This is exactly what we observe in globular proteins: here the right-handed (less restricted) pathway of a connection is the rule, while the left-handed pathway is the exception.

Allow me to remind you of one of these very rare exceptions. I mean the left-handed β-prism discussed a couple of lectures back. Its spatial structure is indeed unique as it consists almost entirely of left-handed connections between β-segments. And what about its primary structure? It appears to be unique as well because it does not look like "random" (that would be typical of globular proteins) but contains ten repeats of an 18-residue peptide forming each turn of the left-handed superhelix (with three β-strands per turn). So, a unique spatial structure is combined with a unique primary structure ...

Let us consider one more problem connected with entropic defects. It concerns not folding patterns but the layered structure of the globule. Which should occur more often, proteins with α- or β-structure in the center?

Figure 16.7 explains why globules with β-strands in the center can be stabilized by many more primary structures than those with α-helices in the center.

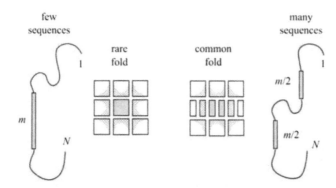

Figure 16.7. A multiple-layer packing with an α-helix in its center should be less probable (and, indeed, is observed more rarely) than a multiple-layer packing with two β-segments in its center. The reason is that the segment occupying the globule's center should contain exclusively hydrophobic residues, and its large length is dictated by the globule diameter. Since an α-helix contains twice as many residues as an extended β-strand of equal length, a central α-helix (with a large block of hydrophobic groups) will require a much less common sequence than is required for the creation of two central β-segments (with two hydrophobic blocks of half the length positioned *somewhere* in the chain). The probability of occurrence of one entirely hydrophobic block of m residues in a *given position* of the random chain is about p^m (where p is the fraction of hydrophobic residues in the chain); there is an equal probability of two hydrophobic blocks of half the length occurring in *two given positions* of a random chain: $(p^{m/2}) \times (p^{m/2}) = p^m$. However, one block can be placed in the N-residue chain in only $\sim N$ different ways, while there are many more ways to position two blocks ($\sim N \times N/2$), i.e. there are many more sequences with two short hydrophobic blocks than with one long block.

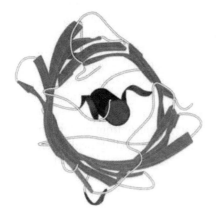

Figure 16.8. Green fluorescent protein has an unusual fold in which the α-helix is surrounded by the β-sheet. The central α-helix is not continuous: actually, it is broken by the chromophore into two approximately equal halves (not shown).

And indeed, the internal α-helix occurs in proteins most rarely (the green fluorescent protein shown in Fig. 16.8 is a striking example of such the most unusual fold), while the internal β-strands are typical (e.g., for "Rossmann folds", see Fig. 16.1).

The same standpoint ("How often is this structural element stabilized by random sequences?") may be used to explain many other structural features observed in globular proteins. Among them, the average domain size for a chain with a given ratio between hydrophobic and hydrophilic groups (this problem was studied by Bresler and Talmud in Leningrad as far back as in 1944, and then by Fischer in the USA), and also the previously discussed average lengths of α-helices, β-segments and irregular loops.

Thus, our analysis shows that the probability of observing the given structural element in stable folds *formed by those random sequences that are able to stabilize at least some 3D structure* should be greater, the lower the element's energy and the greater the number of conformations appropriate for this element.

And since the energy and the number of conformations unite in the free energy, the statistics of occurrence of various structural elements in *randomly created stable globules* should have the form

$$\text{OCCURRENCE} \sim \exp(-\text{FREE ENERGY}/kT_C), \tag{16.10}$$

where T_C is close to (but not equal to, as a thorough analysis shows) the freezing temperature of the globule formed by a random amino acid sequence; in turn, this temperature is close to the protein melting temperature.

And indeed, general protein statistics have the form outlined above, which is typical for statistics of structures built up by random sequences that have been selected only for creation of stable globular structures.

Let me remind you once again of the physical basis of the above relationship. The free energy of a given structural element *exponentially* changes the number of sequences capable of stabilizing the protein whose native structure contains this element. If the element itself is stable, its host protein can stand many even unfavorable mutations, i.e., the protein containing this element can be stabilized by a quite large number of sequences. If the element is unstable, its host protein demands a most thorough selection of its primary structure (stability of its 3D structure can be easily ruined by even a few mutations), and these "thoroughly selected" sequences are rare.

Here it should be stressed that the so-called "prohibited" (not observed or rarely observed) protein structures are *not* impossible in principle but simply improbable because they can only be created by a small number of sequences.

It appears that globular proteins could quite easily originate from random amino acid sequences (to be more accurate, from pieces of DNA coding random amino acid sequences). It would only require slight stabilizing (through a few mutations) of the most stable 3D structure of the initial random polypeptide and "grafting" an active site on its surface (to provide "biologically necessary" interactions with surrounding molecules). Also, it would be necessary to clean the protein's surface of the residues that could involve it in "biologically harmful" associations (like those provoking sickle cell anemia by sticking hemoglobins together).

Inner voice: Are we to understand that you are suggesting that proteins with "new architectures" originated from random sequences rather than through strong mutations of proteins with some "old" architectures? And that the folding patterns whose stability is compatible with many random sequences originated from them many times (as many as there are homologous families in them), while other patterns covering only one homologous family each arose only once?

Lecturer: All we can say with confidence is that more than a negligibly small fraction of random sequences can give rise to proteins, and that the stability of some folds (specifically, of those often observed in proteins) is compatible with a larger number of random sequences than the stability of other, "rare" folds. As to your questions about the historical happenings, I think they cannot be answered at present. In particular, it is impossible to say whether representatives of "popular" folding patterns arose many times or not. Perhaps they originated many times from different random sequences. Perhaps only once (and not from a random sequence but from pieces of some other proteins). It is also possible that later, in the course of evolution, sequences of the same root fell so wide apart within the frames of the preserved folding pattern (since "popular" patterns are compatible with so very different sequences), that all signs of homology and genetic relationships have been wiped out, and we cannot trace them. I only want to stress that the "popular" folding patterns compatible with so many sequences give much more space for any kind of origin and subsequent evolution than the "rare" patterns.

In this connection I cannot but note that contemporary attempts at *de novo* protein design widely use both multiple random mutations of sequences, necessarily accompanied by selection of "appropriate" (say, capable of binding to something) variants, and random shuffling of oligopeptides (accompanied by similar selection). The modern methods of selecting "appropriate" random sequences (I would like to mention perhaps the most powerful of them: it is called phage display; do read about it in molecular biology textbooks) allow examining about 10^8–10^{10} of random sequences. The fact that "protein-like" products were now and then found among these samples shows that the fraction of such "protein-like" chains amounts to 10^{-8}–10^{-10} of all random polypeptides composed of a few dozens of amino acid residues.

I stress that origination from random sequences is especially appropriate for globular proteins because it is their sequences that outwardly resemble "random" (i.e., most abundant) co-polymers (Fig. 16.2). At the same time, *non*-randomness is obvious for primary (and also spatial) structures of fibrous and membrane proteins (but incidentally, the principles of their construction are simple too: repeats of short blocks for fibrous proteins and alternation of hydrophobic and hydrophilic blocks for membrane proteins).

The above analysis emphasizes the fact that evolution-yielded protein structures look very "reasonable" from the physical point of view, just like the DNA double helix and the membrane bilayer. Presumably, at the level of protein domain architectures as well, evolution does not "invent" physically unlikely structures but "selects" them from physically sound ones (i.e., those that are stable and therefore capable of rapid self-organizing, as we will see soon). This is what the sense of "physical selection" consists in.

Part V

COOPERATIVE TRANSITIONS IN PROTEIN MOLECULES

LECTURE 17

Earlier we considered the stability of fixed, "solid" protein structures. However, depending on ambient conditions, the most stable state of a protein molecule may be not solid but molten or even unfolded. Then the protein denatures and loses its native, "working" 3D structure.

Usually protein denaturation is observed *in vitro* as a result of an abnormal temperature or denaturant [i.e., urea, H^+ or OH^- ions (that is, abnormal pH), etc.]. However, decay of the "solid" protein structure and its subsequent refolding can also occur in a living cell; specifically, this is important for transmembrane transport of proteins.

Moreover, even under physiological conditions, not all proteins "by themselves" have a fixed 3D structure. Some of them acquire such a structure only when binding a ligand, another protein, DNA or RNA. (It has been suggested that there is a kind of advantage in binding of a protein with an (individually) unstable 3D structure: this can provide not too stable binding even when the binding surface (and therefore the specificity of binding) is great.) Therefore, studies of the "denatured" (disordered) state of protein chains also directly concern the states of some proteins in the cell.

Denaturation of globular proteins *in vitro* is the subject of intensive studies. This continues to be of interest because of its relationship to the problem of protein folding, that is, to the question as to how a protein chain finds its unique 3D structure among zillions of alternatives. Today, in fact, I will not touch upon the kinetic aspects of protein denaturation and folding; rather, I will concentrate on the thermodynamic and structural aspects of these events.

Water-soluble globular proteins are the best studied in this respect, and I will speak about them.

What does experiment show?

It is well established that denaturation of small proteins is a cooperative transition with a simultaneous abrupt ("S-shaped") change of many (though sometimes not all) characteristics of the molecule (Fig. 17.1). The S-shaped form of experimental curves shows that the plotted characteristics of the molecule change from the values corresponding to the native molecule to those corresponding to its denatured state, and the narrow transition region suggests that the transition embraces many amino acid residues.

Moreover, protein denaturation occurs as an "all-or-none" transition (Fig. 17.2).

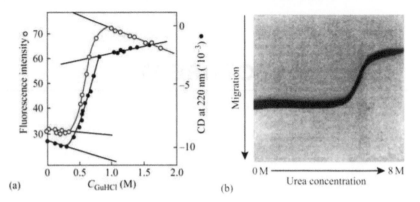

(a)　(b)

Figure 17.1. Protein denaturation is accompanied by an abrupt "S-shaped" change of many characteristics of the molecule. In the given case, protein unfolding is induced by increasing denaturant concentration; the latter is measured in $mol\,l^{-1}$ (often denoted as M). (a) Simultaneous change of CD (at 220 nm) and fluorescence in the process of denaturation of phosphoglycerate kinase in solution with increasing concentration of guanidine dihydrochloride. Adapted from Nojima H., et al., *J. Mol. Biol.* (1977) **116**: 429–442. (b) Electrophoresis of cytochrome c at various urea concentrations. Reproduced, with permission, from Creighton T.E., *J. Mol. Biol.* (1979) **129**: 235–264.

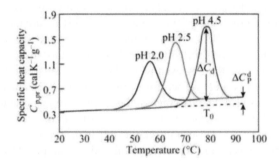

Figure 17.2. Calorimetric study of lysozyme heat denaturation at various pH values. The position of the heat capacity (C_p) peak determines the transition temperature T_0, the peak width gives the transition width ΔT and the area under the peak determines ΔH, the heat absorbed by one gram of the protein. The values ΔT, $\Delta H\times$ (protein's M.W). and T_0 satisfy eqs (17.5) and (17.6), indicating that the denaturation occurs as an "all-or-none" transition. The increased heat capacity of the denatured protein (ΔC_p^d) originates from the enlarged interface between its hydrophobic groups and water after denaturation. Adapted from Privalov P.L., Khechinashvili N.N., *J. Mol. Biol.* (1974) **86**: 665–684.

You may remember that the latter means that only the initial (native) and the final (denatured) states amount to visible quantities (Fig. 17.3), while "semi-denatured" states are virtually absent. (Though, of course, they do exist to a very small extent, since a native molecule cannot come to its denatured state without passing the

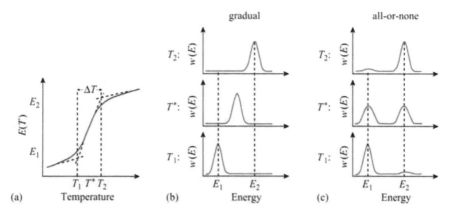

Figure 17.3. With the same temperature dependence of energy E (or any other observable parameter), the cooperative ("S-shaped") transition can be either of the "all-or-none" type (example: protein denaturation), or gradual (example: helix–coil transition in polypeptides). The difference is displayed in the shape of the function $w(E)$ showing the distribution of molecules over energy (or over any other observable parameter) rather than in the shape of the curve $E(T)$. Dashed lines in (a) show a graphical determination of ΔT, the width of the temperature transition.

intermediate forms, and their presence has a crucial effect on kinetics of the transition, which we will discuss later.)

In other words, the "all-or-none" transition is a microscopic analog of the first-order phase transitions in macroscopic systems (e.g., crystal melting). However, unlike true phase transitions, the "all-or-none" transitions in proteins have a non-zero width, since this transition embraces a microscopic system. A little later I will show you how denaturation of small proteins is proved to be of the "all-or-none" type; and now I want to specify that "all-or-none" denaturation actually refers to small proteins and to separate domains of large proteins, while denaturation of a large protein as a whole is the sum of the denaturation of its domains.

One can denature a protein not only by heating (see Fig. 17.2), but sometimes also by cooling (this will be considered later). Besides, proteins denature under the action of too low or too high pH (which annihilates the charges of one sign and thus causes repulsion between remaining charges of the opposite sign). Also, protein denaturation results from addition of denaturants like urea (NH_2—CO—NH_2) or guanidine dihydrochloride ($[(NH_2)_3C]^-Cl^+$): these molecules, with excess hydrogen-bond donors, seem to withdraw the hydrogen bonds of O-atoms of water molecules, disturb the balance of H-bonds in water, and thus force water molecules to break H-bonds in the protein more actively.

(It is not out of place to mention here that some substances like $NaSO_4^-$, on the contrary, work as "renaturants" of protein structures, i.e., they increase the native structure stability.) Proteins can also be denatured by sodium dodecyl sulfate molecules (their hydrophobic tails penetrate into the hydrophobic core of the protein and break it down), by various alcohols, salts, etc., as well as by very high pressure.

It is worthy of note that proteins of organisms living under extreme (for us!) conditions like hot-acidic-salty wells are "extremophilic": they have their solid native form just under such extreme conditions, and denature under "physiological" (for most organisms) conditions. It should be mentioned that the architectures of these "extremophilic" proteins have no striking features that distinguish them from normal ("mesophilic") proteins; rather, adaptation to the extreme conditions is provided through careful fitting of their amino acid sequences aiming to reinforce structure-supporting (in the given environment) interactions. But this story can take us too far ...

Is a denatured protein able to *renature* and to restore its native structure? That is, is denaturation reversible?

Yes, it is (this is known since Anfinsen's experiments in the 1960s that brought him a Nobel prize): the protein can renature if it is not too large and has not been subjected to substantial chemical modifications after *in vivo* folding. In this case, a "mild" (without chemical decay) destruction of its native structure (by temperature, denaturant, etc.) is reversible, and the native structure spontaneously restores after environmental conditions have become normal. However, effective renaturation *in vitro* needs thorough optimization; otherwise, the renaturation can be hindered by precipitation or aggregation (for large multidomain proteins it can also be an intramolecular aggregation of remote chain regions; it is known that the ability to renature usually decreases, and the experimental difficulties increase with protein size).

The reversibility of protein denaturation is very important: it shows that the entire information necessary to build up the protein structure is contained in its amino acid sequence, and that the protein structure itself (to be more exact, the structure of a not-too-modified and not-too-large protein) is thermodynamically stable. And this allows the use of thermodynamics to study and describe de- and renaturation transitions.

The fundamental fact that protein denaturation occurs as an "all-or-none" transition, i.e., that "semi-denatured" states are virtually absent, has been established by P.L. Privalov who worked at our Institute of Protein Research, Russian Academy of Sciences. He studied heat denaturation of proteins, which is usually accompanied by a large heat effect, in the order of one kilocalorie per mole of amino acid residues.

How has it been proven that protein melting is an "all-or-none" transition?

To this end, it is not enough to demonstrate the existence of a heat capacity peak connected with protein melting (or, which is the same, the S-shaped temperature dependence of protein's energy). This only shows an abrupt cooperativity of melting, i.e., that melting embraces many amino acid residues. However, this does not prove that melting embraces the entire protein chain, that is, this does not tell us if the protein melts as a whole or in parts. To prove that melting is an "all-or-none" transition, one has to compare (1) the "effective heat" of transition calculated from its width (i.e., the amount of heat consumed by one independent "melting unit") with (2) the "calorimetric heat" of this transition, i.e., the amount of heat consumed by one melting protein molecule. A coincidence of these two independently measured values proves that the molecule melts as a single unit.

If the "effective latent heat" of a melting unit is less than the calorimetric heat consumed by one protein molecule, then the "melting unit" is smaller than the whole protein, i.e., the protein melts in parts. If the effective heat of transition is greater than the calorimetric heat, then the "melting unit" is greater than the protein, i.e., it is not one protein molecule but some aggregate of them that melts.

This is the van't Hoff criterion for existence (or non-existence) of the "all-or-none" transition. Since it is important, we shall consider it in some detail.

How is the effective heat of transition related to its width?

Let us consider a "melting unit" which can be in two states: "solid", with energy E and entropy S, and "molten", with energy E' and entropy S'. For simplicity, let us assume that E, E', S, S' do not depend on temperature T (consideration of a more general case I leave to the reader . . .).

Since there are only two states of this unit, the probability of its molten state, according to the Boltzmann formula, is

$$P_{MOLTEN} = \frac{\exp[-(E' - TS')/kT]}{\exp[-(E - TS)/kT] + \exp[-(E' - TS')/kT]}$$

$$= \frac{1}{\exp[(\Delta E - T\Delta S)/kT] + 1} \qquad (17.1)$$

where $\Delta E = E' - E$, $\Delta S = S' - S$. The probability of its solid state (for the "all-or-none" transition) is $P_{SOLID} = 1 - P_{MOLTEN}$. The derivative dP_{MOLTEN}/dT shows how steeply P_{MOLTEN} changes with temperature. A simple calculation (using the equation $d(F/T)/dT = d[(E - TS)/T]/dT = -E/T^2$ derived earlier) shows that

$$dP_{MOLTEN}/dT = P_{MOLTEN}(1 - P_{MOLTEN})(\Delta E/kT^2). \qquad (17.2)$$

I think I should give you the details of this calculation: you have to understand rather than just to believe. Thus, let us denote $\Delta F/T \equiv (\Delta E - T\Delta S)/kT$ as X. Then $dX/dT = -\Delta E/kT^2$, $P_{MOLTEN} = 1/(e^X + 1)$, $P_{SOLID} = 1 - P_{MOLTEN} = e^X/(e^X + 1)$, and

$$dP_{MOLTEN}/dT = d[1/(e^X + 1)]/dT$$

$$= -[1/(e^X + 1)^2] \times [de^X/dT]$$

$$= -[1/(e^X + 1)^2] \times e^X \times [dX/dT]$$

$$= -[1/(e^X + 1)] \times [e^X/(e^X + 1)] \times [dX/dT]$$

$$= P_{MOLTEN}(1 - P_{MOLTEN}) \times (-dX/dT)$$

$$= P_{MOLTEN}(1 - P_{MOLTEN})(\Delta E/kT^2)$$

The mid-transition temperature corresponds to $T_0 = \Delta E/\Delta S$. Here $P_{MOLTEN} = P_{SOLID} = \frac{1}{2}$, and here the value $P_{MOLTEN}(1 - P_{MOLTEN})$ has its maximum equal to $\frac{1}{4}$. The point of the steepest change of P_{MOLTEN}, i.e., the maximum of the derivative dP_{MOLTEN}/dT is very close to T_0 when $\Delta E/kT \gg 1$, i.e. when P_{MOLTEN} changes in a narrow temperature range. Here the slope of the curve $P_{MOLTEN}(T)$ has its maximum and equals to

$$(dP_{MOLTEN}/dT)|_{T=T_0} = \tfrac{1}{4}\Delta E/kT_0^2 \tag{17.3}$$

A graphical determination of the transition width ΔT is shown in Fig. 17.3(a). It is done by a linear extrapolation of the maximum slope of the curve $P_{MOLTEN}(T)$ up to its intersection with the base lines corresponding to the native ($P_{MOLTEN} = 0$) and the denatured ($P_{MOLTEN} = 1$) states. In the transition zone, the extrapolated value of $P_{MOLTEN}|^{extrap}$ changes from 0 to 1 (i.e., the change $\Delta P_{MOLTEN}|^{extrap} = 1$), and the temperature changes by ΔT. Thus,

$$(dP_{MOLTEN}/dT)|_{T=T_0} = \Delta P_{MOLTEN}|^{extrap}/\Delta T$$
$$= 1/\Delta T \tag{17.4}$$

That is, ΔT is determined (see Fig. 17.3) by the width of the zone of the steep rise in P_{MOLTEN} (or, more generally: by the width of the steep rise zone of any experimental parameter determining the fraction of the denatured state, e.g., protein helicity).

Finally, the heat consumed by the "melting unit" (ΔE) is connected with the transition width (ΔT) and temperature (T_0) by the relationship

$$\frac{1}{\Delta T} = \tfrac{1}{4}\Delta E/kT_0^2$$

or

$$\Delta E = 4kT_0^2/\Delta T \tag{17.5}$$

The heat ΔE consumed by the "melting unit" (and calculated *only* from the shape of the transition curve) is to be compared with the heat consumed by the whole protein molecule. The latter is calculated as $\Delta H/N$, where ΔH is the heat consumed by all N protein molecules contained in the calorimeter ($N = m/M$, where m is the total mass of the protein taken, and M is its molecular mass). If

$$\Delta E = \Delta H/N \tag{17.6}$$

then melting of the *whole* protein is an "all-or-none" transition. This is the van't Hoff criterion.

If $\Delta E < \Delta H/N$ (i.e., if the transition width ΔT is greater than $4kT_0^2/(\Delta H/N)$), then the "melting unit" is smaller than the whole protein; this means that the protein melts in parts. If $\Delta E > \Delta H/N$, then the "melting unit" is greater than the protein, i.e., the "melting unit" is some aggregate of protein molecules rather than one protein molecule.

Protein melting, i.e., the decay of its structure due to elevated temperature, looks natural; however, a "cold" denaturation of proteins also exists at abnormally low temperatures (Fig. 17.4). It is not observed for all proteins, however, since the water in the tube usually freezes first . . .

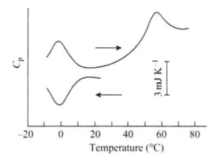

Figure 17.4. Reversible "cold" denaturation of protein (apomyoglobin) at an abnormally low temperature (the lower curve; the arrow shows that temperature decreases during the experiment). The upper curve shows renaturation of the cold-denatured protein (the left heat capacity peak) with increasing temperature and its subsequent melting (the right peak) at high temperature. The curves show the excess of the protein solution heat capacity C_p over that of the solvent; they are not normalized by the amount of protein in solution. Adapted from Griko Yu.V., Privalov P.L., Venyamino S.Yu., Kutyshenko V.P., *J. Mol. Biol.* (1988) **202**: 127–138.

Experimentally, the existence of the cold denaturation of proteins has been shown by Privalov and Griko's group; they have also shown that it is an "all-or-none" transition, like protein melting.

Cold denaturation occurs because of an abnormal temperature dependence of the strength of hydrophobic forces (here I present the arguments of Brandts who predicted cold denaturation long before the experiments by Griko and Privalov). These forces decrease drastically with decreasing temperature. The strong temperature dependence of the hydrophobic effect is manifested by the increased heat capacity of the denatured, less-compact state of the protein (see ΔC_p^d in Fig. 17.2). As a result, the latent heat of protein denaturation increases significantly with increasing temperature – and *decreases* significantly with decreasing temperature (Figs 17.2 and 17.5a). Indeed, it may fall to negative values (Fig. 17.5a)! That is, the energy of the more ordered (native) structure (plus the energy of the surrounding water) is *lower* than that of the less ordered, denatured state (plus surrounding water) at normal temperatures; but it can *exceed* the energy of the denatured state (plus surrounding water) at low temperatures ($\approx +10\,°C$ and below). As a result, the native structure's stability (i.e., the free energy difference between the native and the denatured state) has its maximum at about room temperature, and starts to decrease at low temperatures (Fig. 17.5b). For some proteins the stability decreases so much that the native structure decays at an abnormally low (from the physiological point of view: $\approx 0\,°C$) temperature. However, this event is usually unobserved since water usually freezes at such a low temperature, and all denaturation processes freeze as well (which allows us to store proteins in the cold).

Paradoxically, there is some similarity between the denaturation of proteins and water boiling. As you know, there are two ways to boil water: either to increase temperature or to decrease pressure. The hydrophobic pressure condensing the protein globule decreases significantly at low temperatures, and the protein begins to "boil".

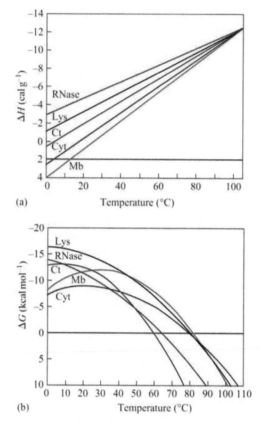

(a)

(b)

Figure 17.5. Temperature dependence of (a) the specific (per gram of protein) difference between the energies of the native and denatured states, and (b) the free energy difference (per mole of proteins) between these states. Notice that the protein melting temperature is 60–80 °C(330–350 K); that the melting heat (at this temperature) is ≈ 8 cal g^{-1} (or ≈ 1 kcal mol^{-1} per residue, since an average residue is about 110 Da); and that the "stability reserve" of the native protein molecule is 10–15 kcal mol^{-1} at room temperature. Adapted from Privalov P.L., Khechinashvili N.N., *J. Mol. Biol.* (1974) **86**: 665–684.

By the way, it really "boils" (meaning a great decrease in its density): the protein chain always unfolds completely, and its volume grows many-fold after cold denaturation, while heat denaturation usually only "melts" the protein, and its volume increases only slightly.

This naturally brings us to the question as to what a denatured protein looks like.

A heated debate on this subject started in the 1950s, continued over the next three decades and only then yielded a kind of consensus.

The problem was that different experimental techniques led to different conclusions.

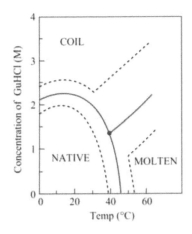

Figure 17.6. Phase diagram of the conformational states of lysozyme at pH 1.7 in guanidine dihydrochloride solution at various temperatures. It shows the regions of existence of the NATIVE state, of the COIL, and of a more compact temperature-denatured state (MOLTEN). The red line corresponds to the mid-transition, the dashed lines outline the transition zone (approximately from the proportion 9:1 in favor of one state to 1:9 in favor of the other). One can see that the COIL–MOLTEN transition is much wider ("less cooperative") than the NATIVE–COIL or the NATIVE–MOLTEN transitions. Adapted from Tanford C., *Adv. Protein Chem.* (1968) **23**: 121–282.

Specifically, numerous *thermodynamic* experiments have shown that there are no cooperative transitions within the denatured state of a protein molecule. Therefore, it was initially assumed that the denatured protein is always a very loose random coil (which is actually true for denatured proteins in "very good" solvents like concentrated solutions of urea).

However, numerous *structural* studies of denatured proteins reported some large-scale rearrangements within the denatured state, that is, on some "intermediates" between the completely unfolded coil and the native state of proteins (Figs 17.6 and 17.7).

Actually, experimental data on the state of proteins after denaturation were rather contradictory. In some cases, proteins seemed to be completely unfolded and "unstructured"; in others, they seemed to be rather structured and/or compact, and they completely unfolded only under the action of highly concentrated denaturants.

This contradictory picture of protein denaturation was clarified only by using a variety of methods. Measurements of protein solution viscosity gave information about the hydrodynamic volume of protein molecules; far-UV CD spectra (circular dichroism in the far-ultraviolet region) revealed their secondary structure; near-UV CD spectra investigated asymmetry, i.e., the ordering of the environment of aromatic side chains in the protein; while infrared spectroscopy; NMR; measurements of protein activity; etc. gave additional information. This whole arsenal of methods has been applied to the study of protein denaturation by the team of O.B.P., and A.V.F. belonged to this group as a theoretician.

It has become clear that, apart from activity, only two protein properties always abruptly change during denaturation. These are: (1) the ordering of the environment of aromatic side chains observed by near-UV CD (Fig. 17.7a) and by NMR, and (2) the rigidity of the globular structure followed (using NMR) by exchange of hydrogens (H) of the protein's polar groups for deuteriums (D) of water, and by acceleration of the protein chain proteolysis.

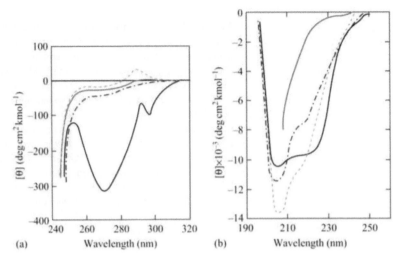

Figure 17.7. CD spectra in (a) the near- and (b) the far-ultraviolet regions for α-lactalbumin in the native (⎯), heat-denatured (⎯..), acid-denatured ( ) and denaturant-unfolded (⎯) forms. Near-UV CD spectra are *always* changed by denaturation. Far-UV CD spectra are significantly changed only by unfolding (see curve ⎯). Reprinted from Dolgikh D.A., Abaturov L.V., Bolotina I.A., Brazhnikov E.V., Bychkova V.E., Bushuev V.N., Gilmansin R.I., Lebedev Yu.O., Semisotnov G.V., Tiktopulo E.I., Ptitsyn O.B. "Compact state of a protein molecule with pronounced small-scale mobility: bovine alpha-lactalbumin." (Figs 3a and 4.) *Eur. Biophys. J.* (1985) **13**: 109–121, ©Springer-Verlag; with permission.

On the other hand, the ordering of the protein main chain (i.e., its secondary structure), the volume of the protein molecule and the density of its hydrophobic core can be virtually preserved in some cases and strongly changed in others, depending on the denaturation conditions (Fig. 17.7b).

These studies have revealed a universal, or nearly universal, intermediate of protein unfolding and folding, which is now known as the "*molten globule*" (Figs 17.8 and 17.9). The most important properties of this intermediate are summarized in Table 17.1.

On the face of it, the properties of the molten globule are contradictory. Its secondary structure is usually nearly as developed as that of the native protein. On the other hand, the molten globule (like the completely unfolded protein) has very little of the ordering of its side chains, which is so typical of the native protein. However, some portion of the native side-chain contacts evidently remains intact in the molten globule; NMR shows that this applies to aromatic side chains and not to aliphatic side chains. The molten globule is a little less compact than the native protein (as hydrodynamic volume measurements indicate), and its core is virtually as compact as that of the native protein (as shown by "middle-angle" X-ray scattering); on the other hand, a rather high rate of hydrogen exchange shows that at least separate solvent molecules easily penetrate into the globule. To some extent, these contradictions may

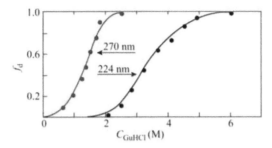

Figure 17.8. Relative changes of α-lactalbumin CD in the near- (270 nm) and far- (224 nm) UV regions at increasing denaturant (guanidine dihydrochloride) concentrations. $f_d = 0$ refers to the signal from the native structure, $f_d = 1$ refers to the signal from the completely denatured molecule. The "intermediate state" exists between the first and the second transitions. At $C_{GuHCl} = 2$ M, all α-lactalbumin molecules are in the "molten globule" state (as shown by subsequent studies). Adapted from Pfeil W., Bychkova V.E., Ptitsyn O.B., *FEBS Lett.* (1986) **198**: 287–291.

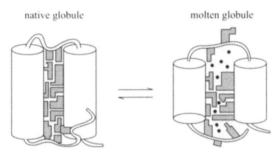

Figure 17.9. Schematic model of the native and the molten globule protein states. For simplicity, the protein is shown to consist of only two helices connected by a loop. The backbone is covered with numerous side chains (shaded). Reinforced by H-bonds, the secondary structures are stable until the globule is "dissolved" by a solvent. Usually, water molecules are unable to do this job without a strong denaturant. In the molten globule, side chains lose their close packing, but acquire freedom of movement, i.e., they lose energy but gain entropy. The water molecules (•) enter into pores of the molten globule (which appear when the close packing is lost), but cause no further decay of the molten globule.

be reconciled by the assumption that the molten globule is heterogeneous: that it has a relatively dense, native-like core and more loose loops. However, the assumption that the molten globule has a completely native core and completely unfolded loops is in clear contradiction to experiment, and specifically with the absence of the ordered environment of even aromatic side chains in the molten globule. The latter is observed by near-UV CD and by NMR spectroscopy.

For very many (though not for all) proteins, the "molten globule" arises from a moderately denaturing impact upon the native protein, and decays (turning into a

Table 17.1. Methods of investigating the "molten globule" state, and conclusions drawn from them.

"Globule" (like native protein)		"Molten" (unlike native protein)	
Investigation	Conclusion	Investigation	Conclusion
Hydrodynamic volume, small- and medium-angle X-ray scattering	Compactness	Near-UV CD, H^1 NMR spectra	No unique packing of side chains
Large-angle X-ray scattering	Presence of core	H $\leftrightarrow$ D exchange+ NMR, proteolysis	Vibrations
NMR (spin echo)	Some aromatic side chains are fixed	NMR (spin echo)	Mobility of aliphatic side chains
Far-UV CD IR spectra, NMR + H $\leftrightarrow$ D exchange	Secondary structure	Scanning and isothermal microcalorimetry	No further melting (as a rule)
Fluorescence	Some Trp residues are not accessible to water	Fluorescence	Some Trp residues are accessible to water
2D NMR	Preservation of some long-range contacts	2D NMR	Decay of most long-range contacts
Chromatography (HPLC)	Possibility of "correct" S—S bond formation		

The molten globule state differs from both the native and the unfolded states in showing enhanced binding of non-polar molecules (demonstrated by enhanced fluorescence of the protein-bound hydrophobic dye ANS)

random coil) only under the impact of a concentrated denaturant. The molten-globule-like state often occurs after temperature denaturation ("melting"), and this melting has been observed always to be an "all-or-none" transition. The molten globule does not usually undergo further cooperative melting with a greater rise in temperature (see Fig. 17.6), but its unfolding caused by a strong denaturant looks like a cooperative S-shaped transition.

However, some proteins (especially small ones) unfold directly into a coil without the intermediate molten globule state; and many other proteins are converted into the molten globule by some denaturing agents (e.g., by temperature or by acid), while other agents (e.g., urea) directly convert them into a coil.

A theoretical explanation was required for the contradiction between the results of structural studies and those of thermodynamic studies of protein denaturation, as well as for the physical nature of the new state of proteins, the "molten globule".

To understand the physics of the molten globule, it is important to take into account that this state arises from the native state by cooperative melting, which is a first-order ("all-or-none") phase transition. This means that the molten globule has a much higher energy and entropy than the native state, that is, that the intra-chain interactions are much weaker and the chain motility is much higher in the molten globule. Since the majority of the protein chain degrees of freedom are connected with small-scale fluctuations of the structure, and predominantly with side-chain movements, it is the liberation of these fluctuations that can make the molten globule thermodynamically advantageous. The liberation of small-scale fluctuations does not require the complete unfolding of the globule; slight swelling would be enough. This swelling, however, leads to a significant decrease of the van der Waals attraction: this attraction strongly depends on the distance, and even a small increase of the globule's volume is enough to reduce it greatly (Fig. 17.9).

Generally, all this is similar to the melting of a crystal, where a small increase in volume reduces van der Waals interactions and liberates the motions of the molecules.

It was unclear, however, why a system so heterogeneous as a protein melts by a phase transition (like a crystal built up from the same or a few kinds of molecules): protein has no regular crystal lattice the decay of which leads to a first-order phase transition at crystal melting.

Studies of the "molten globule" furnished the clue to an understanding of the cooperativity of protein denaturation. This process turned out to be very different from what happens in synthetic polymers, DNA or RNA. The last two have no "all-or-none" denaturation, and the same applies to the helix–coil transition.

But what about the previously studied β-structure formation? As we have seen, the "all-or-none" transition exists there . . . Yes, it does indeed. However, the "non-β-structural" chain is in the coil state: its volume is huge, and it has no secondary structure. And the denatured protein often (in the molten globule state) virtually preserves its volume and the secondary structure content.

Can synthetic polymers serve as an analog?

True, they have globule–coil transitions. However, it turns out that protein denaturation is very different from the globule–coil transition in "normal" polymers, the transition that had been initially assumed to be a close analog of protein denaturation.

The fact is that the events observed in proteins disagree both with experimental studies of synthetic homopolymers (which are extremely difficult because of the threat of aggregation) and with physical theories of homopolymers and of heteropolymers (not of "selected" ones, like proteins, but of random ones).

All of these suggested (Figs 17.10 and 17.11) that the unfolding of a dense polymer globule must start with a gradual increase in its volume, and up to (or nearly up to) its very end, that is, to conversion into a coil, it must go without any abrupt change in the molecule's density. They also suggested that the globule-to-coil transition starts when the globule is already rather swollen, and that the main difference between the globular and coil phases is the difference not in density, but rather in the scale of fluctuations (small in globule, large in coil). This means that the globule-to-coil transition is quite different from the first-order phase transition exemplified by melting, boiling or sublimation, while protein denaturation is essentially similar to melting or sublimation of a solid.

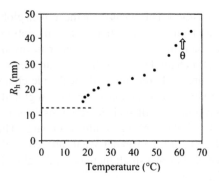

Figure 17.10. Hydrodynamic radius R_h vs. temperature for poly(methyl-methacrylate) chains (M.W. $= 6.5 \times 10^6$ Da). A very diluted isoamyl acetate solution of the polymer is studied. The point of the globule–coil transition (the "θ-point") corresponds to 61 °C (shown by the arrow). The globule exists below 61 °C, the coil above 61 °C. Dashed line: the computed radius of solid poly(methyl-methacrylate) with M.W. of 6.5×10^6 Da. One can see that the radius of the coil is four times larger than that of the solid polymer globule (i.e., the coil is nearly 100 times larger in volume and nearly 100 times less dense), and that after the coil→globule transition (with decreasing temperature) the chain undergoes a gradual compaction down to the swollen globule whose radius is two times larger (i.e., whose volume is 10 times larger and density 10 times less) than those of the solid polymer globule (see the 50–30 °C region). This first compaction is followed (as the temperature decreases) by the second one (see the 25–20 °C region), and only this second compaction leads to a dense globule. Adapted from Kayaman N., Guerel E.E., Baysal B.M., Karasz F., *Macromolecules* (1999) **32**: 8399.

Although the conventional theory of globule–coil transitions in *homo*polymers is *not* applicable to protein denaturation, it would not be out of place to present it to you – of course, in a most simplified form. This will allow me later to emphasize the difference in behavior between the "selected" protein chains and normal polymers. On the other hand, this conventional theory is applicable to the behavior of already denatured protein chains. And, last but not least, I think that acquaintance with the basic physical models is a necessary part of general culture.

I advise that you should "test words by formulas" when reading the following text. It is certainly easier to read words only, but they are often ambiguous, so word verifying by formulas and *vice versa* will help your understanding.

So, I will consider the simplest "polymer" model: many equal monomers *connected by the chain* (like "beads"). And I will compare it with the "cloud of monomers", i.e., with an aggregate of the same N monomers *un*connected by the chain: this is the simplest model, used since van der Waals' time (in the nineteenth century) to describe the gas–liquid transition (Fig. 17.11). We will see that the polymer has no analog of separation into "liquid" and "gas" phases typical of the "cloud".

Preliminary remark. I will assume that the monomers are attracted together (otherwise, it's too dull: no dense phase will form, and that's all), and that the strength of this attraction is

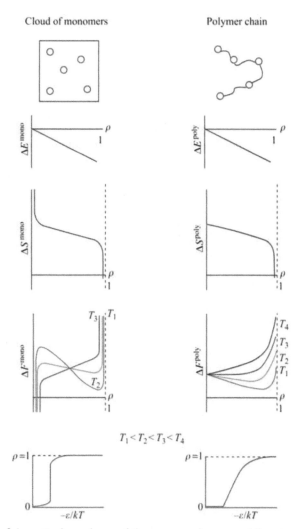

Figure 17.11. Schematic dependence of the monomer's energy ΔE, entropy ΔS, and free energy $\Delta F = \Delta E - T\Delta S$ on the density ρ for two systems: "cloud of monomers" and "polymer chain". The dependence of ΔF on ρ is shown for various temperatures T. A scheme of dependence of the density ρ on the monomer attraction energy (expressed in kT units) is shown in the auxiliary plot for each system (at the bottom). The plots refer to attracting monomers ($\varepsilon < 0$) and to a low pressure (at very high, "supercritical" pressure the loose (with $\rho < 0.5$) states are absent).

temperature-independent, i.e., that it is purely energetic in nature. The latter is often incorrect, since the monomers are floating in the solvent (recall the hydrophobic effect: it increases with temperature); but this allows us to find and consider all the basic scenarios of phase transitions in the simplest way. And later we can complicate the problem by adding the temperature dependence of the interactions and show at what temperature each of these scenarios can take place.

To begin with, I should like to give you the general idea and to explain why the "cloud" can have two phases and the polymer cannot, i.e., why the polymer has to condense gradually.

What is the difference between the life of a monomer in the polymer and in the cloud? In the polymer it cannot go far away from its chain neighbor, while in the cloud it can go as far from any other monomer as the walls of the vessel permit.

Why do monomers form a dense phase, either liquid (if they are not connected by the chain) or globule (if otherwise)? Because they attract each other, and at low temperature, the attraction energy overcomes any tendency for the monomers to scatter and gain entropy: at low temperature, the role of entropy is small (recall: in the free energy $F = E - TS$ the energy E is compared with the entropy *multiplied* by the temperature). Overcomes always *but for* one case: when scattering leads to a virtually infinitely large entropy gain, i.e., when the cloud of monomers is placed in a vast vessel. And just here the monomers have to choose between two phases: either to form a drop (low energy *but* low entropy as well) or to stay in a rarefied gas (negligible energy *but* a very high entropy). The states of intermediate density (e.g., of 10 or 50% of that of the drop) have a higher free energy (their entropy is not as large while their energy is not as low) and therefore are unstable. Thus, the stable state of the "cloud" at low temperature can be either dense, or very rarefied, but not of an intermediate density. In other words, the "cloud" of monomers can exhibit an abrupt evaporation of the dense phase when the temperature rises.

We have dealt with the "cloud". And what about the polymer? Here the situation is quite *different*, because a monomer cannot go far away from its chain neighbor even in the most diffuse coil. This means that the coil entropy cannot exceed a certain limit, and thus the entropy is not able (at low temperatures) to compete with the energy gained by compaction of the polymer. Thus, the polymer has no alternative: at low temperature its monomers cannot scatter, and it has to be a dense globule only, which means that (unlike the "cloud") it cannot jump from the dense to the rare phase. Therefore, a change of conditions can cause only a gradual swelling (or compression) of the polymer globule rather than its abrupt "evaporation".

Now, let us repeat the above in the language of equations.

Let us compare a "cloud" of N monomers confined within the volume V with a polymer of the same N monomers in the same volume V. Assuming that the monomers (of both the cloud and the polymer) are distributed over the volume more or less uniformly, let us add one monomer to each system and estimate the dependence of the free energy of the added monomer (i.e., the *chemical potential* of the system) on the system's density.

The chemical potential is appropriate for analysis of the co-existence of phases. The point is that the molecules pass from a phase with a higher chemical potential to a phase of a lower chemical potential: this decreases the total free energy, tending towards equilibrium. At equilibrium, the chemical potentials of molecules in both phases are equal. This means that an abrupt transition of one phase into another is possible only when the same chemical potential value refers to both phases, to both densities. On the contrary, if the chemical potential only increases (or only decreases) with increasing/decreasing density of the system, an abrupt phase transition is impossible.

It should be noted that phase stability requires an increasing chemical potential with increasing density of the system. Then penetration of each new particle to this phase is more and more difficult. This is the stable state: like a spring, the system increasingly resists the increasing impact.

Let us summarize: the particular state of a system is determined by particular external conditions (that is, by the given volume or the given pressure), but the prerequisite for a possibility of a phase transition is the presence of two ascending (with density) branches of chemical potential values.

In exploring the chemical potential value, let us first consider the energy of the added monomer, or rather, that part of the energy which depends on the system's density, i.e., on the interactions of non-covalently bound monomers.

In both cases the energy of the system ("cloud" or "polymer") changes by the same value when a monomer is added. The change occurs because of the monomer's interactions and is equal to $\Delta E = \varepsilon \rho$, where $\rho = N\omega/V$ is the system's density, i.e., the fraction of its volume V occupied by the N monomers having the volume ω each, and ε is the energy of interactions of this monomer at $\rho = 1$, i.e., in the most dense system ($\varepsilon < 0$, since we have assumed that monomers are attracted together).

Thus, the added monomer's energy behaves alike in both systems; but the added monomer's entropy behaves very differently in the polymer as compared with the "cloud".

The "cloud" leaves the volume V/N at the monomer's disposal. However, since the volume $N\omega$ is already occupied by the monomers, the free volume at the monomer's disposal in the "cloud" is

$$
\begin{aligned}
V_1^{\mathrm{mono}} &= \frac{V - N\omega}{N} \\
&= \frac{V}{N}(1 - \rho) \\
&= \frac{\omega}{\rho}(1 - \rho)
\end{aligned}
\tag{17.7}
$$

Notice that this volume is infinitely large when the density ρ turns to zero.

At the same time, the free volume at the monomer's disposal in the polymer is

$$
V_1^{\mathrm{poly}} = \Omega(1 - \rho)
\tag{17.8}
$$

The volume Ω is limited by the monomer's binding to the previous chain link, and the multiplier $(1 - \rho)$ shows, as above, that other monomers occupy the fraction ρ of the volume.

Then, the entropy change resulting from addition of a monomer to the "cloud" is

$$\Delta S^{mono} = k \ln[V_1^{mono}]$$

$$= k \ln \left[\frac{\omega}{\rho}(1 - \rho) \right], \tag{17.9}$$

and the entropy change resulting from addition of a monomer to the polymer is

$$\Delta S^{poly} = k \ln[V_1^{poly}] = k \ln[\Omega(1 - \rho)]. \tag{17.10}$$

As the functions of the density ρ, ΔS^{mono} and ΔS^{poly} behave alike (see Fig. 17.11) only as $\rho \to 1$ (where the term $\ln(1 - \rho)$ dominates in ΔS, so that both ΔS^{mono} and ΔS^{poly} decrease as $\rho \to 1$), but their behavior is *completely different* as $\rho \to 0$: ΔS^{poly} remains finite, while ΔS^{mono} grows infinitely because of the term ω/ρ.

Thus (see Fig. 17.11), at low temperatures T, the chemical potential $\Delta F = \Delta E - T\Delta S$ of the "cloud" has two ascending branches, i.e., two potentially stable regions *divided* by the first-order phase transition (rarefied gas at a low ρ, owing to the $-T\Delta S$ term, and liquid at a high ρ, owing to the ΔE term). And at high temperatures the "cloud" has only one ascending branch of ΔF value, i.e., only one stable state (gas, owing to the $-T\Delta S$ term).

At the same time, the polymer entropy *does not* grow infinitely even as $\rho \to 0$, and here only *one* ascending branch of the ΔF value exists at all temperatures. At extremely low temperatures, or rather, at high $-\varepsilon/kT$ values, the origin of this branch corresponds to $\rho \approx 1$, that is, to a dense globular state. The temperature increase (or rather, the decrease in the $-\varepsilon/kT$ value) shifts the origin of the ascending branch towards lower densities ρ, and finally, the origin comes to the value $\rho = 0$ and stays there permanently (Fig. 17.11). As you see, the chemical potential of the globule never has two ascending branches.

This means that the globule's expansion does not include a phase separation. And, since there is no phase separation, there is no first-order phase transition. It can be shown (this has been done by Lifshitz, Khokhlov and Grosberg and by De Gennes) that the *second*-order globule-to-coil phase transition occurs at $\rho \approx 0$ (or rather, at the coil state density $\rho \sim N^{-1/2}$), but this is beyond the scope of these lectures.

We can refine the above discussion as follows. Strictly speaking, the general theory states that an abrupt transition of the more-or-less dense globule into the coil, though possible in principle, occurs only under hardly obtainable conditions, e.g., when monomers in the chain are repelled in pairs but attracted to one another when there are many simultaneously interacting particles. This effect is observed in synthetic polymers with very rigid chains. But in contrast to proteins, in these rigid polymers, the jump in energy (per monomer) during coil–globule transitions is very small (as for liquid crystal formation of rigid elongated molecules). A large energy jump requires a very special construction of the chain monomers. Normal polymers do not have such a construction of their links, unlike protein chains. As we will see, protein residues become strongly attracted together when many of them are simultaneously involved in interactions. Why this is so, we will see in the next lecture.

Thus, the conventional theory of coil–globule transitions in normal homopolymers cannot explain protein denaturation. (The same is true for the theory of coil–globule transitions in random heteropolymers, but its consideration is beyond the scope of these lectures.)

Conventional theories state that the globule expands gradually, and that the coil arises from a globule of very low density, and *not* by a first-order phase transition; but denaturation of proteins, these Schrödinger's "aperiodic crystals", occurs at a high density of the globule, often does not lead to random coil formation, and resembles the destruction of a crystal, i.e., a first-order phase transition.

The explanation of these peculiarities of protein denaturation, and of its "all-or-none" phase nature in particular, will be given in the next lecture.

LECTURE 18

As we have learned from the previous lecture, the conventional theory of coil–globule transitions cannot explain protein denaturation. It states that the globule expands gradually, that a coil arises only when the globule's density is very low, and that this does *not* happen by a first-order ("all-or-none") phase transition. On the contrary, denaturation of proteins, these Schrödinger's "aperiodic crystals", occurs at a high density of the globule without its preceding swelling, often does not lead to random coil formation and resembles the destruction of a crystal, that is, it resembles a first-order phase transition.

A "normal" crystal melting also cannot explain protein denaturation: proteins have no periodic crystal lattice, which distinguishes a crystal from liquid or glass (which is nothing but a very viscous liquid). And the origin of crystal melting is just a decay of this lattice.

It is really astonishing that denaturation of protein occurs as a normal sharp first-order phase transition: any protein is a highly heterogeneous system (where, nevertheless, each atom sits at its own place), and heterogeneity is known to slur over the phase transitions in normal molecular systems.

It is not out of place here to draw your attention to an aspect that distinguishes conventional physics from the physics of proteins and of complex systems in general. What I am speaking about are not physical laws: they are certainly the same for both cases; I am speaking about the effects that attract the attention of researchers. The normal subjects of conventional physics are more or less uniform or "averaged": gases, crystals, fields, spin glasses, etc. Therefore, special attention is paid to each heterogeneity arising in these uniform objects; for example, to quasi-particles. Protein physics, on the contrary, deal with a highly heterogeneous object from the very beginning. Protein is full of heterogeneous tensions; it can be said to be full of various frozen fluctuons; but this is just a figure of speech to describe protein heterogeneity. In contrast, the uniform phenomena that deal with the entire heterogeneous system (like denaturation that involves the entire protein) are indeed of special interest for protein physics.

To understand protein denaturation, one has to explain why there exist two equally stable states, i.e., equally stable phases of the protein chain (which, as we already know, is impossible for normal polymers), and why they are separated by a free energy barrier (which, as we also know, is necessary for an "all-or-none" transition). That is, one has to explain why the protein globule cannot decay by gradual swelling, as usual polymer globules do.

In doing so, one has to take into account the main peculiarities of proteins (those which differ them from normal polymers): that each protein has one chain fold distinguished by its peculiar stability; that flexible side groups are attached to a more rigid protein-chain backbone; and that the native protein is packed as tightly as a molecular crystal (although without a crystal lattice): in the protein, as in a molecular crystal, the van der Waals volumes of atoms occupy 70–80% of space, while in liquids (melts) they occupy only 60–65%.

Side groups of the protein chain are capable of rotational isomerization, i.e., jumps from one allowed conformation to another. Each jump requires some vacant volume around the jumping side chain; but the native protein fold is distinguished by tight chain packing (which contributes to the peculiar stability of this fold). Besides, the side chains sit at the rigid backbone, which is especially rigid inside the globule, because here the chain forms the α- and β-structures that are necessary, as we already know, to involve the backbone in the dense globule (Fig. 18.1). These structures are stable at least until the solvent penetrates into the globule (which requires approximately the same free volume as well). Thus, each of these rigid structures has to move as a whole (at least at the beginning of the globule's expansion), with the entire forest of side chains attached. Therefore, expansion of the closely packed globule through the movement of α- and β-structures creates approximately equal free spaces near each side chain, and these spaces are either insufficient for isomerization of each of the side chains (when the globule's expansion is too small), or sufficient for isomerization of many of them at once. This means that liberation of the side chains

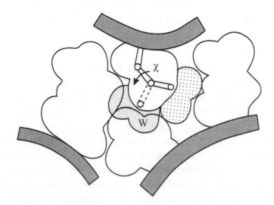

Figure 18.1. A sketch of the side chain packing. Only a small piece of the globule is shown. The dashed region W corresponds to an alternative rotamer of the side chain (χ being its rotational angle); this rotamer is forbidden by close packing. Its appearance requires additional vacant volume W of at least 30 $\overset{\circ}{A}^{3}$ (i.e., the volume of a methyl group), or $\approx 1/5$ of the average amino acid volume. Nearly the same volume is required for H_2O penetration in the core.

(as well as solvent penetration) occurs only when the globule's expansion crosses some threshold, the "barrier" (Fig. 18.2). These two events can make a less dense state of the protein chain as stable as its native state, but only after the density barrier has been passed.

Thus, a small ("pre-barrier") expansion of the native globule is *always* unfavorable: it *already* increases the energy of the globule (since its parts move apart and lose their close packing) but does not increase its entropy because it does not *yet* liberate the rotational isomerization of the side chains. That is, the free energy of the globule increases with a slight expansion. On the contrary, a large ("post-barrier") expansion of the globule liberates the rotational isomerization and leads (at a sufficiently high temperature) to decreasing free energy. As a result, protein denaturation does not occur gradually but as a jump over the free energy barrier, in accordance with the "all-or-none" principle.

Thus, the protein tolerates, without a change, modification of ambient conditions up to a certain limit, and then melts altogether, like a macroscopic solid body. This resistance and hardness of protein, in turn, provides the reliability of its biological functioning.

In other words, the "all-or-none" transition between the native and denatured state is explained by a sudden jump in entropy (mainly side-chain entropy) which occurs only when the globule's expansion crosses a certain threshold (Fig. 18.3). And the latter exists because the side chains sit at the rigid backbone that co-ordinates their positions and cannot be liberated óne by one.

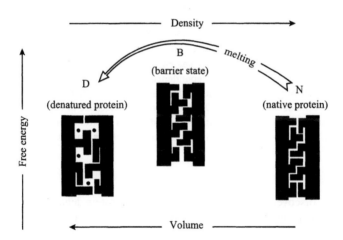

Figure 18.2. Origin of the free-energy barrier between the native and any denatured state of the protein. The "barrier" state B arises at a small expansion of the native, closely packed state N. The pores formed in the state B *already* cause a great increase in the van der Waals energy, but *yet* allow neither side chain liberation nor penetration of the solvent (•) inside the protein. This requires a further expansion of the globule to the state D. Adapted from Shakhnovich E.I., Finkelstein A.V., *Biopolymers* (1989) **28**: 1667–1680.

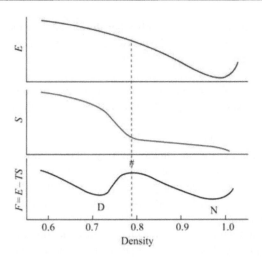

Figure 18.3. Origin of the "all-or-none" transition in protein denaturation. The energy E is at its minimum at close packing (here the globule's density $\rho = 1$). The entropy S increases with decreasing density. Initially, this increase is slow (at $\rho = 1.0$–0.8, till the side chains can only vibrate, but their rotational isomerization is impossible), then rapid (when rotational isomerization becomes possible), and then slow again (when free isomerization has been reached). The non-uniform growth of the entropy is due to the following. The isomerization needs some vacant volume near each side group, and this volume must be at least as large as a CH_3-group. And, since the side chain is attached to the rigid backbone, this vacant volume cannot appear near one side chain only but has to appear near many side chains at once. Therefore, there exists a threshold ("barrier") value of the globule's density after which the entropy starts to grow rapidly with increasing volume of the globule. This density is rather high, about 80% of the density of the native globule, since the volume of the CH_3-group (a typical unit of rotation, see Fig. 18.1) amounts to $\approx 20\%$ of the volume of an amino acid residue. Therefore, the dependence of the resulting free energy $F = E - TS$ on the globule's density ρ has a maximum ("barrier" #) separating the native, tightly packed globule (N) from the denatured state (D) of lower density. Because of this free energy barrier (its width depends on the heterogeneity of the side-chain rotation units) protein denaturation occurs as an "all-or-none" transition, independent of the final state of the denatured protein.

Prior to the discovery of the molten globule state, protein denaturation was usually considered to be a complete decay of the protein structure, i.e., as a transition to the coil. After this discovery, it became clear that the denatured protein can be rather dense as well as loose depending on the solvent's strength and the hydrophobicity of the protein chain. The pores in the molten globule (i.e., the vacant space necessary for the side-chain movements, see Figs 18.1 and 18.2) are "wet", i.e., they are occupied by the solvent because a water molecule inside the protein is still better than a vacuum. Experimentally, the "wetness" of the molten globule is proved by the absence of a visible increase in the protein partial volume after denaturation of any kind.

When the solvent is poorly attracted by the protein core (consisting mainly of hydrophobic groups, although including some polar backbone atoms too), it only occupies the pores that already exist in the molten core to ensure side-chain movements, but it does not create new pores and does not expand the globule (just as water does not expand a sponge, although it occupies its pores). Then the denatured protein remains in the wet molten globule state. Its compactness is maintained by residual hydrophobic interactions.

However, if the residual hydrophobic interactions are weak, i.e., if the solvent is strongly attracted by the protein chain, the solvent starts to expand the pores, and the globule starts to swell. The greater the attraction between the solvent and the protein chain, and the smaller the attraction within the chain, the greater the swelling: up to transition to the random coil.

The coil arises directly from the native state if the temperature is low, i.e. when the role of the entropy of side-chain liberation is not crucial, and the main denaturation effect is produced by the solvent (recall cold denaturation); and at a higher temperature the coil can arise after the swelling of the molten globule (recall the experimental diagram by Tanford; I will show it again later on).

It seems that the swelling of the molten globule can be described by the conventional theory of globule–coil transitions we have considered. Figure 18.4 shows what is predicted to happen: if the intra-chain interactions remain strong (i.e., if the protein chain is hydrophobic enough or/and the solvent is not strong enough), the denatured protein will have the dense molten globule state; otherwise, the globule will swell or even unfold completely – but without any jump in density.

This is a result of the most simplified physical theory of the molten globule state. Actually, it states that the molten globule is separated by an "all-or-none" transition from the native state, but not from the coil. This result does not account for a possible heterogeneity of the molten globule, though. Perhaps a more advanced theory is needed . . . .

And what do experiments show? Is unfolding of the molten globule (or can it be) of the "all-or-none" type?

This is not quite clear yet. Experiments demonstrate that denaturant-induced unfolding of the molten globule is a cooperative, "S-shaped" transition (Fig. 18.5). However, such a form of the experimental curve is compatible both with an "all-or-none" transition and with a gradual (Fig. 18.4) globule-to-coil transition via states of intermediate density.

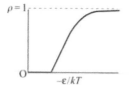

Figure 18.4. Qualitative dependence of the polymer's density ρ on the energy of the monomer interaction ε expressed in kT units. There is no density jump here, i.e., this transition is not of the "all-or-none" type.

True, GuHCl- or urea-induced molten globule unfolding is a narrow transition, not much wider than the denaturation itself (Fig. 18.5), and its width (like the width of denaturation) is shown to be inversely proportional to the protein domain molecular weight, which is in qualitative agreement with the van't Hoff criterion. This led Ptitsyn and Uversky to conclude that the transition in question can be of the "all-or-none" type.

Calorimetric investigations can, in principle, give the final answer; but, molten globules of natural proteins almost never undergo further melting (see Fig. 18.7 below). However, "all-or-none" melting seems to appear in calorimetric studies of some *de novo* designed proteins, which usually have a molten globule rather than a solid form.

Besides, Uversky and Ptitsyn have shown that the molten globule state can be separated from some "less compact" state by an "all-or-none" transition (at least in carbonic anhydrase, see Fig. 18.6). The co-existence of two denatured states is demonstrated by the split peak in the chromatography elution profile of a denatured protein. This split peak shows that the protein molecule exists *either* in one form *or* in another, but the states of intermediate compactness are absent; that is, the split peak is clear evidence of an "all-or-none" transition. However, carbonic anhydrase is a β-structural protein, and the observed transition may refer to decay of the β-structure (which is of an "all-or-none" type in synthetic polypeptides, as we remember from Lecture 9).

At the moment of transition, the "less compact" state is only a little less compact than the molten globule and has half of its secondary structure, but it is far more compact and far more structured than the coil (therefore, it is also called the "pre-molten" globule). It should be noted that structural features of the "pre-molten" and "molten" globules are not that different, and this is the cause of frequently occurring ambiguity in the use of the term "molten globule".

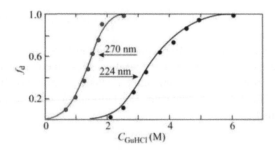

Figure 18.5. Relative changes of the α-lactalbumin CD in the near- (270 nm) and far- (224 nm) UV regions at increasing denaturant (guanidine dihydrochloride) concentrations. $f_d = 0$ refers to the signal from the native structure, $f_d = 1$ refers to the signal from the completely denatured molecule. The "intermediate state" exists between the first and the second transitions, i.e., between protein denaturation and complete unfolding: at $C_{GuHCl} = 2\,M$, the α-lactalbumin molecules have the "molten globule" state. Adapted from Pfeil W., Bychkova V.E., Ptitsyn O.B., *FEBS Lett.* (1986) **198**: 287–291.

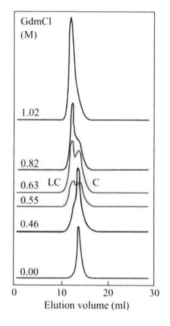

Figure 18.6. Evidence of the co-existence of two denatured states of a protein. Chromatography of carbonic anhydrase demonstrates the co-existence of "compact molten globules" (C) and "less compact globules" (LC) in the region of denaturant (guanidine dihydrochloride) concentrations where the native state is already absent and the molten globule starts to decay. Note the split elution peak at 0.55–0.63 M GdmCl concentrations. Adapted from Uversky V.N., Semisotnov G.V., Pain R.H., Ptitsyn O.B., *FEBS Lett.* (1992) **314:** 89–92.

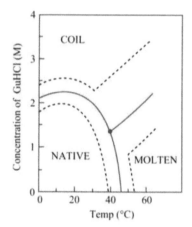

Figure 18.7. Phase diagram of conformational states of lysozyme at pH 1.7 in guanidine dihydrochloride solution at various temperatures. It shows the regions of existence of the NATIVE state, of the COIL state, and of the more compact temperature-denatured state (MOLTEN). The red lines correspond to the mid-transition, the dashed lines outline the transition zone (approximately from the proportion 9:1 in favor of one state to 1:9 in favor of the other). We can see that the COIL–MOLTEN transition is much wider ("less cooperative") than the NATIVE–COIL or the NATIVE–MOLTEN transitions. Adapted from Tanford C., *Adv. Protein Chem.* (1968) **23:** 121–282.

Further unfolding of the "pre-molten" globule seems to be a gradual transition to the coil, as suggested by Fig. 18.4 and the previously discussed theory of globule–coil transitions in "normal" polymers. But let me repeat again that the nature of the molten globule-to-coil transition is yet to be understood, that is, a physical theory of this phenomenon is needed.

Coming back to Tanford's phase diagram (Fig. 18.7), we see that the native state is separated from all the denatured states (both COIL and MOLTEN globule) by an "all-or-none" transition (which is consistent with the narrowness of these transitions in the diagram); that any denatured state differs from the native state in the motility of its side chains, but can be rather compact (in the molten globule form), or more swollen (when the molten globule is about to transform into the coil), or very loose (in the coil); and that denaturant-induced unfolding of the molten globule seems to be a "weak" (wider in the diagram) but still "all-or-none" transition, at least for some proteins.

A short digression. In general, we must not be surprised that different proteins behave differently. And this may not reflect a qualitative difference between proteins, but rather result from the fact that usually we observe each protein through a relatively small "experimental window" with temperature ranging from water freezing to boiling and the action of denaturant is limited by its zero concentration. Therefore, for example, the proteins with less hydrophobic amino acids, i.e., the proteins whose native globule is more susceptible to denaturants, demonstrate (in the "window": see the upper part of Tanford's diagram) only the native state-to-coil transition (since the molten globule is maintained only by residual hydrophobic interactions that are weak in this case). Proteins of this kind (especially if they are "weakened", like, for example, apomyoglobin, which is myoglobin weakened by deprivation of the heme) tend to demonstrate the cold denaturation that leads to the coil. The more hydrophobic, more stable proteins demonstrate (in the "window": see the lower part of Tanford's diagram) the native state-to-molten globule transition, but do not demonstrate cold denaturation: apparently, it can occur only at sub-zero temperatures, when water freezes and thus stops our experiments.

Further consideration of the thermodynamics of protein molecules is done from the following point of view: which physical property distinguishes the protein chain from a random heteropolymer and allows the protein molecule to be capable of "all-or-none" rather than gradual destruction? That is, what makes the protein tolerate, without softening, a change of ambient conditions and then break suddenly? It is clear that this behavior ensures the reliability of the protein function: as long as it works, it works properly . . .

Having plotted the dependence of both the protein entropy S and its energy E on its density ρ (Fig. 18.3), we can plot the S-on-E dependence and then grasp the typical appearance of the spectrum of protein energy (Fig. 18.8a). We want to see how the structures are distributed over the energy; it is of no interest to us now *what* structure has this or that energy, the question is *how many* structures have this or that energy. We can answer it if $S(E)$ is known. Let me remind you that $S(E)$ is proportional to the logarithm of the number of structures with energy E. Let me also remind you that $dS(E)/dE = 1/T(E)$, where $T(E)$ is the temperature corresponding to the energy E.

The distinctive feature of a protein energy spectrum (Fig. 18.8a) is the large gap between the energy of the most stable, i.e., the native protein structure, and the region where the energies of many other, "misfolded" structures appear. (Their energy spectrum is very dense, virtually continuous: the spectrum width is proportional to the number of residues N, while the number of structures is 2^N in order of magnitude, since each residue can have more than one conformation.)

The "energy gap" between the native structure and its numerous high-energy competitors creates a free energy barrier between them. This barrier corresponds to

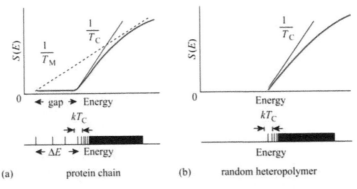

(a)   protein chain   (b)   random heteropolymer

Figure 18.8. Tentative dependence of the entropy S on the energy E for (a) the protein chain and (b) for a random heteropolymer. Typical energy spectra are shown below. Each spectrum line (only a few of them are shown) corresponds to one chain conformation (more precisely, to one energy minimum of the conformational space). The right, high-energy part of each spectrum has plenty of such lines, and they merge into a virtually continuous spectrum. The main peculiarity of the spectrum of protein (the "selected" heteropolymer) is a large energy gap (width $\Delta E \gg kT_C$) between the lowest-energy fold and other "misfolded" structures. The energies of only a few structures get into this "gap"; typically, they are somewhat disordered or nearly folded variants of the lowest-energy fold. The tangent touching the lower part of the $S(E)$ curve to the right of the gap determines the temperature T_C of chain vitrifying (the same temperature, by the way, appears in the statistics of protein structures). The tangent touching also the lowest-energy fold determines the temperature T_M of protein melting (the temperature of co-existence of the lowest-energy structure and a multitude of high-energy ones). Since the slope of the latter is somewhat less than that of the former, $1/T_M < 1/T_C$, that is, $T_M > T_C$. This means that cooling of the protein (a chain with a large "energy gap" ΔE) leads to "all-or-none" freezing of its lowest-energy structure prior to vitrifying of the chain, that is, prior to freezing of misfolded structures corresponding to the high-energy side of the gap. A typical random heteropolymer (a chain having no "gap"), on the contrary, has only a vitrifying transition.

the concave $S(E)$ region. As we have learned, the free energy in such a region is higher than at the ends. It is the "gap" that makes $S(E)$ concave, thereby ensuring the protein's tolerance to heating up to a certain temperature with subsequent melting through an "all-or-none" transition. A few "nearly correctly folded" structures, whose energies get into the gap (and into the concave $S(E)$ region in general), play no role in the thermodynamics of this transition but are important for its kinetics (as we will see later) and for thermal fluctuations.

Two peculiar temperatures are determined by the energy spectrum. These are the melting temperature T_M and the critical vitrifying temperature T_C. The melting temperature T_M is determined by the co-existence of the lowest-energy structure and a multitude of high-energy ones (see Fig. 18.8a). The vitrifying temperature T_C is determined by the rise of the $S(E)$ curve just after the "gap" (Fig. 18.8a), that is, by the abrupt increase in the entropy of "misfolded" structures (or simply, in their number) at the low-energy end of the dense energy spectrum.

If it were not for the jump over the "gap" to the native fold, a temperature decrease would lead to vitrifying of the chain in the ensemble of the low-energy misfolded structures at T_C. This is typical of random heteropolymers that do not have the "gap" (or rather, as the physical theory states, only have a very narrow gap, about kT_C, see Fig. 18.8). Vitrification is not what happens to the protein: its lowest-energy fold is separated from misfolded structures by the energy gap, and as the temperature is decreased, the protein chain jumps to its lowest-energy fold before vitrifying – at the freezing (or melting) temperature T_M which is above T_C (see Fig. 18.8a). The same figure shows that T_M is close to T_C if the "gap" is not too wide, and that $T_M = T_C$ when the gap is absent or close to kT_C.

This means that the "all-or-none" protein melting needs a large energy gap (with a width of $\Delta E \gg kT_C$) between the lowest-energy fold and the bulk of "misfolded" structures, and that this energy gap is a privilege of selected, "protein-suitable" chains only. Only such chains allow a unique stable fold, and this fold is both formed and protected against gradual softening by the "all-or-none" transition.

It seems that natural selection picks up such sequences because of their unique ability to form solid protein globules with fixed 3D structures: only among them can it find those capable of reliable functioning in the organism.

The fraction of random heteropolymers with an energy gap of the given width ΔE is obtained from the extrapolation of the low-energy side of the $S(E)$ curve for random heteropolymers to the still lower energies. The result is:

$$\text{FRACTION}(\Delta E) \sim \exp(-\Delta E / kT_C). \qquad (18.1)$$

This fraction is quite small when the gap is large ($\Delta E \gg kT_C$), which is reasonable: protein creation is not at all easy. However, this fraction is not negligible: if $\Delta E \sim 20kT_C$ (which is plenty to create a reliable barrier between the native fold and its competitors), this fraction amounts to $\sim 10^{-8}$.

Modern experimental methods in molecular biology (for example, the phage display method) are, in principle, already able to select such a small fraction of polypeptides provided they can acquire some structure and become capable of strong binding to something. Therefore, it is no great surprise that some recent experiments with random polypeptides proved to be able to produce "protein-like" molecules. By the way, in principle, these experiments can help to establish the width of the energy gap ΔE using eq. (18.1).

It is noteworthy that the theory yielding eq. (18.1) also predicts that it is much harder to produce a chain that can stabilize two (or more) different folds. The gap ΔE that separates two folds from their competitors occurs in $\sim \exp(-2\Delta E / kT_C)$ of "random" chains only. In other words, to make a chain with one native fold is a wonder, to make a chain with two native folds is a squared wonder, with three a cubed wonder, and so on ... Therefore, it is no surprise that there are almost no proteins having alternative native structures ("almost", because there are at least two proteins of this kind: serpin, slowly passing from its active to its inactive fold without any association of molecules, and prion, an infectious protein, whose extremely slow alternative structure formation is coupled with association of the transformed globules).

The story just presented follows from theoretical studies of random heteropolymers and proteins. It concludes that the values of energy gap width, ΔE, and of vitrifying temperature, T_C, are of crucial importance for protein physics. However, so far, these values have not been established experimentally.

The T_C value does not depend on the gap width, i.e., on the existence of the native fold. It depends on the distribution of many non-native structures over energy. Thus, it has to be the same for random sequences and for protein sequences equal in amino acid content. And it is remarkable that the same T_C appears (cf. eq. (18.1)) in protein statistics that we started to discuss a few lectures ago.

Figure 18.8 shows that T_C has to be only a little lower than the protein melting temperature T_M *if* the ΔE value is relatively low. However, it seems that very wide energy gaps, i.e., very large ΔE values, are not necessary for "protein-like" behavior of the chain. Therefore, natural selection must not seek too large ΔE values, and hence, these have to be highly improbable according to eq. (18.1). It is now a question of experiment, though . . . .

The energy gap is fundamental for protein physics. It is this gap that provides the reliability of a protein's biological functioning: without it, the protein would not be able to tolerate a change of ambient conditions without changing its fold. After a few lectures you will see that the gap is necessary for rapid and reliable folding of a protein structure (and it is quite possible that this determines the appropriate ΔE value). Therefore, an experimental estimate of this gap would be most interesting. Alas, this has not yet been done – and measuring this gap is a challenge to protein physics.

LECTURE 19

In this lecture we will see how protein folding proceeds in time, and consider kinetic intermediates of protein folding. I will talk mainly about water-soluble globular proteins. Folding of membrane proteins is much less studied, and I will touch on them only briefly. Folding of fibrous proteins is better studied, though at the biochemical rather than physical level; we have considered this already when speaking about collagen.

In a living cell, protein is synthesized by a ribosome. The whole life cycle of a bacterium can take twenty minutes only. The biosynthesis takes about a minute and yielding of a "ready" folded protein lasts as long: experiments do not detect any difference. Therefore, it is reasonable to assume that protein chain folding starts on the ribosome before protein chain synthesis is completed.

Unfortunately, reliable experimental data on protein folding *in vivo* (in the cell) are few: it is extremely difficult to notice a structural transformation of a nascent protein chain against the background of thick cell "soup". Usually people have to stop the biosynthesis, to extract the nascent protein and to study it separately. All this takes time, many minutes, during which the 3D structure of the protein chain can change . . . A number of experiments of this kind on large proteins showed that their N-terminal domains are able to fold before the biosynthesis of the whole chain is completed.

I have to stress, though, that these data refer to *multi*-domain proteins; this is an important qualification, since the above mentioned *in vivo* and numerous *in vitro* experiments (see below) show that the "folding unit" is a protein domain rather than the whole protein. A "semi-folded" domain is usually not observed, and we cannot say whether its N-terminal half folds before the C-terminal half.

There are two groups of evidence (provided by *in vitro* rather than *in vivo* experiments) that the folding unit is a domain. First, separate domains are usually capable of folding into the correct structure. Second, single-domain proteins usually cannot fold when as few as ten of their C-terminal amino acids are deleted.

On the other hand, though, it has been shown that a globin chain can bind the heme when only half of its sequence has been synthesized in a cell-free system. (Globin is a single-domain protein from both structural and thermodynamic points of view (the latter means that its denaturation is of the "all-or-none" type). However, its N-terminal half forms a compact sub-domain.)

Some other interesting experimental data on protein folding have also been obtained using biosynthesis in a cell-free system (that is, not entirely *in vitro*, but not entirely *in vivo* as well). This system includes ribosomes, tRNAs, mRNA and other factors necessary for matrix synthesis of a protein.

It is not out of place to mention again here that the terms "*in vivo*" and "*in vitro*" are often understood differently by physicists and biologists. Strictly speaking, between the pure "*in vivo*" and the pure "*in vitro*" there are a number of ambiguous steps. For example, protein folding in a cell-free system (with all its ribosomes, initiation factors, chaperones, etc.) is unequivocally an "*in vivo*" experiment in the physicist's view (for him, "*in vitro*" would be a separate protein in solution; as to the cell-free system... it contains too many complicating life realities). But for a biologist, this is undoubtedly an "*in vitro*" experiment (since "*in vivo*" is referred, for him, to a living and preferably intact organism). However, structural studies of a separate protein in an organism are hardly possible. Therefore, reasonable people compromise by making biologically significant "*in vivo*" events accessible for experimental "*in vitro*" studies.

Unfortunately, it is difficult to see the structure of a nascent protein chain (against the background of the huge ribosome) even in the cell-free system. But, if lucky, one can watch the activity of the synthesized protein.

Luciferase provides such an opportunity. Having folded into its native structure, this protein catalyzes the light-emitting reaction. Therefore, it is easy to observe its appearance: nothing else but the native luciferase emits light in the cell. In their study on luciferase folding and activity, Fedorov, Kolb, Makeev and Spirin from our Protein Research Institute, RAS, have shown that the first active protein appears 10 min after switching on its synthesis in a cell-free system, and that it abruptly stops appearing when its synthesis is switched off (Fig. 19.1). This means that there are virtually no *already* synthesized but *yet not* folded luciferase molecules, that is, folding of this large protein (comprising more than 500 residues) occurs either during biosynthesis or immediately after it.

However, to study not only the result, but also the process of protein folding, one has to carry out "pure" *in vitro* studies (without any ribosomes, chaperones, etc.), that is, to study protein folding in solution.

In about 1960, a remarkable discovery was made: it was shown that a globular protein is capable of spontaneous folding *in vitro* (renaturation) if its chain has not been heavily chemically modified after the initial (*in vivo*) folding. In such a case, a protein that has been gently unfolded by temperature, denaturant, etc. (without damage to the chain) spontaneously restores its activity and structure after solvent "normalization". True, the effective renaturation requires a careful selection of experimental conditions; otherwise, aggregation can prevent the protein from folding.

Inner voice: Is it possible that some features of the native structure are preserved even in the denatured protein, and these residual structures "remembered" from *in vivo* folding direct the *in vitro* protein folding to the right pathway?

Lecturer: No, it is not, if the protein is denatured properly. I mean, if native S—S bonds are disrupted and if the memory of native (*cis-* or *trans-*) states of prolines

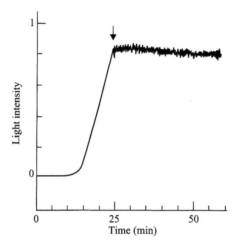

Figure 19.1. Light emission by luciferase synthesized in a cell-free system. Time "0": biosynthesis is switched on. Light emission stops increasing as soon as biosynthesis is switched off (denoted by arrow). Adapted from Kolb V.A., Makeev E.V., Spirin A.S., *EMBO J.* (1994) **13**: 3631–3637.

is erased (which occurs in a few minutes after unfolding), and if the denaturant is strong enough to destroy the secondary structure and to disrupt other non-covalent contacts, i.e., to convert the chain into the coil so that its volume increases many-fold. All this can be checked experimentally. And the protein still renatures . . .

Inner voice: However, there are reliable data on preservation of some native contacts even in a quite unfolded protein chain!

Lecturer: There are such data, though they only concern contacts between residues which are more-or-less close in the chain (if proteins with preserved S—S bonds are not taken into account). The question is, however, are these contacts inherited from the native protein fold and "frozen", or are they occasionally formed spontaneously in the unfolded chain? Unfortunately, NMR (which is the main source of information about these contacts) cannot distinguish "frozen" contacts from those formed spontaneously from time to time. The point is that the NMR signal decreases as R^{-6} with increasing contact distance R. Therefore, the same signal is produced by a contact having $R = 4.5\,\text{Å}$ and existing for 100% of the time, and a contact having $R = 3.0\,\text{Å}$ and existing for 10% of the time. I do not see any physical forces which can freeze some contacts between non-neighbor residues in unfolded protein chain (if preserved covalent S—S bonds are not considered). I think that the contacts we are discussing are sometimes formed spontaneously between those unfolded chain regions that attract each other (and let me remind you that a native protein structure typically contains just those contacts that give sufficient stability by themselves). The same consideration refers, I think, to the "residual" secondary structure sometimes observed by CD.

The phenomenon of a spontaneous protein folding was discovered by Anfinsen's team in 1961. This phenomenon was first established for bovine ribonuclease: its biochemical activity and its correct S—S bonds were restored (after complete unfolding of the protein and disruption of all its S—S bonds) when the protein chain was placed again into the "native" solvent. Later, numerous research groups dealing with a great many other proteins experimentally confirmed this discovery. Furthermore, it was demonstrated (by Gutte and Merrifield) that a protein chain synthesized chemically, without any cell or ribosome, and placed in the proper ambient conditions, folds into a biologically active protein.

The phenomenon of spontaneous folding of protein native structures allows us to detach, at least to a first approximation, the study of protein folding from the study of its biosynthesis.

Protein folding *in vitro* is the simplest (and therefore, the most interesting to me as a physicist) case of pure *self*-organization: here there is nothing "biological" (but for the sequence!) to help the protein chain to fold. As a matter of fact, self-organization of the unique 3D structures of proteins (and RNAs) is a physical phenomenon having no close analogs in inanimate nature. However, this self-organization resembles the formation of crystals (but very peculiar crystals: with no regular space lattice, very complicated and very small).

From the mathematical point of view, this self-organization belongs to the "order from order" class (according to Prigogine's classification): the protein structure emerges as a 3D "aperiodic crystal" (in Schrödinger's wording) from the order of amino acid residues in its chain. Notice that the self-organization of 3D structures of proteins (and RNAs, and crystals) results from the tendency to thermodynamic stability, which distinguishes this self-organization from the more widely discussed self-organization of the "order from disorder" class (existing in oscillating Belousov–Zhabotinsky chemical reactions, in ecological "predator–prey" systems, etc.), which occurs in non-equilibrium systems at the cost of energy flow.

However, before beginning to consider the physics of protein folding *in vitro*, I would like to remind you briefly of the machinery that is used by a cell to increase the efficiency of protein folding under the conditions faced by a nascent protein chain in the cell.

A ribosome makes a protein chain residue by residue, from its N- to its C-end, and not quite uniformly: there are pauses, temporary pauses in the synthesis at the "rare" codons (they correspond to tRNAs which are rare in the cell, and these codons are rare in the cell's mRNAs, too). It is assumed that the pauses may correspond to the boundaries of structural domains, that can help a quiet maturation of the domain structures.

Some enzymes, like prolyl peptide isomerase or disulfide isomerase accelerate *in vivo* folding. The former catalyzes slow, if unaided, *trans↔cis* conversions of prolines; in some cases this is the rate-limiting step of the *in vitro* folding. The latter catalyzes the formation and decay of S—S bonds.

The next point to consider is that in the cell, a protein chain folds under the protection of special proteins, chaperones. Chaperones are the cell's trouble-shooters;

their main task is to fight the consequences of aggregation (which would be only natural in a thick cell "soup"). "Small" chaperones, like hsp70 (heat shock protein, of 70 kilodaltons), bind to a nascent protein to protect it against aggregation, and then they dissociate from the protein (at the cost of ATP consumption). "Large" chaperones like GroEL/GroES or TriC work mainly with multi-domain proteins, and especially with proteins whose domains are composed of remote chain regions. GroEL forms a kind of micro-test-tube (a few nanometers in diameter, with GroES as a lid). The nascent protein or rather its domain (initially covered with hsp70 and/or hsp40 chaperones) seems to come into this tube (though this has yet to be finally proved). This "tube" (sometimes called the "Anfinsen cage") protects the nascent protein against aggregation and against the action of the cell "soup" with all its proteases, etc., and lets the protein fold. GroEL undergoes some conformational changes increasing and decreasing its hydrophobic surface, i.e., the "test-tube" "shakes" from time to time, and the GroES lid closes and opens (all this at the cost of ATP consumption). The chaperone lets the protein go only when it is already folded and has ceased sticking to the "tube". Thus, chaperons work as "incubators" of protein folding; however, as shown by recent studies of Spirin's group, they can work also as "freezers" that postpone folding up to the time when the protein is to be assembled into a quarternay structure or is transported to a proper place of the cell.

Many chaperones are produced in response to heat shock, since a rise in temperature intensifies the hydrophobic forces causing aggregation.

Protein aggregation in a cell often leads to the formation of "inclusion bodies" where the proteins do not have their native structures (specifically, their S—S bonds are formed randomly). However, the "correct" native protein can be obtained after dissolving the inclusion bodies and subsequently allowing protein renaturation *in vitro*.

In a living cell, folding takes place in a highly crowded molecular environment. Over evolutionary time, this has resulted in a range of mechanisms to avoid the problems this raises. These mechanisms include the emergence of various families of molecular chaperones and the increased importance of co-translational folding relative to post-translational folding, especially in eukaryotic cells. However, there is no reason to assume that anything other than the amino acid sequence determines protein conformation in the cellular environment, despite some propositions to the contrary.

One of the most exciting ideas to emerge from recent studies is the "evolutionary shift" from post-translational to co-translational folding as eukaryotic cells appeared from combinations of prokaryotic cells, and larger and more complex proteins developed. (A typical eukaryotic protein consists of four–five domains, and a prokaryotic protein of two domains.)

Co-translational folding, that is, the folding of a nascent chain while it is still attached to the ribosome, is of a special interest at present because of the tremendous progress being made in defining the structure of the ribosome at atomic resolution. The folding of nascent peptides may possibly occur in the sheltered environment of a visible ribosomal "tunnel", in which (many researchers believe) they are protected from aggregation and degradation. Moreover, there is some (incomplete) evidence that ribosomes and some of their components, notably the large subunit and, in particular, its 23S RNA, can function as molecular chaperones to mediate and accelerate the refolding of denatured proteins. This would provide an explanation for the fact that the refolding of denatured proteins *in vitro* (without ribosomes) often appears to be much slower than the folding *in vivo* that occurs simultaneously with biosynthesis on the ribosome.

However, experiments on *in vitro* protein folding show that the work of the whole cellular machinery can be replaced by careful selection of the experimental conditions (a low protein concentration, a suitable redox potential, etc.). This substitution does not change the result of the chain folding: if the protein folds rather than precipitates *in vitro*, then the resultant native structure is the same as *in vivo* folding yields. True, this can take more time (or much less, for some small proteins) than *in vivo* folding, but the *result* is the same. Moreover, as I have already mentioned, it has been shown that the chain of a small protein can be synthesized in a test-tube by purely chemical methods, and a large fraction of these chains fold into a correct native (and active) 3D structure. All this proves that all the information necessary to build up the 3D protein structure is inscribed in its amino acid sequence.

It looks as though the biosynthetic machinery of the cell (ribosomes + chaperones + · · ·) not only ensures synthesis of the protein chain but also serves as a kind of incubator that helps folding of 3D protein structures. This "incubator" does not determine the protein structure, though, but rather provides "hothouse" conditions for its maturation – just like a normal incubator, which helps a nestling to develop but does not determine whether a chicken or a duckling will be developed.

Inner voice: All that you have just said refers mostly to small water-soluble globular proteins or to the separate domains of large proteins. They usually renature easily, that's true. However, to be frank, you have to say that one faces far greater difficulties when dealing with whole large proteins, especially with those from higher organisms: far from all of them renature spontaneously. As for the spontaneous renaturation of membrane and fibrous proteins, this only happens with some of them; usually their complete renaturation cannot be obtained . . .

Lecturer: Concerning the proteins that are "difficult" for renaturation, I would suggest that aggregation is the common major obstacle (for example, some membrane proteins renature in detergents and do not fold in water because of aggregation). Multi-domain proteins can also experience "aggregation" of remote chain regions which is absent *in vivo* since ribosomal synthesis allows folding of the N-terminal domains prior to synthesis of the C-terminal domains. Also, refolding difficulties caused by post-folding modifications should not be neglected, especially for eukaryotic proteins.

Let us agree that for now I am considering relatively small water-soluble proteins. Let us understand the folding of these first . . .

Returning now to the main topic, the fundamental physical problem of spontaneous folding of proteins (and RNA) has come to be known as the Levinthal paradox. It reads as follows. On the one hand, the same native state is achieved by various folding processes: *in vivo* on the ribosome, *in vivo* after translocation through the membrane, *in vitro* after denaturation with various agents ... The existence of spontaneous renaturation and the correct folding of chemically synthesized protein chains suggests that the native state is thermodynamically the most stable state under "biological" conditions. On the other hand, a chain has zillions of possible conformations (at least 2^{100} for a 100-residue chain, since at least two conformations are possible for each residue), and the protein can "feel" the correct stable structure only if it is achieved exactly this conformation, since even a 1 Å deviation can strongly increase the chain energy in a closely packed globule. Thus, the chain needs at least $\sim 2^{100}$ picoseconds, or $\sim 10^{10}$ years to sample all possible conformations in its search for the most stable fold.

So, how can the chain find its most stable structure within a "biological" time (minutes)?

The paradox is that, on the one hand, the achievement of the same (native) state by a variety of processes is (in physics) clear-cut evidence of its stability. On the other hand, Levinthal's estimate shows that the protein simply does not have enough time to prove that the native structure is the most stable among all possible structures!

Then, how does the protein choose its native structure among zillions of others, asked Levinthal, and answered: It seems that there exists a specific folding pathway, and the native fold is simply the end of this pathway rather than the most stable chain fold. Should this pathway be narrow, only a small part of the conformational space would be sampled, and the paradox would be avoided.

In other words, Levinthal suggested that the native protein structure is under kinetic rather than under thermodynamic control, i.e., that it corresponds not to the global but rather to the easily accessible free energy minimum.

The question as to whether the protein structure is under kinetic or thermodynamic control is not a purely speculative question. It is raised again and again when one faces practical problems of protein physics and engineering. For example: when trying to predict a protein structure from its sequence, what do we have to look for? The most stable or the most rapidly folding structure? When designing a *de novo* protein, what do we have to do? To maximize the stability of the desired fold or to create a rapid pathway to this fold?

A discussion on protein folding mechanisms started immediately after solution of the first 3D protein structures and the discovery of spontaneous folding. It seems that the first proposed hypothesis was that by Phillips, who suggested that the folding nucleus is formed by the N-end of the nascent protein chain, and that the remaining part of the chain wraps around it. In various forms this appealing hypothesis is present in some works up to now. However, this hypothesis has been refuted experimentally (as far as single-domain proteins are concerned). The elegant work of Goldenberg and Creighton has shown that the N-terminus has no special role in *in vitro* folding. It was demonstrated that it is possible to glue the ends of the chain of a small protein (trypsin inhibitor) with a peptide bond, and it folds into the correct 3D structure, nevertheless. Moreover, it is possible to cut this circular chain so as to make a new N-end at the

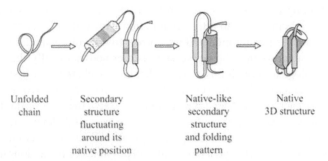

| Unfolded chain | Secondary structure fluctuating around its native position | Native-like secondary structure and folding pattern | Native 3D structure |

Figure 19.2. Framework model of sequential protein folding according to Ptitsyn (1973). The secondary structures are shown as cylinders (α-helices) and arrows (β-strands). Both predicted intermediates have already been observed; the first is now known as the "pre-molten globule" and the other as the "molten globule".

former middle of the chain; and it still folds to the former native structure. Nowadays, protein engineering routinely produces circularly modified proteins.

In an effort to solve the folding problem, O.B.P. proposed in 1973 a model of sequential protein folding (Fig. 19.2). Later given the name "framework model", this hypothesis stimulated further investigation of folding intermediates. It postulated a sequential involvement of different interactions in protein structure formation, stressing the importance of rapidly folded α-helices and β-hairpins in the initial folding steps, the gluing of these helices and hairpins into a native-like globule, and the final crystallization of the structure within this globule at the last step of folding.

The cornerstone of this concept was the then hypothetical and now well known folding intermediate, the "molten globule". The theoretically predicted molten globule was experimentally discovered and studied by Dolgikh, Semisotnov, Gilmanshin, Bychkova and others at the laboratory of O.B.P. in the 1980s – first as the equilibrium state of a "weakly denatured" protein (I mentioned this in my previous lectures), and then as a kinetic folding intermediate.

The molten globule is assembled in the course of the folding of many proteins. In the experiments shown in (Fig. 19.3), the initial coil state is obtained by incubation of a protein in solution with a high denaturant concentration. The renaturation is achieved by rapidly diluting this solvent with water. That is, the folding process starts from the coil and ends with the native protein. However, different properties of the native protein have two quite different rates of restoration, which is evidence for the accumulation of some "intermediate" state of the protein molecule at the beginning of the folding process.

Figure 19.3 shows that the "native" values of intrinsic viscosity (which characterize the globule's volume) and ellipticity at 222 nm (i.e., CD in the far-UV region, which characterizes the secondary structure) are restored, or, rather, nearly restored in the course of renaturation much faster than the ellipticity at 270 nm (i.e., CD in the near-UV region, which characterizes the side-chain packing) or protein activity

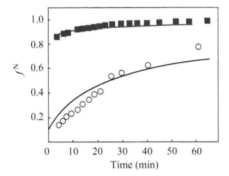

Figure 19.3. Kinetics of recovery of a "degree of nativity" (f_N) in the process of carbonic anhydrase B renaturation. Transition of the protein from its fully unfolded state (existing at 5.45 M GuHCl concentration) to the native state (at 0.97 M GuHCl) is followed by intrinsic viscosity (■■■), by ellipticity at 222 nm (__) and at 270 nm (__), and by enzymatic activity (o o o). Adapted from Dolgikh D.A., Kolomiets A.P., Bolotina I.A., Ptitsyn O.B., *FEBS Lett.* (1984) **164**: 88–92.

(which is a general indication of the native state of the protein). This indicates the presence of a metastable (i.e., quasi-stable, since it lives long but then disappears) kinetic intermediate, accumulated in a large quantity. This intermediate is called the "molten globule" since its compactness and secondary structure are close to those of the native protein, although it has neither native side-chain packing nor the enzymatic activity of the native "solid" protein.

The molten globule is an early folding intermediate in the *in vitro* folding of many proteins. It is noteworthy that it can be observed under physiological conditions. This intermediate takes a few milliseconds to form, while the complete restoration of the native properties of a 100–300 residue chain can take seconds for some proteins and up to hours for others. Thus, the rate-limiting folding step is the formation of the native "solid" protein from the molten globule rather than the formation of the molten globule from the coil (Fig. 19.3).

The molten globule is not the only intermediate observed in protein folding. A state with a partially formed secondary structure and a partially condensed chain (that also fits the "framework model") has been observed to precede formation of the molten globule. This intermediate was discovered with the use of recently developed ultra-fast (sub-millisecond) measuring techniques. This state of the folding chain resembles the "pre-molten" globule. In addition, proteins with disulfide bonds allow the trapping of various intermediates indicative of the order of formation of S—S bonds, etc.

As a matter of fact, the "kinetic control" hypothesis initiated very intensive and numerous studies of protein folding intermediates. Actually, it was clear almost from the very beginning that the stable intermediates are not obligatory for folding (since the protein can also fold near the point of equilibrium between the native and denatured states, where the transition is of the "all-or-none" type, which excludes any stable intermediates). The idea was, though, that the intermediates, if trapped, would help to trace the folding pathway, just as intermediates in a complicated biochemical reaction trace the pathway of this reaction. This was, as it is now called, "chemical logic". However, this logic worked only in part when it came to protein folding. Intermediates (like molten globules) were found for many proteins, but the main question as to how the protein chain can rapidly find its native structure among zillions of alternatives remained unanswered.

Progress in understanding was achieved when the studies involved small proteins (of 50–100 residues). Many of them fold *in vitro* without any observable intermediates accumulating in the experiment, and have only two observable states: the native fold and the denatured coil. The absence of folding intermediates refers not only to the vicinity of the denaturation point (where the two-state kinetics can be expected *a priori* because of the "all-or-none" thermodynamics of denaturation, which means the absence of a visible quantity of intermediates). For these small proteins, this absence also refers to "physiological" conditions where folding intermediates of larger proteins are usually observed.

The investigation of small proteins having no S—S bonds or *cis*-prolines (features that were widely used previously to trap the folding intermediates) has led to a striking result: it turned out that, in the absence of such "unnecessary complications" as folding intermediates, these proteins fold very rapidly: not only much faster than larger proteins (which would not be a surprise), but also at least as rapidly as small proteins that have folding intermediates that would be expected to accelerate their folding...

Some small proteins with *no* observable folding intermediates fold within milliseconds (Fig. 19.4) or even faster under nearly physiological conditions. The absence

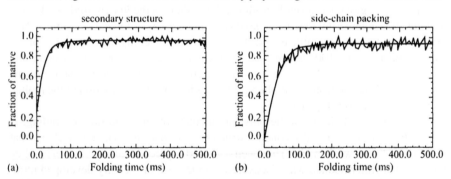

Figure 19.4. Two-state renaturation of ACBP (acetyl coenzyme A binding protein), registered by the restoration of ellipticity (a) at 225 nm and (b) at 286 nm. The former reflects restoration of the secondary structure, the latter restoration of the tertiary structure (specifically, restoration of the aromatic side group packing). The experimental signal (broken line) and the result of its smoothening and extrapolation (smooth line) are shown. Note the absence of changes within the measurement dead-time: this provides evidence that no intermediates accumulate within this time. Reprinted with permission from Kragelund B.B., Robinson C.V., Knudsen J., Dobson C.M., Poulsen F.M., *Biochemistry* (1995) **34**: 7217–7224. © 1995, American Chemical Society.

of folding intermediates was established by the equal rates of restoration of secondary structure, side-chain packing, hydrogen exchange, etc.

However, what can these experiments provide to shed light on the nature of protein folding? What are we supposed to study if there are no intermediates to single out and investigate?

The answer is: just here one has the best opportunity to study the *transition state*, the bottleneck of folding.

I think it is important to emphasize from the very beginning that the transition state, which is of crucial importance for folding kinetics and the folding rate, is, by definition, the *most unstable* state along the protein folding pathway. Thus, it is *not* a kind of previously discussed "folding intermediate". The "intermediates" correspond to free energy minima; that is, they are stable at least for some time and therefore can accumulate during folding and can be directly observed. On the contrary, the transition state corresponds to the free energy maximum; it never accumulates and therefore is never observed directly. It can *only* be followed by its influence on the folding rate.

Let us postpone a detailed consideration of the transition states observed in protein folding till the next lecture and just see what results they produce. Transition state studies outlined the "folding nucleus", i.e., a part of the chain that is folded already in the transition state (Fig. 19.5). A relatively small fraction of residues are involved

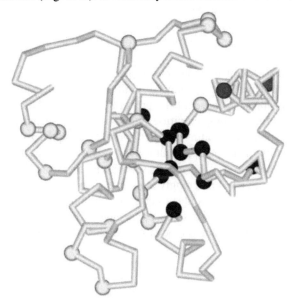

Figure 19.5. Experimentally outlined folding nucleus for CheY protein according to E. López-Hernándes and L. Serrano, *Folding & Design* (1996) 1: 43–55. The residues studied experimentally are shown as beads against the background of the native chain fold. The residues forming the folding nucleus are shown in dark blue. The yellow beads indicate the residues that are not involved in the nucleus. The red beads show two residues that are difficult to interpret.

in the nucleus; there, they already have their native-like (i.e., the same as in the native protein) conformations and contacts. Other residues are not involved in the nucleus and remain in the unfolded state. All this will be discussed in detail in the next lecture.

Thus far we have considered water-soluble globular proteins. True, most studies of protein folding deal with these proteins. But nearly a third of proteins in the genome of any organism are membrane proteins, which can only achieve their functional state in the presence of a lipid environment. There is some evidence that the folding of helical membrane proteins occurs through intermediate states that share many features of the "molten globules" of water-soluble proteins, i.e., have a native-like secondary structure, but disordered tertiary interactions. The presence of exposed hydrophobic residues within a condensed globular form makes them attractive in the context of folding in a membrane. A number of experiments suggest that different helices in membrane proteins fold in a relatively autonomous manner, consistent with the flexibility anticipated for the molten globule. The analogy between the folding of membrane-bound and water-soluble proteins may extend to the presence of a structural "core" that might serve as a nucleus for folding. The other membrane proteins such as the porins have β-barrel structures, and it appears that they fold much more co-operatively than helical membrane proteins, that is, without observable stable intermediates and visible folding steps.

As well as satisfying our natural curiosity about the folding of this group of proteins, progress in protein folding studies will contribute to our ability to produce membrane proteins efficiently and increase our understanding of the regulation of membrane processes.

In conclusion, it seems to be not out of place to suggest some analogy between the folding of proteins and assembling of viral particles. For many simple viruses, the coat proteins appear to have all the properties needed for the formation of the intact particles. For more complex viruses, however, there are known to be assembly pathways that involve "scaffolding proteins" or other components, which are essential to the formation of the particles but are not incorporated in the mature virions. These "molecular chaperones" are clearly very fascinating from both a structural and a mechanistic point of view. Intriguingly, as simple viruses assemble without the accumulation of intermediates (as simple proteins fold without populating partially folded states), we know more about the assembly pathways of the more complicated species, where partially structured states can be observed under favorable circumstances.

LECTURE 20

As I mentioned in the previous lecture, in the understanding of protein folding a key role was played by the simplest folding event, in which no metastable intermediates accumulate during the process.

Only the native and the completely unfolded state are seen in such a "two-state" transition. The absence of folding intermediates is established by the equal rates of restoration of the secondary structure, of the side-chain packing, etc.

Notice that two-state kinetics can be *a priori* expected around the mid-point of the de/renaturation, i.e., close to the point of co-existence of the native and denatured states of the protein molecule – since the thermodynamics of denaturation are known to be of the "all-or-none" type. The latter means that no intermediates are present in any visible amount around the mid-point of this transition. Less trivial is the fact that two-state folding occurs (in many small proteins) also under physiological conditions (which are far from the equilibrium mid-point) when the folding is most rapid, and when the majority of larger proteins display accumulating folding intermediates like the molten globule.

This same two-state folding, which occurs within a wide range of conditions (and has no such complications as accumulating intermediates), provides the best opportunity to study the *transition state*, that is, the bottleneck of folding that determines the folding rate.

As you may remember, the transition state corresponds, by definition, to the free energy *maximum* on the pathway from one stable state (the free energy *minimum*) to another. Let me also remind you that the kinetics of the simplest transition between two stable states (A and B in Fig. 20.1) are satisfactorily described by the *transition-state theory*.

The essence of this theory is that the rate of the process is limited by the *low occupancy* of the transition state, i.e., the state having the highest free energy and therefore the least populated species along the reaction pathway (as such, it does not accumulate and cannot be observed directly). In other words, the kinetics of the A $\rightarrow$ B (and B $\rightarrow$ A) transitions is determined by the *height* of the free energy maximum along the reaction pathway.

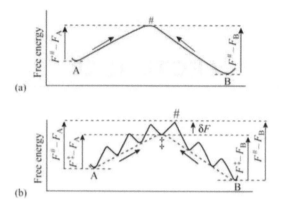

Figure 20.1. Overcoming of the free energy barrier on the pathway from the stable state A to the stable state B (the left arrow in each scheme), and from B to A (the right arrow in each scheme). F_A, F_B and $F^{\#}$ are the free energies of the stable states A, B and the transition state # (which is the "barrier", i.e., the free energy maximum on the pathway). (a) Elementary reaction; (b) multi-step processes consisting of several elementary transitions via free energy minima of a high free energy (compared with that of the A and B states). δF, the free energy barrier for one step; ‡, the free energy minimum of the highest free energy $F^{‡}$.

The rates of the A $\rightarrow$ B and of the B $\rightarrow$ A transitions are given, let me remind you, by the equations

$$k_{A \rightarrow B} = k_0 \exp[-(F^{\#} - F_A)/RT]$$
$$k_{B \rightarrow A} = k_0 \exp[-(F^{\#} - F_B)/RT]$$

(20.1)

Here F_A, F_B and $F^{\#}$ are the free energies of the states A, B and # (the transition state, see Fig. 20.1), and k_0 is the rate of an "elementary step" of the process.

When the elementary (one-step) reaction is considered (Fig. 20.1a), the value k_0 is usually taken as $k_{0,el} = RT/h$ (where h is the Planck constant); RT/h is equal to 10^{13} s^{-1} at $T = 300$ K: this is the frequency of attacks on the barrier under the action of thermal vibrations.

The rate of a multi-step process (Fig. 20.1b) is also determined only by the highest free energy barrier on the pathway (when the number of intermediate minima is not too large and their free energies are higher than both F_A and F_B). Considering a multi-step process, it is often convenient to express its rate through the rates of one-step transitions between intermediate free energy minima and the heights of these minima. To this end, we can

1. take the rate of transition between two adjacent minima as the rate of an elementary step:

$$k_{0,step} = k_{0,el} \exp(-\delta F/RT)$$

(20.2)

where δF is the free energy barrier that is to be overcome during one step;

2. take the maximum free energy of an intermediate metastable state, $F^{\ddagger}$, as the barrier free energy for the whole process; then:

$$k_{A \to B} = k_{0,el} \exp[-(F^{\#} - F_A)/RT]$$
$$= k_{0,el} \exp[-(F^{\ddagger} + \delta F - F_A)/RT]$$
$$= k_{0,step} \exp[-(F^{\ddagger} - F_A)/RT] \qquad (20.3)$$

This equation looks like eq (20.1) with $k_{0,step}$ in place of k_0 and $F^{\ddagger}$ in place of $F^{\#}$.

The time for an A $\to$ B transition is

$$t_{A \to B} = 1/k_{A \to B}$$
$$= (1/k_{0,step}) \exp[+(F^{\ddagger} - F_A)/RT]. \qquad (20.4)$$

The same substitutions and explanations are, of course, equally true for $k_{B \to A}$ and $t_{B \to A}$.

With these refinements, eqs (20.1) are applicable to multi-step processes that (a) take much more time than $1/k_{0,step}$, the time of an elementary step; (b) include not too many intermediate free energy minima; and (c) proceed without accumulating intermediates. When considering the two-state protein folding, we are dealing with just such processes.

It is noteworthy that the temperature dependence of the A $\to$ B reaction rate allows us to estimate the transition state energy relative to that of state A. According to Arrhenius, we obtain

$$\frac{d[\ln(k_{A \to B})]}{d(1/T)} = \frac{d[\ln(k_{0,el}) - (F^{\#} - F_A)/RT]}{(-T^{-2} dT)}$$
$$\approx -\frac{(E^{\#} - E_A)}{R} \qquad (20.5)$$

Here I have used the known relationship $d(F/T)/dT = -E/T^2$ and neglected the weak dependence of the elementary step rate $k_{0,el}$ on temperature T. The latter is usually acceptable in chemical reactions, and thus should be at least as acceptable in proteins where the energies $E^{\#}$ and E_A, determined by many interactions, are large.

Figure 20.2 shows the dependence of lysozyme folding and unfolding rates on the value of $1/T$. The plot shows that the rates of both processes are close to $e^{-2.5} \approx 0.1 \, s^{-1}$ (i.e., their times are close to 10 s) at the mid-transition (where the folding rate $k_{u \to N}$ is equal to the unfolding rate $k_{N \to u}$, so that the curves for $\ln(k_{u \to N})$ and $\ln(k_{N \to u})$ intersect).

This plot also shows that the denaturation accelerates as it gets deeper into the "denaturation region", and the renaturation accelerates as it gets deeper into the "renaturation region".

A most interesting conclusion from this plot is as follows.

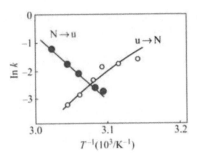

Figure 20.2. Arrhenius plots for the rates of lysozyme de- and renaturation vs. the reciprocal temperature value (T^{-1}); the plot is extracted, with some abridgement, from Segava S., Sugihara S., *Biochemistry* (1984) **23**: 2473–2488. The rate constants (k) are measured in sec^{-1}. Renaturation: rate $k_{u \to N}$ (experimental points ○ and the thin dark blue interpolation curve); denaturation: rate $k_{N \to u}$ (experimental points ● and the bold red interpolation line). The mid-transition is the temperature point where $k_{u \to N} = k_{N \to u}$, i.e., where the curves intersect (at about $1000/3.08 = 325$ K). Folding prevails in the "renaturation region" $(u \to N)$ at low temperatures (i.e., at T^{-1} values to the right of the intersection point); unfolding prevails in the "denaturation region" $(N \to u)$ at high temperatures (i.e., at T^{-1} values to the left of the intersection point).

The unfolding rate $k_{N \to u}$ *decreases* with T^{-1} (i.e., it grows with temperature T, which is typical of physicochemical reactions), while the folding rate $k_{u \to N}$, *on the contrary, increases* with T^{-1} (i.e., it drops with temperature T, which is *a*typical of physicochemical reactions). According to eq. (20.5), this means that $E^{\#} - E_N > 0$ and $E^{\#} - E_u < 0$. In other words, $E_u > E^{\#} > E_N$, i.e., the barrier energy $E^{\#}$ is above the native state energy E_N, but *below* the unfolded state energy E_u; the latter is *a*typical of chemical reactions where the barrier energy is higher than *both* the initial and the final state energy. Additional analysis of this plot (taken now in the form $T \ln(k/k_{0,el})$ vs T) shows that the same relationship, $S_u > S^{\#} > S_N$, is valid for the entropies of the native, transition and denatured states. This means that the barrier between the native and denatured states looks like a normal energy barrier when viewed from the native state, but from the denatured state it looks like an *ab*normal entropy barrier.

However, just this could be expected from our previous analysis of the free energy barrier between the native and denatured states and of the reasons for the "all-or-none" protein melting (see Fig. 20.3 which reproduces the previously discussed scheme of energy, entropy and free energy changes during expansion of the globule).

Let us return to eqs (20.1) once again. They show that the ratio between the rates of the direct and reverse reactions, $k_{A \to B}/k_{B \to A}$, is simply the constant of equilibrium between the final (B) and initial (A) states:

$$K_{B:A} = \frac{k_{A \to B}}{k_{B \to A}}$$

$$= \exp[-(F_B - F_A)/RT] \tag{20.6}$$

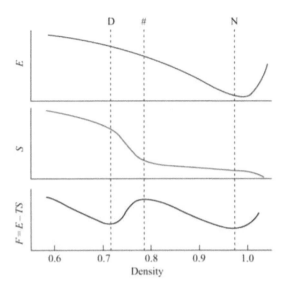

Figure 20.3. Change of the energy E, the entropy S and the free energy $F = E - TS$ with changing density ρ. N, the native globule; #, the "barrier", i.e., the free energy maximum on the path of uniform swelling of the globule (from N to D); D, the denatured state (the molten globule in this case, since its density is rather high).

The $K_{B:A}$ value gives the ratio between the equilibrium numbers (n_A^∞, n_B^∞) of molecules in the states B and A, which is achieved as a final result of the transition process under given conditions (temperature, etc.), i.e., when (in equilibrium) the flow A → B equilibrates the flow B → A (so that $n_A^\infty k_{A\to B} = n_B^\infty k_{B\to A}$):

$$K_{B:A} = \frac{k_{A\to B}}{k_{B\to A}}$$

$$= \frac{n_B^\infty}{n_A^\infty} \tag{20.7}$$

What is the rate of approach to this equilibrium? To answer this question, we have to solve the differential equation

$$\frac{dn_A(t)}{dt} = -k_{A\to B}n_A(t) + k_{B\to A}n_B(t) \tag{20.8}$$

where $n_A(t)$, $n_B(t)$ are the numbers of molecules in the states A and B, respectively, at time t. (There is no need to solve the second equation, for dn_B/dt, since $n_A(t) + n_B(t) \equiv n_0$, and the full number of molecules n_0 is constant; i.e., $dn_B/dt \equiv -dn_A/dt$.)

Substituting $n_B(t) = n_0 - n_A(t)$ into (20.8), one obtains $dn_A(t)/dt = -(k_{A\to B} + k_{B\to A})n_A(t) + k_{B\to A}n_0$. Now, using $n_A^\infty k_{A\to B} = n_B^\infty k_{B\to A}$ (see 20.7) and $n_A^\infty + n_B^\infty = n_0$, so that $n_0 \times k_{B \to A} = n_A^\infty (k_{A \to B} + k_{B \to A})$, we have:

$$\frac{dn_A(t)}{dt} = -(k_{A \to B} + k_{B \to A})\left[n_A(t) - n_A^\infty\right] \tag{20.9}$$

The final answer (I recommend you to verify the calculation yourselves) is:

$$n_A(t) = \left[n_A(0) - n_A^\infty\right] \exp[-(k_{A \to B} + k_{B \to A})t] + n_A^\infty \tag{20.10}$$

This means that the apparent rate of approach to the equilibrium is

$$k_{app} = k_{A \to B} + k_{B \to A} \tag{20.11}$$

i.e., it is equal to the sum of the rates of the forward and reverse reactions. The value k_{app} depends only on the conditions of the process rather than on the initial numbers of folded and unfolded molecules.

The most rapid reaction dominates in the k_{app} value. When the conditions (solvent, temperature, etc.) stabilize the native state, the folding rate $k_{u \to N}$ dominates in $k_{app} = k_{u \to N} + k_{N \to u}$. When the denatured state is more stable, then the unfolding rate $k_{N \to u}$ dominates in k_{app}.

The rate of approach to the equilibrium, k_{app}, can be measured (unlike $k_{u \to N}$ and $k_{N \to u}$ separately) both when the native state is more stable and when it is less stable than the denatured one; this is an important advantage of the k_{app} value.

Values of k_{app} are commonly presented as a function of the denaturant's concentration, in the form of so-called "chevron plots" ("chevron", because its shape resembles the military long-service stripe) (Fig. 20.4). This plot shows that the rates of folding ($k_{u \to N}$) and unfolding ($k_{N \to u}$) are oppositely dependent on the denaturant's concentration. We have already seen a similar case when considering the temperature dependence of the folding and unfolding rates. As in that case, the opposite slopes of the two branches indicate that the transition state is intermediate in properties between the native and denatured states.

Indeed, the ability of a denaturant to unfold the protein means that it is attracted to the denatured state more than to the native state. The denaturant-induced decrease in the folding rate of the initially unfolded protein (see the left part of Fig. 20.4) means that the denaturant is more strongly attracted to the unfolded state than to the transition state (i.e., that the denaturant destabilizes the transition state relative to the initial unfolded state). The denaturant-induced increase in the unfolding rate of the initially native protein (see the right part of Fig. 20.4) means that the denaturant is more strongly attracted to the transition state than to the native state. This shows that the denaturant's contact with the transition state is stronger than with the native state, but weaker than with the denatured state. Thus, in terms of protein–solvent contact, the transition state is intermediate between the native and denatured states. The plot shows even a little more: since the $k_{u \to N}$ slope is a little steeper than the $k_{N \to u}$ slope, the compactness of the transition state is somewhat closer (in this case) to that of the native state than to the compactness of the denatured state.

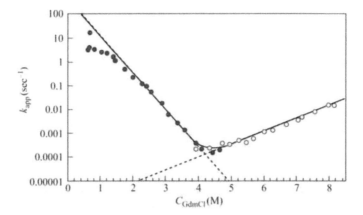

Figure 20.4. "Chevron plot" for the apparent rate (k_{app}) of approach to the equilibrium between the native and unfolded forms of hen egg white lysozyme vs. the concentration of guanidine dihydrochloride. The experiment was carried out in the presence of native S—S bonds. The filled circles are obtained when diluting a concentrated GdmCl solution of the denatured protein, i.e., they correspond to refolding; $k_{app} \approx k_{u \to N}$ in this region (since here the refolding rate $k_{u \to N}$ is greater then the unfolding rate $k_{N \to u}$). The open circles are obtained when adding GdmCl to aqueous solution of the native protein, i.e., they correspond to unfolding; $k_{app} \approx k_{N \to u}$ here (since $k_{u \to N} < k_{N \to u}$ upon unfolding). At the lowest part of the chevron, where $C_{GdmCl} \approx 4.2\,M$, $k_{u \to N} \approx k_{N \to u} \approx k_{app}/2$. Note the overlapping of the open and filled circles in this region. It shows that here, at the mid-transition, the protein folds at the same rate as it unfolds. The dotted lines show extrapolation of the $k_{u \to N}$ and $k_{N \to u}$ values to the chevron bend region and beyond. The solid line shows the $k_{u \to N}$ extrapolation to the low-GdmCl region. The points that deviate from the solid line at low GdmCl concentrations (i.e., *far* from the mid-transition) indicate either some rearrangement of the transition state or the appearance of some additional metastable intermediates (possibly molten globules); these may be on, as well as off the main renaturation pathway. Note that these rearrangements and/or intermediates do not increase the renaturation rate compared with what could be expected in their absence: although they correspond to the highest observed renaturation rate, they are *below* the extrapolation line. Adapted from Kiefhaber T., *Proc. Natl. Acad. Sci. USA* (1995) **92**: 9029–9033.

Figure 20.4 (and also Fig. 20.2) refers to lysozyme, the protein whose GdmCl-induced denaturation leads directly to the coil state, rather than to the molten globule state. Its folding takes about an hour at moderate GdmCl concentrations. However, in "almost pure" water the folding time is as short as ≈ 0.1–0.3 s.

Inner voice: Does this mean that the protein has no such characteristic as the folding rate?

Lecturer: That's true. The protein does not have a strictly defined "folding rate". Indeed, Fig. 20.4 shows that lysozyme folding takes about 0.1 s under native conditions and about 10 000 s when close to the denaturant-induced equilibrium between its native and denatured forms. And at high temperature (but without denaturant) its folding takes about 10 s at equilibrium (see Fig. 20.2). Therefore, when discussing the protein's folding rate, one has either to refer to the entire observed range of its folding rates, or specify the ambient conditions (for example, those which refer to the protein folding in the cell, or in water at 20–25°C, as is also often done).

Coming back to lysozyme folding in water (*far* from the denaturant-induced equilibrium between the native and the coil forms), we see that here the folding rate is at a maximum, and is virtually independent of denaturant concentration (this is called a "rollover" of the chevron). This is typical of the folding of middle-sized and larger proteins at low denaturant concentrations; therefore, the in-water folding time is often taken as the characteristic folding time of a given protein.

It has been shown that here, in pure or almost pure water, lysozyme folding has a compact intermediate that does not appear at higher GdmCl concentrations. Thus, lysozyme exhibits "three-state" folding far from the equilibrium of the native and denatured states, and "two-state folding" in the vicinity of this equilibrium. This means that there is no physical difference between "three-" and "two-state" folding proteins, if we consider the vicinity of mid-transition only, and their unfolded states are the same.

However, the study of proteins exhibiting only two-state folding is more straight-forward: their in-water folding transition state is the same as that at other denaturant concentrations. Another advantage is technical: the extrapolation of the long chevron branches is easier. Therefore, the "two-state folders" played a key role in investigations of transition states.

The appearance of compact intermediates always makes the folding rate dependence more shallow (or even changes the sign of the slope) because the rate-limiting step of the folding now starts from a more compact intermediate, i.e., from the state whose properties (such as interaction with the denaturant) are closer to those of the transition state than to the properties of the former starting state, i.e., the unfolded state. It is known also that the rate of folding that starts from the molten globule rather than from the coil state (e.g., carbonic anhydrase folding) shows a very shallow dependence on the denaturant concentration. This means that the compactness of the transition state in this folding is close to that of the molten globule state.

In other words, the "intermediate" character of the transition state properties makes them similar to the properties of the molten globule. However, this similarity refers only to the averaged properties of these two states (such as energy or compactness), and *does not* mean that the transition state *is* the molten globule: the transition state is (by definition) unstable, while the molten globule is stable or at least metastable.

Experiments (to be described later) show that the transition state is far less uniform than the molten globule. As far as the "coil → native protein" transition is concerned, the transition state can be imagined as similar to a piece of the native protein, while the rest of the chain remains in the unfolded state. The appearance of transition states in the "molten globule → native protein" and "coil → molten globule" transitions has not yet been determined experimentally, but it is plausible that they also include a piece of the more structured state, while the rest of the chain remains in a less structured state.

The nature of the transition state is established using many mutants of the protein (ideally, each separate chain residue is mutated), and analyzing changes in the chevron plots (see Figs 20.4 and 20.5). This allows one to find out, at the cost of hard work, which residues are involved in the native-like part of the transition state (their mutations change the folding rate substantially), and which are not (their mutations do not change the folding rate, but change the rate of unfolding). This protein-engineering method was developed by A. Fersht in England. It can be applied to all "all-or-none" transitions. However, up to now it has mostly been used for proteins whose denatured state is the coil rather than the molten globule.

To estimate the involvement of a residue in the native-like part of the transition state ("folding nucleus"), one estimates: (a) the mutation-induced shift of the folding rate and (b) the mutation-induced shift of the native protein stability (Fig. 20.5).

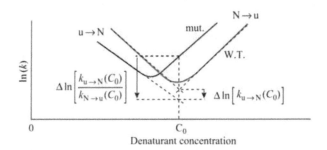

Figure 20.5. A scheme illustrating a mutation-induced change of the chevron plot: k, apparent rate of transition. Green line, the initial, "wild type" protein (W.T.); red line, the same protein with one residue mutated (mut.). C_0, the denaturant concentration at the mid-transition for the W.T. protein. The dotted lines show extrapolation of the folding and unfolding rates, $k_{u \to N}$ and $k_{N \to u}$, to the W.T. chevron bend region. Two measured mutation-induced shifts are shown. The first is the change in the height of the free energy barrier on the pathway from the unfolded to the native state: $\Delta(F^\# - F_u) = -RT \Delta \ln[k_{u \to N}]$. The other is the change in protein stability, i.e., in the free energy difference between the native and unfolded states: $\Delta(F_N - F_u) = -RT \Delta \ln[k_{u \to N}/k_{N \to u}]$. The ratio $\Delta(F^\# - F_u)/\Delta(F_N - F_u)$ is called Φ_f for the mutated residue. In the shown case $\Phi_f \approx 1/4$, i.e., when in the transition state, the mutated residue has only a quarter of its native interactions. (A mnemonic rule. If the left branches of the W.T. and mut. chevrons are closer ·to each other than the right branches, Φ_f is closer to 0 than to 1. If otherwise, Φ_f is closer to 1 than to 0.) The values shown in the plot refer to the C_0 denaturant concentration, where the necessary extrapolations (and therefore, extrapolation errors) are minimal. However, usually one makes the extrapolation to $C = 0$ and determines the change in protein stability, $\Delta(F_N - F_u)$, the change in the barrier height, $\Delta(F^\# - F_u)$, and the Φ_f value referring to pure water rather than to the denaturant concentration C_0.

The folding (u $\rightarrow$ N) rate is determined (see eq. (20.1)) by the free energy difference between the transition state (#) and the initial unfolded (u) state of the protein (i.e., by the $F^{\#} - F_u$ value).

The stability of the native (N) relative to the unfolded (u) form of the protein is determined by their free energy difference, $F_N - F_u$, which is in turn determined (see eq. (20.4)) by the ratio between the folding and unfolding rates. (These values depend on the ambient conditions. Since it is in-water folding that is usually of interest, the rates of protein unfolding (and folding) are usually extrapolated to zero denaturant concentration. It should be noted, though, that a smaller and therefore more accurate extrapolation (Fig. 20.5) is needed to determine the folding nucleus at mid-transition.)

Mutations influence both the folding rate and the stability of the protein. This provides an opportunity to use them in outlining the folding nucleus experimentally. The interpretation is done under the assumption that the rate-limiting step of protein folding is the nucleation of its native structure, and that the residues involved in the *folding nucleus* (the globular part of the transition state) are positioned there in the same way as in the native protein.

The nucleation mechanism is typical of the first-order phase transitions in conventional physics (such as crystal freezing), and therefore it is highly plausible that it should occur in protein folding: it is an "all-or-none" transition, which, as we know, is a microscopic analog of a first-order phase transition in macroscopic systems.

The experimental check of this assumption will be described shortly. First, I would like to tell you how one can interpret the mutation-induced shifts of the $F_N - F_u$ and $F^{\#} - F_u$ values to outline, under the above assumption, the residues involved in the nucleus.

If the residue's mutation changes the transition state stability value $F^{\#} - F_u$ and the native protein stability value $F_N - F_u$ *equally*, this means that the residue in question is involved in the folding nucleus and has there the same contacts and conformation as in the native protein.

If, on the contrary, the residue's mutation changes *only* the native protein stability value $F_N - F_u$, but *does not* change the folding rate (and thus does not change the transition state stability value $F^{\#} - F_u$), this means that the residue in question is *not* involved in the nucleus and comes to the native structure only after the rate-limiting step.

And, lastly, if the residue's mutation affects the transition state stability to a lesser degree (but with the same sign) than the native protein stability, this means that the residue in question either belongs to one of a few alternative folding nuclei or forms only a part of its native contacts within the nucleus (i.e., that this residue is at the surface of the nucleus).

This is how the folding nucleus is outlined (Fig. 20.6): one estimates the value

$$\Phi_f = \Delta(F^{\#} - F_u)/\Delta(F_N - F_u) \equiv \Delta \ln[k_{u \rightarrow N}]/\Delta \ln[k_{u \rightarrow N}/k_{N \rightarrow u}] \qquad (20.12)$$

for each mutated residue. If its Φ_f is close to 1, the residue is interpreted as participating in the folding nucleus; if its Φ_f is close to 0, as non-participating. If $0 < \Phi_f < 1$,

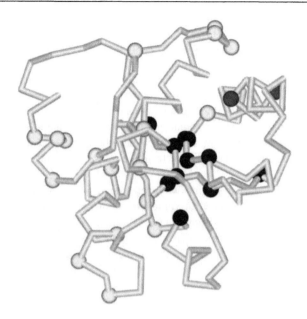

Figure 20.6. Folding nucleus for CheY protein according to López-Hernándes E., Serrano L., *Folding Design* (1996) **1**: 43–55. The residues studied experimentally are shown as beads against the background of the native chain fold. Dark blue beads show the residues involved in the nucleus, i.e., in the folded part of the transition state (they have $\Phi_f > 0.3$, i.e., each of them forms more then 30% of its native contacts there). Yellow beads show residues having $\Phi_f < 0.3$; they are not involved in the folded part of the transition state. Red beads show two residues that are difficult to interpret experimentally. They have such low $\Delta(F^{\#} - F_u)$ and $\Delta(F_N - F_u)$ values (the latter is more important, since this value is the denominator in eq. (20.11)), that the errors in their determination exceed the measured values themselves.

the residue is interpreted as either a surface residue of the nucleus, or as a member of one of a few alternative folding nuclei.

It is significant that only a negligible part of residues cannot be interpreted in such a way (and *this is experimental evidence for the nucleation folding mechanism*). In other words, only a negligible fraction of mutations have their Φ_f values beyond the range 0–1. That is, almost no mutations influence the folding rate only leaving stability unaffected (these would have $\Phi_f \gg 1$ or $\Phi_f \ll -1$), or stabilize the native fold but decrease the folding rate (these would have $\Phi_f < 0$), and even these exceptions are mostly connected with unreliably measured (too small) differences $\Delta(F^{\#} - F_u)$ and especially $\Delta(F_N - F_u)$.

This allows us to believe that the nucleation mechanism is basically correct, i.e., that the residues, if involved in the folding nucleus, are positioned there in the same way as in the native protein.

Inner voice: Actually, you are saying that any mutation affecting the protein folding rate affects its stability as well. But there are experiments showing the opposite . . .

Lecturer: All that I have said refers only to "two-state" transitions, i.e., to those without stable folding intermediates. If such an intermediate is present (which is possible only far from the mid-transition, see the upper left part of Fig. 20.4), the mutation may affect it without affecting either the final native structure or the initial coil (as, for example, mutation of a surface residue (in the native fold) can affect the stability of the intermediate molten (or pre-molten) globule, not to mention possible transient association of these intermediates, but affects neither the native nor the coil form stability). Then the effect you mentioned (which, incidentally, is always observed far from the transition point) becomes possible.

Figure 20.6 shows that in this way detected residues of the folding nucleus (i.e., residues having the highest Φ_f values, those of key importance for the protein folding) form a small compact region, a small compact folding nucleus positioned at the periphery rather than in the center of the protein globule.

A similar picture is observed in other proteins already studied experimentally (though not yet very many): usually, the folding nucleus is compact and does not coincide with the protein's hydrophobic core.

LECTURE 21

In this lecture we shall continue to discuss protein folding.

All the experimental data we discussed, though very interesting by themselves, cannot answer the main question as to how a protein manages to find its native structure among zillions of others within those minutes or seconds that are assigned for its folding.

And the number of alternative structures is vast indeed: it is at least 2^{100} but may be even 10^{100} for a 100-residue chain, because at least 2 (but more likely 10) conformations are possible for each residue. Since the chain cannot pass from one conformation to another faster than within a picosecond (the time of a thermal vibration), their exhaustive search would take at least $\sim 2^{100}$ picoseconds, or $\sim 10^{10}$ years. And it looks like the sampling has to be really exhaustive, as the protein can "feel" that it has come to the stable structure only when it hits it precisely, because even a 1 Å deviation can strongly increase the chain energy in the closely packed globule.

Then, how does the protein choose its native structure among zillions of possible others, asked Levinthal (who first noticed this paradox), and answered: It seems that the protein folding follows some specific pathway, and the native fold is simply the end of this pathway, no matter if it is the most stable chain fold or not. In other words, Levinthal suggested that the native protein structure is determined by kinetics rather than stability and corresponds to the easily accessible local rather than the global free energy minimum.

The difficulty of this problem is that it cannot be solved by direct experiment. Indeed, suppose that the protein has some structure that is more stable than the native one. How can we find it if the protein does not do so itself? Shall we wait for $\sim 10^{10}$ years?

On the other hand, the question as to whether the protein structure is controlled by kinetics or stability arises again and again when one has to solve practical problems of protein physics and engineering. For example, in predicting a protein's structure from its sequence, what should we look for? The most stable or the most rapidly folding structure? In designing a protein *de novo*, should we maximize the stability of the desired fold, or create a rapid pathway to this fold?

However, is there a real contradiction between "the most stable" and the "rapidly folding" structure? Maybe, the stable structure *automatically* forms a focus for the "rapid" folding pathways, and therefore it is *automatically* capable of fast folding.

Before considering these questions, i.e., before considering the *kinetic* aspects of protein folding, let us recall some basic facts concerning protein *thermodynamics* (I will talk about single-domain proteins only, i.e., chains of 50–200 residues). These facts will help us to understand what chains and what folding conditions we have to consider. The facts are as follows:

1. Protein unfolding is reversible, and it occurs as an "all-or-none" transition. The latter means that only two states of the protein molecule, native and denatured, are present (close to the denaturation point) in a visible quantity, while others, semi-native or misfolded, are virtually absent. Such a transition requires an amino acid sequence that provides a large energy gap between the most stable structure and the bulk of misfolded ones.
2. The denatured state, at least that of small proteins unfolded by a strong denaturant, is often the random coil.
3. Even under normal physiological conditions the native state of a protein is only more stable than its unfolded state by a few kilocalories per mole (and these states have equal stability at mid-transition, naturally). The native structure is stable because of its low energy, i.e., because of strong interactions within this structure, and the coil state is stable because of its high entropy, that is, because of the vast number of unfolded conformations. (It is essential that you note that, as is usual in the literature, the term "entropy" here means this transition only for the conformational entropy: this "entropy" does not include the solvent entropy; and the term "energy" here means "free energy of interactions" since, for example, the hydrophobic and other solvent-mediated forces are connected with the solvent entropy. This terminology is commonly used to concentrate on the main problem of sampling the protein chain conformations.)

Thus, to solve the "Levinthal paradox" and to show that the most stable chain fold can be found within a reasonable time, we could, to a first approximation, consider only the rate of the "all-or-none" transition between the coil and the most stable structure. And we may consider this transition only for the case when the most stable fold is as stable as (or only a little more stable than) the coil, all other forms of the chain being unstable, i.e., close to the "all-or-none" transition midpoint. Here the analysis can be made in the simplest form, without accounting for accumulating intermediates. True, the maximum folding rate is achieved when the native fold is much more stable than the coil, and then observable intermediates often arise. But let us first consider the situation when the folding is not the fastest but the simplest...

Since the "all-or-none" transition requires a large energy gap between the most stable structure and misfolded ones (Fig. 21.1), we will assume that the considered amino acid sequence provides such a gap. I am going to show you that the "gap condition" provides a rapid folding pathway to the global energy minimum, to estimate

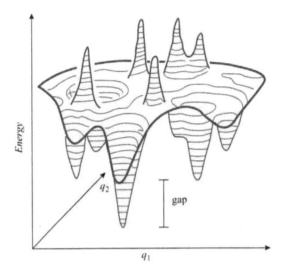

Figure 21.1. Schematic representation of the bumpy energy landscape of a protein chain. Only two coordinates (q_1 and q_2) can be shown in the drawing, while the protein chain conformation is determined by hundreds of coordinates. A wide energy gap between the global and other energy minima is necessary to provide the "all-or-none" type of decay of the stable protein structure. This, in turn, makes the protein function specific: like a light bulb, either the protein works or it does not.

the rate of folding and to prove that the most stable structure of a normal size domain can fold within seconds or minutes.

To prove that the most stable chain structure is capable of rapid folding, it is sufficient to prove that at least one rapid folding pathway leads to this structure. Additional pathways can only accelerate the folding since the rates of parallel reactions are additive. (One can imagine water leaking from a full to an empty pool through cracks in the wall between them: when the cracks cannot absorb all the water, each additional crack accelerates filling of the empty pool. And, by definition of the "all-or-none" transition, all semi- and mis-folded forms together are too unstable to absorb a significant fraction of the folding chains and trap them.)

To be rapid, the pathway must consist of not too many steps, and most importantly, it must not require overcoming of a too high free energy barrier. An N-residue chain can attain its lowest-energy fold in N steps, each adding one fixed residue to the growing structure (Fig. 21.2). *If* the free energy went downhill along the entire pathway, a 100-residue chain would fold in $\sim$ 100–1000 ns, since the growth of a structure (e.g., an α-helix) by one residue is known to take a few nanoseconds.

Protein folding takes seconds or minutes rather than a microsecond because of the free energy barrier, since most of the folding time is spent on climbing up this barrier and falling back, rather than on moving along the folding pathway.

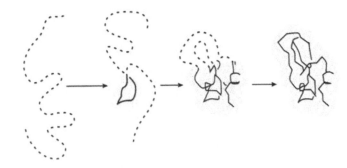

Figure 21.2. Sequential folding pathway. At each step one residue leaves the coil and takes its final position in the lowest-energy structure. The folded part (shaded) is compact and native-like. The bold line shows the backbone fixed in the already folded part; the fixed side chains are not shown for the sake of simplicity (the volume that they occupy is shaded). The broken line shows the still unfolded chain.

You should remember that, according to conventional transition state theory, the time of the process is estimated as

$$TIME \sim \tau \times \exp(+\Delta F^{\#}/RT) \tag{21.1}$$

where τ is the time of one step, and $\Delta F^{\#}$ the height of the free energy barrier. For protein folding, τ is about 1–10 ns (according to the experimentally measured time of the growth of an α-helix by one residue).

As for $\Delta F^{\#}$, this is our main question: how high is the free energy barrier $F^{\#}$ on the pathway leading to the lowest-energy structure? Folding of this structure decreases both the chain entropy (because of an increase in the chain's ordering) and its energy (because of the formation of contacts stabilizing the lowest-energy fold). The former increases and the latter decreases the free energy of the chain.

If the fold-stabilizing contacts start to arise only when the chain comes very close to its final structure (i.e., if the chain has to lose almost all its entropy *before* the energy starts to decrease), the initial free energy increase would form a very high free energy barrier (proportional to the *total* chain entropy lost). The Levinthal paradox claiming that the lowest-energy fold cannot be found within any reasonable time since this involves exhaustive sampling of all chain conformations originates exactly from this picture (loss of the entire entropy *before* the energy gain).

However, this paradox can be avoided if there is a folding pathway where the entropy decrease is immediately or nearly immediately compensated for by the energy decrease.

Let us consider a *sequential* (Fig. 21.2) folding pathway. At each step of this process, one residue leaves the coil and takes its final position in the lowest-energy 3D structure.

Inner voice: This pathway looks a bit artificial ... How can the residue know its position in the lowest-energy fold?

Lecturer: As I told, we have to trace *one* pathway leading to the native fold. And the outlined pathway does so. As for the impression of its being artificial: look, it is exactly the pathway that you expect to see watching the movie on unfolding, but in the opposite direction.

Inner voice: And why cannot the protein use one way to unfold and quite another to fold?

Lecturer: It cannot, since the direct and reverse reactions must follow the same pathway(s) under the same conditions (and we have already agreed to consider the mid-point of the folding–unfolding equilibrium). This is the *detailed balance* law of physics. If the direct and reverse reactions followed different pathways, a perpetual circular flow would arise. And you could use it to rotate a small turbine. That is, your suggestion would lead to a device (called a *perpetuum mobile* of the second order) that converts surrounding heat into work. And, as you know, or should know, the second law of thermodynamics, the law of maximum possible entropy, states that such a *perpetuum mobile* is impossible. (Being more specific: the direct and reverse reactions must follow the same pathway under the same conditions; under different conditions the pathways can be different, of course. That is, folding in water need not follow the same pathway as unfolding in concentrated denaturant; but at the same denaturant concentration (e.g., at mid-transition) they must use the same pathway.)

Inner voice: Still, a movie about explosion of a building, even watched in the opposite direction, is quite different from a movie about building construction...

Lecturer: Both construction and explosion proceed at the expense of a huge energy (or, to be more precise, free energy): fuel, manpower, explosives... On the contrary, protein folding and unfolding do not consume any "fuel", and, as the chevron plots show, they can occur around the equilibrium point; this fact (I stressed it many times) is very significant for an understanding of protein folding. Here, near the equilibrium point, the free energy difference between the folded and unfolded states (and thus the free energy expenditure in the course of both processes) is about $k_B T$. And, at the very mid-transition (i.e., the point where unfolding is in equilibrium with folding and the free energy difference is completely absent), these processes occur simultaneously under the same ambient conditions, and here the direct and reverse reactions must follow exactly the same pathway.

Thus, let us consider the energy change ΔE, the entropy change ΔS and the resultant free energy change $\Delta F = \Delta E - T \Delta S$ along the *sequential* (Fig. 21.2) folding pathway.

When a piece of the final globule grows sequentially, the interactions that stabilize the final fold are restored sequentially as well. If the folded piece remains compact, as in Fig. 21.2, the number of restored interactions grows (and their total energy decreases) approximately in proportion to the number n of residues that have taken their final positions (Fig. 21.3a).

At the beginning of folding, though, the energy decrease is a little slower, since the contact of a newly joined residue with the surface of a small globule is, on average,

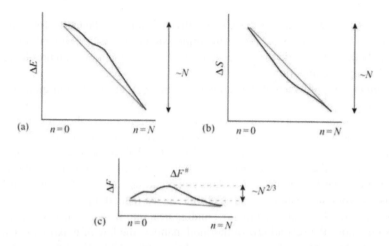

Figure 21.3. The change of energy (a), entropy (b) and free energy (c) along the sequential folding pathway close to the point of thermodynamic equilibrium between the coil ($n = 0$) and the final structure ($n = N$: all the N chain residues are folded). The full energy and entropy changes, $\Delta E(N)$ and $\Delta S(N)$, are approximately proportional to N. The blue lines show the linear (proportional to n) parts of $\Delta E(n)$ and $\Delta S(n)$. The non-linear parts of $\Delta E(n)$ and $\Delta S(n)$ result mainly from the boundary between the folded and unfolded halves of the molecule. The maximum deviations of the $\Delta E(n)$ and $\Delta S(n)$ values from linear dependence is only $\sim N^{2/3}$. As a result, the $\Delta F(n) = \Delta E(n) - T\Delta S(n)$ value also deviates from linear dependence (the blue line) by a value of $\sim N^{2/3}$. Thus, at the equilibrium point (where $\Delta F(0) = \Delta F(N)$) the maximum free energy $\Delta F^{\#}$ along the pathway is proportional to $N^{2/3}$ as well.

smaller than its contact with the surface of a large globule. This results in a non-linear *surface* term (proportional to $n^{2/3}$) in the energy ΔE of the growing globule. Thus, the maximum deviation from a linear energy decrease is proportional to $N^{2/3}$, while the total energy decrease is proportional to the total number N of residues.

The entropy decrease is also approximately proportional to the number of residues that have taken their final positions (Fig. 21.3b). At the beginning of folding, though, the entropy decrease can be a little faster owing to disordered but closed loops protruding from the growing globule (Figs 21.2 and 21.4). Their number is proportional to the interface between the folded and unfolded phases, and the free energy of a loop is known to have a very slow, logarithmic dependence on its length. This again results in a non-linear *surface* term in the entropy ΔS of the growing globule, and the maximum deviation from the linear entropy decrease again is proportional to $N^{2/3}$, while the entropy decrease itself is proportional to N.

Both linear and surface constituents of ΔS and ΔE enter the free energy $\Delta F = \Delta E - T\Delta S$ of the growing globule. However, when the final globule is in thermodynamic equilibrium with the coil, the large linear terms *annihilate* each other in the difference $\Delta E - T\Delta S$ (since $\Delta F = 0$ both in the coil (i.e., at $n = 0$) and in the final globule (at $n = N$)), and only the surface terms remain: $\Delta F(n)$ would be *zero* all along the pathway in the absence of surface terms.

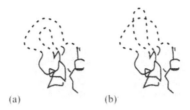

(a)   (b)

Figure 21.4. (a) Compact semi-folded intermediate with protruding unfolded loops. Its growth corresponds to a shift of the boundary between the folded (globular) and unfolded parts. Successful folding requires correct knotting of loops: the structure with incorrect knotting (b) cannot change directly to the correct final structure: it first has to unfold and achieve the correct knotting. However, since a chain of ~ 100 residues can only form one or two knots, the search for correct knotting can only slow down the folding two-fold or at most four-fold; thus, the search for correct chain knotting does not limit the folding rate of normal size protein chains.

Thus, the free energy barrier is connected *only* with the relatively small surface effects at the coil–globule boundary, and the height of this barrier on a sequential folding pathway (Fig. 21.2) is proportional *not to N* (as Levinthal's estimate implies), but to $N^{2/3}$ only (Fig. 21.3c).

As a result, the time of folding of the most stable chain structure grows with the number of the chain residues N *not* "according to Levinthal" (i.e., not as 2^N, or 10^N, or any exponent of N), but as $\exp(\lambda N^{2/3})$ only. The value $N^{2/3}$ arises from the separation of the "native" and "unfolded" phases, and it is much smaller than N. A thorough estimate of the coefficient λ shows that $\lambda = 1 \pm 0.5$, the particular value depending on the distribution of strongly and weakly attracted residues within the lowest-energy structure, and in the main, on the *topology* of the lowest-energy structure (λ is large when this structure is such that its folding requires intermediates with many closed loops protruding from the native-like part, and λ is small if such loops are not required).

The observed protein folding times (for the coil → native globule transition at the point of equilibrium between these two states) are indeed (Fig. 21.5) in the range $10 \times \exp(0.5N^{2/3})$ ns to $10 \times \exp(1.5N^{2/3})$ ns, in accordance with the estimate obtained.

The reason for the "non-Levinthal" estimate obtained,

$$TIME \sim \exp[(1 \pm 0.5)N^{2/3}]\,\text{ns} \tag{21.2}$$

is that the entropy decrease is almost immediately compensated for by the energy gain along the sequential folding pathway, and the free energy barrier occurs owing to the surface effects only.

It is noteworthy that the sequential folding pathway does not require any rearrangement of the globular part (which could take a lot of time): all rearrangements occur in the coil and therefore are rapid.

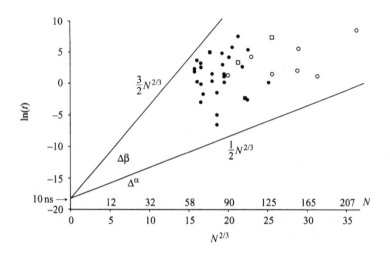

Figure 21.5. Observed folding time (t sec) at the point of equilibrium between the unfolded and the native states, vs $N^{2/3}$ (N being the number of residues in the chain). The circles and squares refer to all 36 proteins listed in Jackson S.E., Folding Design (1998) **3**: R81–R91, whose folding time at the mid-transition can be calculated from the available data. Circles, proteins without S—S bonds; squares, with S—S bonds. Filled symbols, no stable folding intermediates are observed; open symbols, the intermediates are observed (but far from the mid-transition). The triangles refer to α-helical and β-hairpin peptides (Thompson P.A., Eaton W.A., Hofrichter J., Biochemistry (1997) **36**: 9200; Muñoz V., Thompson P.A., Hofrichter J., Eaton W.A., Nature (1997) **390**: 196). The theoretically predicted region of ln(t) is limited by the lines $\ln(10^{-8}\,\text{s}) + 0.5N^{2/3}$ and $\ln(10^{-8}\,\text{s}) + 1.5N^{2/3}$. As seen, all the experimental points are within this range (except for that of the α-helix, which is a one-dimensional rather than three-dimensional object). Adapted from Galzitskaya O.V., Ivankov D.N., Finkelstein A.V., FEBS Lett. (2001) **489**: 113–118.

Anyhow, the estimate obtained, eq. (21.2), illustrated by Fig. 21.5 shows that a chain of 100 residues will find its most stable fold within minutes even near the mid-transition. It also explains why a large protein should consist (according to the "divide and rule" principle) of separately folding domains ("foldons"): otherwise, chains of more than 300 residues would fold too slowly.

Four more things remain to be said:

1. Having found the free energy of the transition state (Fig. 21.3), one can further estimate the size of its globular part. This estimate shows that the nucleus must be as large as half the protein. This is compatible with experiment (Fig. 21.6) which shows a crudely equal (but of opposite sign) dependence of the folding and unfolding rates on the denaturant concentration. This means that the solvent-accessible area of the chain in the transition state is between the solvent-accessible areas of the unfolded and the globular states. Such a large folding nucleus implies that there cannot be many alternative folding nuclei, which means that consideration of only one folding pathway can give a rather accurate estimate of the folding time.

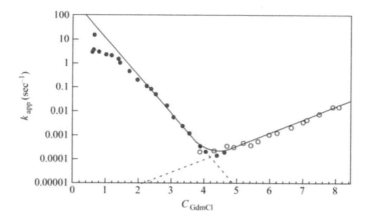

Figure 21.6. Refolding and unfolding rates of hen egg white lysozyme vs. the concentration of guanidine dihydrochloride. The experimental points show the apparent rate k_{app} of approach to the equilibrium between the native and denatured forms of the protein. The filled circles correspond to refolding obtained upon diluting a concentrated GdmCl solution of the denatured protein. The open circles correspond to unfolding occurring after GdmCl addition to aqueous solution of the native protein. Adapted from Kiefhaber T., *Proc. Natl. Acad. Sci. USA* (1995) **92**: 9029–9033.

2. Our estimate, eq. (21.2), refers to the point of equilibrium between the unfolded and the native states where the observed folding time is at a maximum and can exceed by orders of magnitude the folding time under native conditions (Fig. 21.6); the latter depends on the mid-transition folding time and on the native (relative to the unfolded) state stability under the native conditions.
3. The influence of protein chain topology upon the folding time (the factor (1 ± 0.5) in eq. (21.2)) can be estimated using a phenomenological "contact order" parameter (CO%) of Baker and Plaxco. CO% is equal to the average chain separation of the residues that are in contact in the native protein fold, divided by the chain length. A high CO% value reflects the existence of many long closed loops in the protein fold; therefore, CO% is approximately proportional to the value of λ.
4. The "quasi-Levinthal" search over intermediates with different chain knotting (Fig. 21.4) can, in principle, be a rate-limiting factor, since knotting cannot be changed without a decay of the globular part. However, since the computer experiments show that one knot involves about a hundred residues, the search for knotting can only be important for extremely long chains, which cannot fold rapidly (according to eq. (21.2)) in any case.

So far, we have only considered the folding rate close to the mid-point of the folding phase transition, when only one fold (the "native") competes with the coil, and all other globular forms, even taken together, are unstable (Fig. 21.7a).

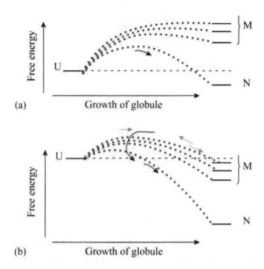

Figure 21.7. Folding under different conditions. Thin lines show the free energies of the unfolded chain (U), of the most stable ("native") fold (N), and of the mis- or semi-folded structures (M). Dotted lines schematically show the behavior of the free energy along the folding pathways leading to different structures (ignoring small irregularities); their maxima correspond to the transition states on these pathways. The main folding scenarios are given. (a) Fold N is more stable than the coil U, and the coil is more stable than all misfolded states M taken together. Misfolding and kinetic traps do not hinder rapid folding to the native state. The chain virtually does not explore the pathways leading to misfolded states, since on these pathways the free energy rises too high. (b) Both native and some of the mis- folded folds are more stable than the coil (which is the phase where a fast rearrangement occurs). The chain rapidly comes to some misfolded form and then slowly undergoes tran- sition to the stable state N via the partly unfolded state. Arrows show the mainstream of the folding process. Adapted from: Finkelstein A.V., Badretdinov A.Ya., *Folding Design* (1997) **2**: 115–121.

How will the folding rate change when the native state becomes more stable then the coil? (The opposite case is not interesting: then no folding will happen, and that's all.)

The initial growth of native fold stability has to increase the folding rate, since the transition state is stabilized as well, and the competing misfolded structures are still unstable (relative to the coil) owing to the energy gap between them and the most stable fold (Fig. 21.7a). This acceleration is indeed observed (see the left chevron limb in Fig. 21.6). However, the acceleration proceeds up to a certain limit only (see the plateau at the top of the left chevron limb in Fig. 21.6). It seems that the maximum folding rate is achieved when the "misfolded" states become as stable as the unfolded state. After that, the further increase in stability of the folded states leads to a rapid misfolding followed by a slow conversion into the native state that occurs via the unfolded state (Fig. 21.7b).

Actually, all this resembles crystallization: at the freezing temperature a perfect large single crystal (the lowest-energy structure) arises, although extremely slowly. As the temperature is lowered a little the single crystal grows faster; and a further temperature decrease leads to rapid formation of a multitude of small crystals rather than of a perfect large single crystal.

I should like to mention that the similarity of the thermodynamic aspects of protein folding to those of first-order phase transitions has been experimentally proved by Privalov, and the similarity of the kinetic aspects of these two events has been outlined by the experiments of Fersht and his co-workers and by the computer experiments of Shakhnovich and Gutin.

Shakhnovich and his team used simple computer models of protein chains to explore the region of fastest folding. They showed that the folding time grows with the chain length N in this region much more slowly than predicted by eq. (21.2) for the mid-transition. Specifically, in this region, the folding time grows with N not as $\exp(N^{2/3})$ but as N^6 for "random" chains and as N^4 for the chains selected to fold most rapidly (i.e., having a large energy gap between the most stable fold and the other ones). This emphasizes once again the dependence of the folding rate on the experimental conditions and on the difference in stability between the lowest-energy fold and its competitors.

It would be only natural if here you asked me the question as to what happens if two structures instead of one are separated from others by a large energy gap.

The answer is: If these two structures are stable relative to the denatured state, the one with the better folding pathway (with a lower barrier) will be the first to fold. However, if this "more rapidly folded structure" is even a little less stable than the other one, a very slow transition to the more stable structure will follow. The transition will be slow since unfolding of the first stable structure will be required (Fig. 21.7b). This transition is similar to a polymorphous transition in crystals (recall the "tin disease", i.e., the transition of white tin into gray tin: in the sixteenth to eighteenth centuries this "disease" is known to have destroyed whole stores of tin buttons during unusually severe frosts). It seems, though, that "polymorphous" proteins must be rare: as I told you, theoretical estimates show that the amino acid sequence coding one stable chain fold is a kind of wonder, but the sequence coding two of them is a squared wonder . . .

There is evidence, though, for polymorphous transitions in some proteins. I mean serpins (a family of **ser**ine **p**rotease **in**hibitors) and prions (the "mad cow disease" proteins, in particular).

In the last case, the polymorphous transition is coupled with association of the transformed (infected) prion globules, which either facilitates or is facilitated by the new, β-structural form of the protein. Anyway, the association enlarges a folding unit, and this, according to the estimates given above, can make the initiation of a new phase extremely slow. This may underlie the extremely long incubation periods of prion diseases like scrapie or mad cow disease. However, when the nucleus of the new phase has been formed, subsequent transitions of other prion molecules to the "infected" form is fast (just as in the "tin disease" case).

However, there is a more simple, "peaceful", and well-studied example of the same kind. The water-soluble polypeptide poly(Lys), at pH > 10 and 20–50 °C, undergoes a very fast coil to α-helix transition followed by a very slow conversion of the α-helix into the β-structure. The latter can be accompanied by association and can take hours or weeks; we have already discussed it when considering the kinetics of β-structure formation.

In conclusion, let us return to the energy landscape (Fig. 21.1) again to visualize the fast folding pathways that automatically lead to the global energy minimum, provided it is much deeper than other energy minima.

Each low-energy structure, each energy minimum is surrounded by a rosette of more or less smooth tracks going to the minimum through the hilly, even rocky landscape (the "rocks" correspond to powerful but short-range repulsions at collisions of chain fragments). Some of these tracks correspond to the sequential (Fig. 21.2) pathways of folding to the given energy minimum. No rearrangements of the already folded part of the globule take place on the sequential pathways, and therefore they have no large pits and bumps (and small ones pose no obstacle when the temperature is not too low). When moving along these pathways to the energy minimum, the molecule gains energy but loses entropy (Fig. 21.3), since its structure becomes more and more fixed. The deeper the energy minimum, the greater the energy gain – and the easier it is to overcome the "entropic resistance". The strength of this resistance is proportional to temperature. At too high a temperature the resistance is strong, and it does not allow the chain to fold into any fixed structure. At too low a temperature the resistance is too small, and the chain falls in any neighboring energy well; then it spends a lot of time to get out of it just to fall in a local minimum again – and thus it cannot approach the global energy minimum for a very long time (Fig. 21.7b). And at last, when the temperature is good for folding, the entropic resistance is overcome only on the pathways to the deepest energy minimum (which is much deeper than other minima: just recall the "energy gap"), and the chain gets there rapidly (Fig. 21.7a).

The scheme given above of entropy-by-energy compensation along the folding pathway and the conclusion that it can solve the Levinthal paradox are applicable to formation of the native protein structure not only from the coil but also from the molten globule or from another intermediate. However, for these scenarios, all the estimates would be much more cumbersome, while these processes do not show (experimentally) any drastic advantage in the folding rate. Therefore, I will not go beyond the simplest case of the coil-to-native globule transition.

All the above emphasizes the significance of development of the "new view" on folding, the view based on the nature of the energy surfaces or "landscapes" that define the folding reaction rather than on discrete accumulation of folding intermediates. This process differs from the more familiar reactions of small molecules by the following: (1) a particular protein structure is the result of many weak interactions; (2) a large change in configurational entropy occurs during folding; (3) the native and denatured phases separate during folding.

One of the most satisfying features of recent conceptual advances is the emergence of a "unified mechanism" of folding and the realization that apparently disparate patterns of experimental behavior result from the interplay of different contributions to the overall free energy under different circumstances. This unification of ideas provides a framework for joint work by experimentalists and theoreticians in order to obtain a more detailed understanding of the intricate folding process.

In particular, special attention is now paid to the link between the fundamental process of folding and the multiplicity of folding protein states that can exist in a cellular system, which leads to the need to consider their intermolecular as well as intramolecular interactions, such as the interaction of the folding nuclei with chaperones or other proteins, RNA and DNA.

The understanding of protein folding is much more than just a challenge to the skills and intellects of protein scientists. Folding (or self-organization) represents a bridge between the laws of physics and the results of the evolutionary pressure under which the character of biology has developed. For many years we could only discuss this issue in very general terms, emphasizing the biological importance of this complex physical process. Now, we are beginning to understand that folding is a physical process taking place only in systems that have been biologically selected to ensure its efficiency.

Protein physics is grounded on two fundamental experimental facts: (1) protein chains are capable of forming their native structures spontaneously in the appropriate environment, and (2) the native state is separated from the unfolded state of the chain by an "all-or-none" phase transition. The latter ensures the robustness of protein action and minimal populations of partially folded and therefore aggregation-prone species.

It appears that biological evolution selects only those sequences that fold into a well-defined three-dimensional native structure. This is necessary for the protein to work reliably. Such a well-defined structure can be formed only by a sequence with a free energy that is much lower than that of alternative structures. And this is precisely the sequence that is capable of "all-or-none" folding and unfolding. The enhanced stability of the native structure seems to be due to its tight packing (even though it is not yet clear exactly what constraint is placed on protein chains by their capability of tight packing). Interestingly, these selected sequences appear to meet the requirement of correct folding into the stable structure simultaneously with the ability to fold quickly. It is probable that the latter helps incompletely folded polypeptide chains to avoid the competing process of intermolecular aggregation. The ability of stable structures to fold quickly solves the long-standing contradiction between the kinetic and thermodynamic choice of the native fold.

The sequences of globular proteins look rather "random", and their secondary and tertiary structures bear many features typical of all more or less stable folds of random co-polymers. However, the ability of polypeptides to fold rapidly and reproducibly to definite tertiary structures is not a characteristic feature of random sequences. The main feature of "protein-like" amino acid sequences, the feature that determines all their physical properties, is the enhanced stability of their native fold, i.e., the existence of a large gap between the energy of the native fold and the minimum energy of misfolded globular structures. Although the size of this energy gap is not yet known, some theoretical estimates show that even a rather narrow gap (of a few kilocalories per mole) can cause "protein-like" behavior of a protein domain. And it is already clear that the necessary reinforcement of the lowest-energy fold can be achieved by a gradual evolutionary selection among many random point mutations.

Protein folding is perhaps the simplest example of a biological morphogenesis that can only take place in a system evolutionarily designed to allow it to happen. Nevertheless, protein folding has the typical physical characteristics associated with a complex process obeying the laws of statistical mechanics. In particular, folding of globular proteins involves a nucleation mechanism generally typical for "all-or-none" (i.e., first-order) phase transitions.

Folding pathways seem to be not unique: various pathways can lead to one target, although their rates may be different and may depend on the folding conditions. Folding is the result of statistical fluctuations within the unfolded protein chain, which (under appropriate temperature and solvent conditions) result in its transition to the native state, the structure of the lowest free energy. In this way a single well-defined structure can emerge from the statistical ensemble of unfolded or partially folded species.

All this once again emphasizes a connection between protein structure, stability and ability to fold spontaneously and the role of natural selection in the introduction and maintenance of these qualities.

It is easy to imagine and it is possible to show by computer simulations (but much more difficult to prove experimentally) that selection could start from the slightly increased stability of some structure of a polypeptide coded by a "random" piece of DNA, provided this structure can do something even marginally useful for a cell. And that then the pressure of evolution results, by random mutations and selection of the more and more reliably folding sequences, in the emergence of a sequence with the rare quality of spontaneously folding into a definite fold, which also allows specificity of function to be achieved, by additional selection, in a manner susceptible to feedback control and regulation in a complex and crowded cellular environment.

However, this way of originating new proteins, origination from random sequences, seems to play no role at observable (i.e., not the earliest) stages of bio-logical evolution. As I have already mentioned, it is commonly accepted that now "new" proteins originate in two ways. One is gene duplication, which allows the "old" protein to keep working, while the sequence of the "new" one is free to mutate and to drift (with the aid of selection) towards some "new" function. The other is the merging of separate domains and small proteins into a multi-domain protein capable of performing more complicated functions and thus more susceptible to reg-ulation because of the physical interactions of these domains. It is worth mentioning that multi-domain proteins are more typical of higher organisms than of bacteria and unicellular eukaryotes.

However, this does not revoke the privilege of some ("defect-free") protein archi-tectures, whose stability is compatible with a greater variety of sequences to ensure more freedom for evolution and selection. Of course, selection is capable of creating even most improbable structures if they give an advantage to a species (the eye and brain being examples). However, protein function has only a little connection with its architecture, as we had seen and will see again. Therefore, it is not really surprising (although it is remarkable) that protein structures often look like those to be expected for stable folds of random sequences; that is, the "multitude principle" still seems to work in biological evolution.

Part VI

PREDICTION AND DESIGN OF PROTEIN STRUCTURE

LECTURE 22

When it became known that the amino acid sequence of a protein chain determines its 3D structure, the "protein structure prediction problem" arose.

Prediction of protein structures from their amino acid sequences is interesting for two reasons: intellectual and practical.

Intellectual interest in the protein structure prediction problem is aroused not only because it is a challenge (can we do it or not?) but mostly, I think, because we well remember the very great importance of the DNA structure prediction for molecular biology as a whole.

The practical interest is obvious. Experimentally, it is much more difficult to determine a 3D protein structure than an amino acid sequence. The flow of new sequences is immense: the genome projects produce hundreds of sequences a day. Now (in 2001; all the numbers given double every two years) the number of known amino acid sequences approaches a million. Many of them are only read from DNA or RNA, and the functions of these proteins are not yet established experimentally. The flow of new 3D structures is less by two orders of magnitude "only" many thousands of them have been already established. And any understanding of the protein action mechanism, the search for protein's inhibitors and activators (i.e., the search for potential drugs) – all this requires a knowledge of the protein's 3D structure. Thus, any tips on the 3D folds from the sequences are valuable.

What can be said about the 3D structures of "new" proteins (not yet studied by NMR or X-ray) when only their amino acid sequences are known?

The first thought is to predict the 3D structure of a "new" sequence on the basis of its generic similarity ("homology") to a sequence of some "old" (X-ray- or NMR-solved) protein.

Establishing sequence homology (Fig. 22.1) is indeed a very powerful method for elucidating structural and functional similarity (and not only for proteins – for DNA fragments and RNAs as well; but I will speak on proteins only).

It may be worth mentioning that the term "homology of sequences" is somewhat ambiguous. Here is an illustrative example. Suppose, one additional base is included in the protein gene just after the initiation codon. Then the new and the old DNA (and RNA) sequences are virtually the same; one will find them highly "homologous" (which is correct, in the strict sense of

	1	10	20
Human, chimpanzee		**GDVEKGKK**I F**IMKCSQCHTV**...	
Pig, bovine, sheep		**GDVEKGKK**I F<u>V</u>Q **KCAQCHTV**...	
Chicken, turkey		**GD**<u>I</u>**VEKGKK**I<u>V</u>Q **KCSQCHTV**...	
Puget sound dogfish		**GDVEKGKK**<u>V</u>**F**<u>V</u>Q **KCAQCHTV**...	
Screw-worm fly		GV P A**GDVEKGKK**I F<u>V</u>Q <u>R</u>**CAQCHTV**...	
Rust fungus		G F E**DGDAKKGA**<u>R</u>I FK T <u>R</u>**CAQCHT**<u>L</u>...	
Rape, cauliflower		ASFDE A P P**GNSKAGE**K I FK T **KCAQCHTV**...	

Figure 22.1. Homologous amino acid sequences of the N-terminal fragments of cytochrome c from various eukaryotic mitochondria and chloroplasts. The sequence alignment is taken from [4]. Bold letters indicate residues identical to the residue having the same position in the human protein; the residues similar to human protein residues are underlined. Proteins with such a high similarity of sequences (if it is observed throughout the chain) are known to have nearly identical 3D structures.

the word "homology", since their origin is the same). At the same time, the new and the old protein sequences are completely different due to the frame shift, one will find no "homology" between them (which is, strictly speaking, a wrong conclusion because they have the same genetic origin; but this conclusion is instrumental as far as it simply means that amino acid sequences (and 3D folds) of the proteins have nothing in common).

Further on, I will use the term "homology" in a narrow sense, in the sense of similarity of amino acid sequences only, having in mind that our aim is to use "homology" in finding similarity between 3D structures.

Experience tells us that even a moderate sequence similarity is sufficient for a high similarity of 3D structures (Fig. 22.2). It is often said that the 3D structure is much more conservative than the sequence (or that "coding of the 3D structure by sequence is degenerate"). Establishing sequence homology only leads to a straightforward and precise reconstruction of 3D structures when the sequence similarity is high enough (this case is illustrated in Fig. 22.1).

The case when a family of sequences has diverged significantly is more difficult (Fig. 22.3). The sequence similarity is not obvious now, and finding the correct alignment (that is, finding residues equally positioned in the common 3D structure) is far from easy.

Usually, sequence alignment is done by computer. There are many programs searching for homology and alignment; one can use them via the Internet. The most popular programs are "BLAST", "PSI-BLAST", "HMMer", and "Smith-Waterman". All of them align sequences trying to achieve the highest similarity of matched sequence regions. To this end, sequence gaps are often introduced (see "–" in Fig. 22.3). The program estimates the similarity of aligned sequences and reports on (1) whether they are homologous (i.e., whether they are genetically related), and (2) what the best alignment of the sequences looks like. It should be noted that even the correctly established homology (in the case of a low sequence similarity) could not guarantee that the alignment found is structurally correct (see below).

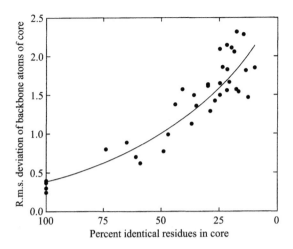

Figure 22.2. The root mean square deviation (Å) between the positions of equivalent backbone atoms in superposed 3D folds of homologous proteins vs. the amino acid identity of residues that occupy the equivalent positions. The equivalent positions include more than 90% of protein chain residues when the identity exceeds 40%, but can be as low as about 50% of the residues when the identity is as small as 10–25%. The exponential curve that shows a general tendency is fitted to the experimental points. The initial variation (at 100% identity) in the r.m.s. deviation reflects small variations of the 3D structures obtained under different conditions or/and by different refinement procedures for the same proteins. The plot reflects conservation of homologous 3D protein structures up to a rather low sequence identity. Reproduced from: Lesk A.M., Chothia C. "The response of protein structures to amino acid sequence changes." (Fig. 2). *Philos. Trans. R. Soc. London* (1986) **A317**: 345–356 (University Press, Cambridge), with permission.

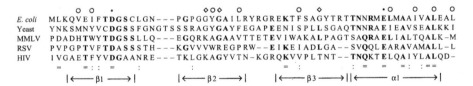

Figure 22.3. Amino acid sequences of the N-terminal halves of ribonucleases H from bacteria (*E. coli*), eukaryote (yeast), and three viruses. In this case the alignment was done ("by hand", without a computer) so as to rule out sequence gaps ("– – –") within inner regions of α- and β- structures (these are shown in the very bottom). The residues that are identical in three or more sequences are shown in bold. The active site residues are marked with black dots above the sequences. The open circles and rhombs (above the sequences) indicate the residues involved in two hydrophobic cores of the protein. Below the alignment, I have indicated the identical (=) and similar (:) residues of the RSV and HIV sequences. Adapted from [10].

Various programs have various estimates as to the cost of identity of compared residues, the cost of their similarity and mismatch, and the penalties for gap initiation, elongation and termination. The costs of matches, mismatches and similarities are derived from the statistics of amino acid substitutions in evident homologs, while the gap penalties are fitted using various methods and criteria to have a satisfactory recognition of already known homologous proteins (and discrimination of non-homologous ones). Using a program, inexperienced people ("users") often have only a general idea as to what is "good" and "bad" according to the program, and express the result as follows: "the homology level found is 25%", based on the information that 25% of aligned residues are identical.

Does a 25% coincidence of residues mean that the sequences are homologous? Generally speaking, this depends on the length of the sequence alignment (the longer, the more probable), on the number and the length of gaps introduced to align the sequences (the more, the less probable) and on other details. Good programs normally report a probability of finding the observed similarity for randomly chosen sequences. However, it is always instructive to compare obviously *non*-homologous sequences (Fig. 22.4) using the same program. Then one can see that such sequences (especially the short ones) often have a "homology level" of 15–20% and even 25% (especially with gaps introduced into the sequences to match the most similar regions).

These values are somewhat different for different programs. Experience, however, leads to the following conclusions. If a "good" (according to common opinion) program reports a 30–35% sequence identity, this means that homology of the sequences is reliable (with the proviso that the alignment includes at least 50, better 100, residue pairs). Another reservation: at the 30–35% homology level, only 70–80% of the sequence alignment obtained correctly reproduces the alignment of 3D structures, while the remaining 20–30% of the sequence alignment is structurally wrong; the sequence alignments are more than 95% correct only when the sequence identity exceeds 40–50%.

If the "homology level" is below 10–15%, the similarity is at the level of "random noise", and homology cannot be detected (which, however, does not prove its non-existence; I will deal with this later on). And the "twilight zone" covers the range between 15 and 25–30% of sequence identity: there, the sequences look homologous, but who can guarantee this? (And the sequence alignment is often rather different from the correct structural alignment in this case.)

rop   MTKQEKTALNMARF I RSQTLTLLEKLNELDADEQAD I CESLHDHADELYRSCLARFGDDGENL---
mjc   SGKMTG I VKWFNADKGFGF I TPDDGSKDVFVHFSA IQNDGYKSLDEGQKVSFT I ESGAKGPAAGNVTSL

Figure 22.4. Alignment of sequences of *unlike*, non-homologous proteins often shows 10–15% coincidence of amino acid residues (here: α-helical RNA-binding protein *(rop)* and β-structural cold shock protein *(mjc)* have 14.5% residues coinciding: 10 (shown in bold) out of 69). The alignment was done using the BLAST program. No gaps inside the sequences are used. Thus, alignment of definitely unrelated *rop* and *mjc* shows approximately the same sequence identity and even needs fewer gaps than alignment of the N-terminal halves of RSV and HIV ribonucleases shown in Fig. 22.3.

Unfortunately, as I have said, all these thresholds are not exactly equal for various programs (and operating modes), and many programs do not let one know these values (they can be usually found in original papers, but who cares to read them . . . ; however, some programs, like BLAST, report on the statistical reliability of any similarity found). In any case, I would recommend you, prior to a search for homology, to check the program and the operating mode you use against a set of known proteins (both homologous and *non*-homologous) of approximately the same length as the protein you are interested in, and to gain some experience (another variant: to read the original papers . . .).

The drift of estimates of the reliability and *un*reliability levels is connected mostly with different gap penalties used by different programs. If the gap penalty is so high that it forbids the gaps, the homology of randomly chosen 20-letter (protein) sequences is $\approx$ 5–7% only. If the gap penalty equals zero, that is, if gap insertion is free of charge, then a homology level as high as 30–35% is typical of randomly chosen protein (and all 20-letter) sequences (and the 60–65% level is typical of the 4-letter DNA and RNA sequences). Experience suggests that optimal discrimination of homologous and non-homologous sequences is achieved when the gap-initiation penalty is equal to the price of 2–3 residue matches, and the penalty for gap elongation by one residue is about 1/20 to 1/100 of the gap-initiation penalty.

I told you nothing about the very interesting mathematics underlying the algorithms of the search for homology. This would take us too far. But I want to mention the key word: "dynamic programming". This is the most powerful method used to optimize one-dimensional systems (and sequence alignment is a one-dimensional system where shifts and gaps are used to optimize the similarity).

Is it possible to extend homology recognition below the level of 30%, that is, to the "twilight zone" and beyond it? Yes, it is possible – but this requires comparison of *many* sequences. Involvement of many sequences is a feature that distinguishes the most advanced programs like PSI-BLAST and HMMer from, say, "simple" BLAST working with separate sequences.

The general scheme is as follows. Firstly, obvious homologs (at a level above $\approx$ 40% similarity) are selected for each of the two compared sequences. And then these two bunches of sequences (these two "sequence families") are compared, with special attention paid to the most conserved positions in the families. If both families are shown to be similar, or if some sequences are members of both families, the compared sequences are proved, though indirectly, to be similar. Apart from the conservative positions, the HMMer (Hidden Markov Model) program uses conserved correlations between the neighboring positions in alignments. In this way one can lower the homology recognition level to 10% similarity between the "seed" sequences – *if* the similarity includes the positions and correlations which are most conservative in the families.

Figure 22.3 shows that ribonucleases H of HIV and RSV viruses are similar only at the "noise level" (9 identical residues out of 60: 15% sequence identity). However, this similarity is localized in the key regions where various ribonucleases H are most similar. This increases the value of the similarity found. Furthermore, provided one takes into account that all active site residues are identical, and that the similarity is concentrated in secondary structures and hydrophobic cores where the sequence identity is 30% (rather than 15%), then recognition of homology is proven.

```
                                   M
              K                    F
              I                    K                        E
              V F V   G            V G    L                 C
    S D       A T R   A T          I A    E                 G
    K K       S S K   R S          S N    Q                 A
    N Q       E A N   K N T        T Q    N T       G Y
    E E     S D D T   D D V        E E Q  S A       A I
    A A     N N E S   Q R L        D S R A G Q      S L
    P P G D P K A G E K I F K T K C A E C H T V
                                   = _ _ = =
    −1+1                  10                      20
```

Figure 22.5. Amino acid contents for various positions at the N-terminus of cytochrome c from eukaryotic mitochondria. The most important, i.e., most conservative residues are those where no variations are observed. The underlined site Cys–X–X–Cys–His is responsible for heme binding in the majority of cytochrome species (not only c, and not only from eukaryotes). The alignment is taken from [4].

It is convenient to use "consensus sequences" (Fig. 22.5) underlining the conservative features of protein families already studied. When a consensus sequence is supplied with residue frequencies for each position, it is called the family's "sequence profile".

To establish a functional similarity, one has to pay attention also to the "functional sites" that are already known for many functions. The "sites" are relatively short, highly conservative sequences ensuring these functions (see Fig. 22.5: the site Cys–X–X–Cys–His binds heme in cytochromes). There are libraries of these sites, and there are special programs (e.g., PROSITE) used to search for them in sequences.

When a "new" protein structure is predicted from its homology with an "old" one(s), it should be kept in mind that the similarity of 3D structures cannot be extended to regions where the sequences have diverged too far. These are mainly irregular chain regions (see Fig. 22.3). Here, *homology modeling* methods are used, so far with only moderate success; I will not talk on them. Besides, experience tells us that even a complete identity of two short sequence fragments cannot guarantee identity of their 3D structures. For example, identical 6–10-residue sequences can be involved in an α-helix of one protein and β-structure of another.

Now, let us turn to methods of structure prediction for "new" sequences having no detectable homology with "old" proteins, already solved by X-ray or NMR methods.

More than half the "new" sequences belong to this class; their structures cannot be recognized from sequence similarity. Thus, the problem is to predict these structures, and then perhaps their functions as well, from the amino acid sequences and the physics of protein folding: if a protein is capable of spontaneous folding, all the necessary information is to be coded in its sequence (Fig. 22.6).

It has to be said from the very beginning that to date there are no perfectly precise and sufficiently reliable methods to predict a protein's structure from its sequence.

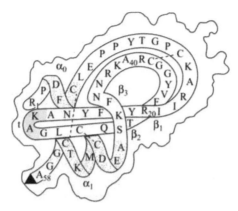

Figure 22.6. Scheme of the primary and 3D structure of a small protein (bovine pancreatic trypsin inhibitor). The backbone (with the one-letter-coded sequence) is drawn against the background of the molecule's contour. The α-helices, β-strands, β-turn (t), and S–S bonds (- - -) are outlined. When a protein is capable of spontaneous folding, all these details can be predicted, *in principle*, from its amino acid sequence alone. The side chains are not shown here, but their conformation, *in principle*, can also be predicted from the sequence.

It seems the reason is two-fold: (1) the limited accuracy of the energy or energy-like estimates underlying theoretical computations of protein structures; and (2) a relatively small energy difference ("gap") between one "correct" and many possible "wrong" folds of a protein chain: the small gap can be easily erased by a low accuracy of energy estimates. I have to emphasize the latter: the small gap radically discriminates between the protein (and RNA) situation and the DNA situation (where the gap is large) – to the deep regret of people dealing with protein structure predictions. In DNA, the complementary base pairing exists over the whole double helix, since the sequence of one DNA strand reproduces the sequence of the other. This strict complementarity of the two sequences forming the DNA molecule leads to formation of a continuous double helix and ensures the great energy advantage of the "correct" double helix over all "wrong" structures. The advantage is so great that even a very crude estimate of base pairing was sufficient for the correct prediction of the double helix. The same "base-pairing code" exists in RNA as well, but here the complementarity of sequences is much less strict and covers much shorter regions; therefore, prediction of the RNA structure is much more difficult. And the proteins have neither an unambiguous "code" for interaction of residues nor any strict complementarity of long contacting chain regions and accordingly, they lack the huge advantage of the "correct" fold . . .

However, the existing methods provide fairly rich and quite reliable information on the probable structure (or rather, on plausible structures) of a protein, and specifically on the main structural elements of its 3D structure.

Pragmatically, the protein structure prediction problem is usually set as follows: is the 3D structure of a given protein sequence similar to any of the already known 3D protein structures? If "**yes**", then **how** can the sequence in question be superimposed onto the known 3D fold? If "**no**", what features of the spatial organization of the sequence in question can be established?

I know many cases, including several from my own practice, when answers to these questions facilitated the experimental determination of the 3D protein structure and/or its function, helped to plan protein-engineering experiments, etc. However, I should say that it would be of great interest to me to read a good review on practical applications of protein structure predictions . . .

Now let us turn to the *methods* of protein structure prediction. When considering these methods, I will emphasize the underlying physical ideas.

There are two strategies of protein structure prediction: (1) to seek for the structure resulting from the kinetic folding process; and (2) to seek for the most stable (or, equally, the most probable) chain structure.

In principle, both these strategies can lead to a correct answer, since the most stable structure of a protein chain has to fold rapidly (recall the previous lecture).

However, it is important that protein structure prediction requires neither consideration nor reproduction of the folding mechanisms, and it is enough to consider only the stability of the folded structure. Moreover, the kinetic approach requires more difficult computations, needs additional (kinetic) parameters, and still does not lead to a substantial success. And all the speculations on the specific (say, hierarchic) rules of protein folding (which looked so attractive and could indeed, if correct, facilitate the search for the native structure enormously) appear to be wrong (at least as far as single-domain proteins are concerned).

The second strategy turned out to be easier and more successful. The evaluation of stability (or, equally, of probability) of various chain folds is based either on physics-derived potentials of interactions, or (what is more common now) on the frequency of occurrence of various structural elements and interactions in proteins. This makes no major difference since, as we already know, the quasi-Boltzmann statistics of protein structures are determined by the potentials of interactions. Thus, I shall speak of the "potentials", even if their evaluation originates from protein statistics rather than from physical experiments. The interactions with water molecules are usually not taken into account explicitly; instead, the water-mediated interactions of the residues (like hydrophobic interactions) are used.

Let us start with secondary structure predictions. α-Helices and β-strands are important elements of the protein; as you remember, they determine many features of its architecture (Fig. 22.6).

Let us forget for a while that the protein chain is packed into a solid globule: it is too difficult to predict the secondary structure simultaneously with the tertiary one. Is it possible to predict, on the basis of stability, the secondary structure of the chain from its amino acid sequence *prior* to the tertiary structure? The answer is: yes, usually it is possible, though the resulting prediction is not absolutely precise.

First of all: what residues stabilize a separate secondary structure, an α-helix, for example, and what residues destroy it?

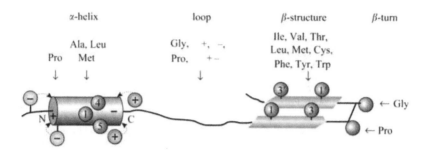

Figure 22.7. "Templates" for an α-helix, a loop, a β-strand and a β-turn. The amino acid residues that stabilize these elements or their separate parts are singled out; "+" denotes all positively charged residues; "−" denotes all negatively charged residues; all residues with a dipole in the side chain are marked with "+−". The patterns of alternating hydrophobic side chains shown (see the numbered groups) stabilize the α- and β-structure. Such an alternation also leads to the formation of continuous non-polar and polar surfaces at opposite sides of the α-helices and β-strands.

Experiment gives a direct answer to this question. I mean, first of all, the immense work on determining the α- and β-forming propensities of amino acid residues in polypeptides and proteins performed by the groups of Scheraga, Fasman, Baldwin, Fersht, Serrano, DeGrado, Kim and by a number of other groups (and in our laboratory by O.B.P. and Bychkova). Also, a wealth of information (well consistent with and extending physicochemical experiments) is given by the statistics of α- and β-structures in proteins. Figure 22.7 summarizes the most important results (those worth remembering) obtained by all these methods.

It is necessary to stress that all the rules formulated for protein structures are of a probabilistic nature. Despite many attempts, no strict "code" of protein structures has been found; i.e., in proteins there is nothing like the strict A–T and G–C pairings of nucleotides, which is so typical of nucleic acids. However, I have to say that the nucleotide pairing is not all that strict in RNAs either, with their diverse (in contrast to DNA) repertoire of 3D structures . . .

The majority of the experimental and statistical data obtained can be easily understood from the physics and stereochemistry of amino acid residues. We have touched on this question already. For example: Pro likes to enter neither the α-helix (but for its N-terminal turn), nor the β-structure. Why? Because Pro has no NH-group, and therefore it cannot form the corresponding H-bonds in the α- and β-structures. Therefore, Pro destabilizes these structures – and it does not like to enter them. On the other hand, the NH-groups of residues forming the N-terminal turn of the α-helix (and NH-groups in some positions at the β-sheet edges) are not involved in H-bonding; thus, Pro loses nothing here, and it does not avoid these positions. Moreover, Pro often enters the N-ends of helices, since its ϕ angle is already fixed in the α-helical conformation by the proline ring (which stabilizes the α-helix with Pro at its N-end). This is also a reason for the frequent occurrence of prolines at the N-ends of the β-turns.

Another example: Ala stabilizes α-helices in polypeptides (and it is most abundant there in proteins), while Gly destabilizes both α- and β-structures and facilitates the formation of irregular regions. What is the origin of this difference? Gly lacks a side chain, and therefore it has a much larger area than Ala of possible ϕ, ψ angles; that is, while the α- and β-conformations of both residues are essentially the same, the possible irregular conformations of Gly are much more numerous and therefore more probable.

For the same reason, the C^β-branched residues (Val, Ile, Thr) stabilize the β-structure, where their side chains have all three possible rotamers, and destabilize the α-helix and irregular regions, where only one, on average, side-chain rotamer is allowed for any backbone conformation.

Hydrophobic groups generally prefer α- and β-structures, where they can stick together "for free" (the price has been already paid by H-bonds) in a hydrophobic cluster (see Fig. 22.7), and do not like the coil, where they cannot stick together "for free". On the other hand, polar side chains, and especially short side chains, prefer irregular regions where they can form additional irregular H-bonds to the backbone, and do not like α- and β-structures where all the backbone's H-bond donors and acceptors are already occupied.

The influence of amino acids on secondary structure formation can be not only explained but also theoretically predicted. For example, in 1970, prior to experimental evidence, we (A.V.F. and O.B.P.) predicted that a negatively charged residue would stabilize the C-end of the α-helix because of attraction to the N-end's positive charge. We predicted also that a negatively charged residue would destabilize the C-end of the α-helix because of repulsion from the C-end's negative charge; and that a positively charged residue must act in the opposite direction. The potential of such "charged residue–helix terminus" interaction was predicted to be about 1/4 or 1/3 kcal mol^{-1}, which is consistent with present-day experimental estimates.

When we know what residues stabilize the middle of the α-helix and its N- and C-ends, we obtain a kind of "template" for recognition of helices in amino acid sequences. The α-helical "template" can be roughly described as follows. A sequence fragment forms the α-helix, when: four or five positions near its N-end are enriched with negatively charged groups and include a Pro in addition; the middle of the fragment is enriched with Ala, Leu, Met rather than with Gly, and does not include any Pro at all; and its C-end is enriched with positively charged groups and avoids negatively charged residues. In addition, the α-helix is stabilized by alternation of hydrophobic groups in the sequence resulting in their sticking together within the helix (Fig. 22.7), and the same order of side chains in the sequence is necessary for incorporation of the helix into the globule. The importance of side-chain ordering for secondary structure formation in globular proteins was demonstrated by V.I. Lim. The better the amino acid sequence satisfies this template, the higher the probability that this sequence forms an α-helix.

Other templates have been described for β-strands, for β-turns, and for loops. Moreover, the templates can be used to describe sequences forming more complicated structures, for example, the β–α–β units consisting of two parallel β-strands connected with an α-helix (Fig. 22.8). These "supersecondary structures" are typical

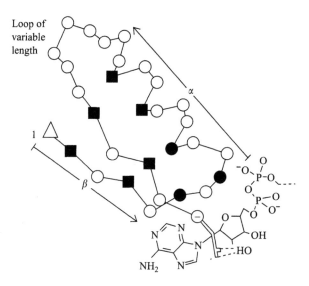

Loop of
variable
length

Figure 22.8. Template of the nucleotide-binding β–α–β unit. Squares indicate positions usually occupied by relatively small hydrophobic residues (Ala, Ile, Leu, Val, Met, Cys): they form a hydrophobic core of the β–α–β superhelix. Filled circles are the key positions (sharp turns) occupied by Gly only. The open triangle indicates the beginning of the β–α–β motif; it is usually occupied by a basic or a dipolar side chain. The last (–) motif's position includes a nucleotide-binding Asp or Glu. Adapted from Wierenga R.K., et al., *J. Mol. Biol.* (1986) **187**: 101–107.

of nucleotide-binding domains. A special role is played by the so-called "key residue positions" of the templates that can be occupied only by strictly defined amino acids – for example, by Gly which is the only residue that can have a conformation with $\phi \approx +60°$: this conformation is forbidden for other amino acids.

It is noteworthy that the "templates" can include not only structural, but also functional information (see the last residue of the nucleotide-binding β–α–β unit in Fig. 22.8).

Let us come back to secondary structure predictions. To begin with, let us completely ignore interactions between different secondary structures and consider an "unfolded" polypeptide chain (with separate secondary structures).

When the increments of separate interactions to the α-helix and β-hairpin stability are known, one can compute the free energies of these structures in any part of the sequence. Then it is possible to compute the probability of occurrence of α- and β-structures in each chain region and the average content of α- and β-structures in an "unfolded", i.e., non-globular, chain (at a given temperature, pH, and ionic strength of the solution). For nearly 20 years our program ALB (in the operating mode "unfolded chain") has been used to this end. The results can be compared with experimental (e.g., CD spectra-based) data on the secondary structure content in non-globular polypeptides. Figure 22.9 shows that the theoretical estimates are in reasonable agreement with experiment.

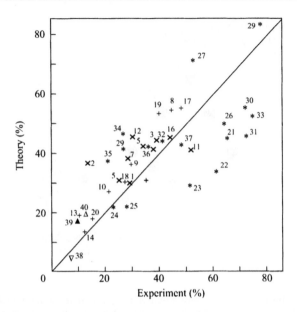

Figure 22.9. Theoretical (computed by the program ALB-"unfolded chain") and experimentally observed helicity of several tens of peptides at 0–5 °C and various pH and ionic strengths of the solution. Adapted from Finkelstein A.V., Badretdinov A.Ya., Ptitsyn O.B., *Proteins* (1990) **10**: 287–299.

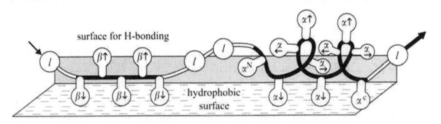

Figure 22.10. Fluctuating secondary structure of a protein chain (β, β-strand; l, loop; α, α-helix) on the surface mimicking the protein globule (the "floating logs" model). The surface consists of a hydrophobic "lake", the "shore" of which is capable of H-bonding to β-strands. The model takes into account different alternations of side chains directed to ($\downarrow$), from ($\uparrow$) and along ($\rightarrow$, $\leftarrow$) the surface in different secondary structures, as well as the N- and C-terminal effects in α-helices (these effects are summed up and attributed to the N- and C-terminal residues of the helix, α^N and α^C). Adapted from Ptitsyn O.B., Finkelstein A.V., *Biopolymers* (1983) **22**: 15–25.

When predicting the secondary structures of water-soluble globular proteins, we should take into account the interaction of each secondary structure with the rest of the globule. This can be done only approximately, since 3D structure of this globule is not known. One knows, though, that the secondary structures are attached somehow to its hydrophobic core. This interaction can be approximated by interactions of secondary structures with a "hydrophobic lake" (Fig. 22.10).

The strength of hydrophobic interactions is known experimentally. The stereochemistry of α- and β-structures tells us which of their residues face the same direction and can stick to the "hydrophobic lake" simultaneously. Thus, we can estimate the energy of secondary structure adhesion, add it to the internal secondary structure energy, calculate the Boltzmann probability of formation of α- and β-structures in each chain region, and use these probabilistic estimates as a basis for the prediction of secondary structures of globular proteins. This is exactly what the program ALB performs in the operating mode "globular chain" (the Internet address: http://indy.ipr.serpukhov.su/~rykunov/alb/).

The question arises as to what temperature should be used to calculate the above mentioned Boltzmann probabilities. If we take an extremely low temperature (0 K), the secondary structures forming only *one*, the lowest-energy (according to our calculations with all their errors) chain fold is singled out. If we take a higher temperature, the secondary structures common for *many* low-energy folds (again, according to our calculations with all their errors) are singled out. What temperature is to be preferred?

On the one hand, we are interested in a stable structure of the protein chain, which may suggest that only the secondary structures forming the lowest-energy fold should be singled out, i.e., that we should compute the probabilities for 0 K. On the other hand, the protein exists at approximately room temperature ($\approx$300 K), and its structure is maintained by the temperature-dependent hydrophobic interactions; this implies that room temperature should be used.

There is another strong reason to use a temperature like $\approx$300 K rather than 0 K.

When the protein-stabilizing interactions are known only approximately, this temperature ($\approx$300 K) is most suitable to single out probable states of the chain's elements. More strictly: the most suitable temperature is our good old T_C, which is somewhat lower than the protein melting temperature, which, as you may remember, is about 350 K.

I will try to explain rather than to prove this difficult point.

In essence, *any* protein structure prediction is based on only some of the interactions operating in the chain. Perhaps some interactions are neglected (like close packing in secondary structure predictions), while others are known only approximately (which means that only a part of their energy is properly taken into account). This refers, in particular, to powerful hydrophobic interactions of secondary structures with the rest of the globule (since we do not know its shape).

Then, making *any* prediction is akin to predicting the inner/outer residue position exclusively from the hydrophobicity of this residue. In this case only a minor part of the interactions within the protein are taken into account. Nevertheless, we know that this problem has a solution, though probabilistic rather than precise. The solution can be found in the statistics of the distribution of residues between the interior and the surface of proteins. We already know that the statistics look like

PROBABILITY TO BE "IN" / PROBABILITY TO BE "OUT"

$\sim \exp(\text{HYDROPHOBIC FREE ENERGY}/kT_C)$ (22.1)

where T_C is a characteristic temperature of protein statistics, which does not depend on the kind of interaction taken into account.

The above example suggests that the same characteristic temperature T_C should be used for *any probabilistic* protein structure prediction, and that only a probabilistic prediction can be made when we know only a part of the interactions.

If the "known" part is small, our prediction will be rather vague, "highly fluctuating". On the other hand, if we obtain a prediction without any or with only small uncertainty – this means that the interactions taken into account are sufficient to fix the structure in question at temperature T_C . . .

Figure 22.11 presents the result of a secondary structure calculation performed by the program ALB-"globular chain" for a small protein. This calculation takes into account the interactions within secondary structures and the interactions of secondary structures with a "hydrophobic lake" that models the protein globule. The solidity of the protein and the close packing of its side chains are not taken into account in these calculations. Thus, one can say that this secondary structure prediction refers rather to the molten globule than to the native, solid protein; but we already know that the secondary structures of these two are similar. The resulting plot shows that the calculated secondary structure probability, even in the most structured (according to the calculation) chain regions is far from 100%. However, in accordance with the X-ray 3D structure of the protein, the predicted α-helix prevails at the chain's C-end, and the β-structure (actually, the β-hairpin) in its middle. Thus, the peaks of probability of the α-helical and β-structural states (if they exceed some empirically established level) can be used for protein secondary structure prediction. This is a

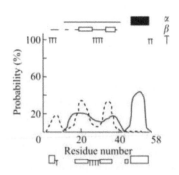

Figure 22.11. Secondary structure calculation performed by the program ALB-"globular chain" for the sequence of bovine pancreatic trypsin inhibitor. Abscissa, the residue number in the sequence; ordinate, the computed probability of the α-helical (—) and β-structural (- - -) states for each chain residue. Above the plot: prediction of probable α-helices (α), β-strands (β), and chain turns (T). Filled rectangles, the most probable secondary structures; open rectangles, probable (and also predicted) secondary structures; lines, possible but not predicted secondary structures. Below the plot: α-helices (deep rectangles), β-strands (shallow rectangles), and turns (T) in the native protein structure.

probabilistic prediction, of course, though many secondary structure elements can be predicted with confidence.

Actually, secondary structure prediction is always probabilistic in nature. This refers not only to the program ALB, which is explicitly a probability-based one. This refers to all other methods, even to those that are aimed, as is usually done, to single out only one, the "best" (read: the most probable from the program's point of view) position of α- and β-structures in the protein chain.

When the protein's secondary structure is predicted from its amino acid sequence *alone*, the precision of recognition of α, β, and an "irregular" state is about 65%, which means that the secondary structure is correctly predicted for 65% of protein chain residues.

Is it possible to improve the accuracy of secondary structure prediction? Yes, it is. The Internet-accessible "PHD" method of Rost and Sander predicts the protein's secondary structure *not* from *its* sequence alone, but (whenever it is possible) from a *set* of sequences that includes the sequence in question *and* its homologs. This approach leads to the partial annihilation of random errors made in considering each of the sequences, and the prediction is made with more confidence (reaching an average level of 72–75% instead of 63–65%). The recently developed method of Jones (who uses homologs as well) demonstrates even somewhat better results.

Secondary structure prediction has become, in fact, a routine procedure in the analysis of protein sequences, even though its accuracy is still not perfect. Another problem, however, that of predicting the 3D protein fold from its sequence, is much more difficult and much less routine. This problem will be considered in the next lecture.

LECTURE 23

Let us turn now to attempts to predict three-dimensional folds of protein chains.

Prediction of 3D folds is often based on the previously predicted secondary structure. Although this way is not, so to speak, self-consistent, since tertiary structure can affect the secondary structure, it is sometimes successful, since a stable 3D fold must consist mostly of stable secondary structures.

Figures 23.1 and 23.2 illustrate prediction of 3D folds for interferon; this prediction has been done in 1985. The secondary prediction (Fig. 23.1) shows a predominance of α-helices in the interferon chains, especially in its N-terminal part (where the functional domain had been localized by functional studies). The α-helices of interferon have been predicted very definitely (which is far from being often), and we dared to sculpture the N-terminal domain from the predicted helices. For the C-terminal domain, where the secondary structure prediction is more ambiguous, we were not able to give an unambiguous prediction of its 3D fold.

Figure 23.2a shows that the N-terminal domain is predicted to be built up from three large helices and one tiny helix. The X-ray structure of interferon β was solved (Fig. 23.2b) five years after this prediction and showed a rather accurate agreement with the predicted structure. This interferon N-terminal domain fold was one of the first successful *a priori* predictions of a protein structure from its sequence.

I would like to emphasize the factors contributing to this success: the prediction of α-helices was unambiguous, and predictions were very similar for several remote homologs, which allowed us to believe them. (However, such definite and similar predictions for remote homologs are relatively rare. Less definite and less consistent predictions are more common. This is exemplified by the secondary structure prediction made for the interferon C-terminal domain (Fig. 23.1); an unambiguous fold prediction is not possible in this case.) And one more thing was crucial for the successful prediction of the interferon fold. When looking for the folding pattern, we could perform a rational and exhaustive search using the *a priori* classification of α-helical complexes developed by Murzin and A.V.F.

The problem of protein structure prediction can be posed as a problem of choice of the 3D structure best fitting the given sequence among many other possible folds.

However, what can be the source of "possible" structures?

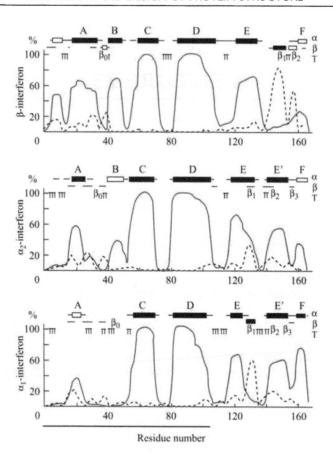

Figure 23.1. Secondary structure calculation for interferons. The calculation was performed by the program ALB-"globular chain". Abscissa, residue number in the sequence; ordinate, the computed probability of the α-helical (—) and β-structural (- - -) states for each chain residue. Above the plot: prediction of probable α-helices (α), β-strands (β) and chain turns (T). Filled rectangles, the most probable secondary structures; open rectangles, probable (and also predicted) secondary structures; lines, possible, but not predicted secondary structures. The chain region forming the N-terminal domain is underlined. Adapted from Ptitsyn O.B., Finkelstein A.V., Murzin A.G., *FEBS Lett.* (1985) **186**: 143–148.

One answer is: an *a priori* classification; this way was used in predicting the interferon fold. Another, a more practical answer is: the Protein Data Bank (PDB) where all the solved 3D structures are collected. In this case, actually, we will deal with "recognition" rather than "prediction": a fold cannot be recognized if PDB does not contain its already solved analog. This limits the power of recognition. However, it has an important advantage: if the protein fold is recognized among the PDB-stored structures, one can hope to recognize also the most interesting feature of the protein, its function – by analogy with that of an already studied protein.

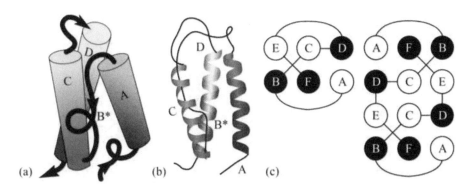

Figure 23.2. (a) Prediction of the chain fold for the N-terminal domain of interferon (from Ptitsyn O.B., Finkelstein A.V., Murzin A.G., *FEBS Lett.* (1985) **186**: 143). Three large α-helices (A, C, D) are shown as cylinders, the tiny helix B* as a separate helix turn. (b) X-ray structure of the N-terminal domain of interferon β (the C-terminal domain is not shown) solved in 1990 (Senda T., Matsuda S., Kurihara H., Nakamura K.T., Shimizu G.K.H., Mizuno H., Mitsui Y., *Proc. Jpn. Acad. Sci., Ser.* B (1990) **66**: 77). The structure is given in the same orientation as the predicted model, and the helices are lettered identically. The region B* is helix-like, but it is not an α-helix in interferon β; however, a short α-helix exists in this place in the closely related interferon γ (Ealick S.E., Cook W.J., Vijay-Kumar S., Carson M., Nagabhushan T.L., Trotta P.P., Bugg C.E., *Science* (1991) **252**: 698). (c) Topologies of β (on the left) and γ (on the right) interferons according to Ealick *et al.* Interferon γ consists of two subunits. Notice that these subunits "swap" the C-terminal regions (helices E and F).

Certainly, not all protein folding patterns have been collected in PDB yet; however, it hopefully already includes half of all the folding patterns existing in nature. This hope, substantiated by Cyrus Chothia, is based on the fact that the folds found in newly solved protein structures turn out to be similar to already known folds more and more frequently. Extrapolation shows that perhaps about 1500 folding patterns of protein domains exist in genomes, and we currently know half of them (including the majority of the "most popular" folds).

To recognize the fold of a chain having no visible homology with already solved proteins, one can use various superimpositions of the chain in question onto all PDB-stored 3D folds in search of the lowest-energy chain-with-fold alignment (see Fig. 23.3). This is called the "threading method". When a chain is aligned with the given fold, it is threaded onto the fold's backbone until its energy (or rather, free energy) is minimized, including both local interactions and interactions between remote chain regions. The threading alignment allows "gaps" in the chain and in the fold's backbone (the latter are often allowed for irregular backbone regions only).

The threading approach was suggested by A.V.F. and B.A. Reva in 1990–91, and, independently, by David Eisenberg's group in 1991 (the latter variant, although less "self-consistent", has the advantage of a simple form and practical convenience). Since then, various threading methods have become popular tools for recognizing the folds of "new" proteins by their analogy with "old" ones.

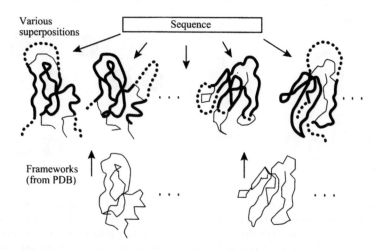

Figure 23.3. Scheme illustrating the idea of "threading" the studied sequence onto the PDB-stored backbone structures. The bold line shows the regions where the chain is aligned with the backbone of a "known" protein structure; the broken line shows the non-aligned region of the studied chain.

In principle, threading is similar to a homology search; the difference is that only sequences are aligned in a homology search, while threading aligns a "new" sequence with "old" folds.

Before coming to the results, I would like to stress some of the principal problems of threading methods (actually, no method of protein structure prediction is free of these problems in one form or another).

First, the conformations of the gapped regions remain unknown, together with all interactions in these regions. Second, even the conformations of the aligned regions and their interactions are known only approximately since the alignment does not include side-chain conformations (which differ in "new" and "old" structures). Estimates show that threading takes into account, at best, half of the interactions operating in the protein chain, while the other half remain unknown to us. Thus, again, the protein structure is to be judged from only a part of the interactions occurring in this structure. Therefore, this can only be a probabilistic judgement.

The next problem: how do we sort out all possible threading alignments and to find which is the best? The number of possible alignments is enormous... To cut a long story short: powerful mathematical tools have been already developed to this end. I will only name these methods: perhaps they will be of use to some of you (not to the majority, though). Dynamic programming (not to be confused with molecular dynamics! – they are absolutely different...) and its variant, statistical mechanics of one-dimensional systems; they are used to sort out the alignment variants. Self-consistent molecular field theory is used to calculate the field acting at each residue in each position of the fold. Stochastic Monte-Carlo energy minimization; various branch-and-bound methods including dead-end elimination, etc., are also in use. And

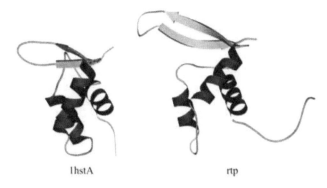

Figure 23.4. 3D structures of chicken histone H5 (1hstA) and replication terminating protein (rtp). In rtp, the last helix is not shown since it has no analog in the histone. The prediction is made by threading (Flöckner H., Braxenthaler M., Lackner P., Jaritz M., Ortner M., Sippl M.J., *Proteins* (1995) **23**: 376–386) in the course of a "blind" assessment of protein structure prediction methods (CASP-1994). The root mean square deviation between 65 equivalent C^{α} atom positions in the presented structures is 2.4 Å.

one more: an intuitive estimation of the variants! It seems to be old-fashioned in the computer age, but just this method allowed Alexei Murzin to predict many protein structures and to beat even the best computer programs in this kind of chess game.

Now, let us turn to results. As an example, I will show you the structure of replication terminating protein (rtp) predicted by threading. Having threaded the rtp sequence onto all PDB-stored folds, Sippl and his Salzburg group showed that the rtp fold must be similar to that of H5 histone (Fig. 23.4). This *a priori* recognition turned out to be correct.

However, it also turned out that the alignment provided by threading deviates from the true alignment obtained from superposed 3D structures of rtp and H5 histone (Fig. 23.5). On the one hand, this shows that even a rather inaccurate picture of residue-to-residue contacts can lead to an approximately correct structure prediction. On the other hand, this shows once again that all the mentioned flaws (errors in or insufficient knowledge of the interaction potentials, uncertainty in conformations of non-aligned regions, of side chains, etc.) allow us to single out only a more-or-less narrow set of plausible folds rather than *one* unique correct fold. The set of the "most plausible" folds can be singled out quite reliably, but it still remains unclear, which of these is the best. The native structure is *a member* of the set of the plausible ones, it is more or less close to the most plausible ("predicted") fold, but this is all that one can actually say even in the very best case.

I have certainly illustrated this lecture only with examples of "good", successful predictions. Examples of wrong predictions are more numerous, and the reason is evident: one has to pick out only one variant (or a few of them) from zillions of others. Therefore, it is really a great success if even the vicinity of this infinitely small target is hit in half the attempts (which is the level that the best "predictors" work at now, in 2001).

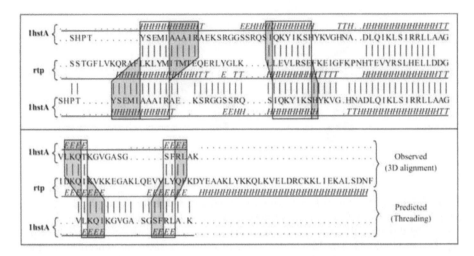

Figure 23.5. Histone H5 (1hstA) sequence aligned with that of replication terminating protein (rtp) by superposing their 3D structures (the "Observed" pair), and the rtp and 1hstA sequences aligned by threading (the "Predicted" pair). Ideally, these two alignments should coincide. However, the two 1hstA-with-rtp alignments are shifted (see the gray zones indicating the regular secondary structures; *H*, α-helix; *E*, β-strand; *T*, turn). The gaps in sequences, introduced to optimize the alignments, are shown with dots. The similarity of folds has been recognized by threading in spite of a very low homology of the sequences. Adapted from the paper "Protein structure prediction by threading methods: evaluation of current techniques" by Lemer C.M.-R., Rooman V.J., Wodak S.J., *Proteins* (1995) **23**: 337–355, where the CASP-1994 results are summarized.

Threading methods became a tool for a tentative recognition of protein folds from their sequences. The advantage of these methods is that they formulate a recipe – do this, this and this, and you will obtain a few plausible folds, one of which has a fairly high chance of being correct.

The success of protein structure prediction by modeling its folding is generally less pronounced, although a few most promising works have appeared recently. Specifically, I mean David Baker's work. He starts by collecting known 3D structures for each short, usually nine-residue long, piece of the sequence whose fold is to be predicted. This models the formation of local structures at the initial folding step. Then the alternative local structures found are used as blocks to build up many possible 3D structures of the whole chain. At the next step, the "best" of these chain folds are singled out by energy calculations. In this way Baker succeeded (in the course of CASP-2000, the last "blind" test of protein structure prediction methods) not only in "recognizing" the folds of many new sequences (i.e., in finding the PDB-stored analogs to their folds), but also in *ab initio* predicting some folds with no analogs in PDB. As might be expected, Baker's method is more successful in predicting folds with many local interactions than in predicting folds supported mainly by interactions between remote chain regions.

Is it possible to compensate somehow for the harmful action of errors in estimates of the interactions in protein chains and thus to improve protein structure recognition and prediction? Yes, it is possible, if we use a set of homologs and try to predict not the best structure for a single sequence, but the best common structure for the set of them. This approach has already increased the quality of secondary structure predictions of protein (and earlier of RNA) chains. It is also used now in recognizing protein folds. Chains with a homology of $\sim$ 40% seem to be most useful for the "set of homologs" used in predictions, since they still have a common 3D structure, although coded differently; this allows detection of the "signal" (structure) in spite of the "noise" of errors. It should be noted, though, that this approach is not aimed at predicting *the* 3D structure for an individual protein chain, but rather a common (that is, approximate for each sequence) fold for a protein family.

So far this lecture has been mainly about globular proteins. And what about fibrous and membrane proteins?

Sequences of fibrous proteins are so regular that one usually needs no computer to recognize their secondary structures. And these can form a basis for predicting a few variants of secondary structure packing into a fibril.

Prediction of membrane protein structures is less developed than prediction of structures of water-soluble globular proteins, since only a handful of membrane protein structures have been experimentally solved so far. It is fairly easy, however, to recognize the intra-membrane parts of proteins in the sequence: these are nearly continuous hydrophobic blocks; they form intra-membrane α- and β-structures, whose length is dictated by the membrane thickness. However, the principles of packing of these blocks are not yet developed enough (although the main principle is clear: membrane proteins do not have a hydrophilic interior, like water-soluble proteins, but a hydrophobic one; it is the external surface that is hydrophobic and is in contact with the membrane's lipids). Thus, predictions of membrane protein structures are just starting to evolve.

Let me summarize.

Prediction of the 3D fold of a protein chain, from a physicist's point of view, is a search for the most stable structure for this chain. Such a prediction is possible in principle, and is sometimes successful. The main problem is, however, that we still have insufficient information on the repertoire of possible 3D structures of protein chains and, most importantly, on the potentials of many interactions operating in these chains.

As a result, an unambiguous and reliable prediction of a protein chain structure from its sequence is hardly possible, but one can single out a very narrow set of plausible structures and, importantly, rule out a great many "impossible" folds. And the predictions can be improved by additional information concerning homologous sequences, active sites, etc.

Pragmatically, the protein structure prediction problem is usually set now as a task of recognition: is the 3D structure of a given sequence similar to some already known 3D protein structure? to which exactly? and how can the sequence in question be superimposed on the known 3D fold?

The recognition strategy is restricted. It is restricted to known protein folds (since unknown ones cannot be recognized). However, there is one clear pragmatic advantage of "recognition" over the "*a priori*" (unrestricted by the set of known folds) prediction. A protein fold once recognized among the PDB-stored structures allows easy analogy-based recognition of the function, active site, etc. of this protein. The *a priori* prediction by itself provides no hints of the function, which is usually the most interesting feature of the protein.

You should realize that physics-based methods form only one branch of protein structure prediction (or rather, recognition) methods. The mainstream is formed by methods using similarities of the sequence of interest, and often of its genome environment as well, to sequences of already identified proteins.

The latter is completely based on sequence and structure data banks. The amount of data stored in these banks is so tremendous that people try to retrieve from there all the information necessary for predictions, i.e., data on the repertoire of possible structures, energies of various interactions, homologs, functions, active sites, etc.

This is called "bioinformatics". This branch of science is evolving very rapidly. A huge amount of money is allocated to its development, as well as to the collection of new data and replenishment of the databases. All of you know about the Human Genome project aimed at collecting the sequences of all the human (and not only human) chromosomes and genes. As for human genome sequencing, this project has been accomplished already. The situation with gene identifying and annotating is a little trickier. Identification of a gene in the eukaryotic chromosome is not a trivial task. Recall splicing, and especially alternative splicing: one has to identify all the pieces that compose a gene, and a protein-encoding gene in particular. Hopefully, this problem can be also solved within a few years.

Other large-scale projects in the field of bioinformatics have been already launched.

The "Proteomics" project aims to proceed from identification of all genes in the genome to identification of the biological functions and roles of their protein products. It has to establish the protein's biochemical function (does this gene encode an oxireductase? or nuclease? or?), to recognize the protein's role in the cell (is it involved in the nucleotide metabolism? or in regulation? or?), as well as to elucidate relationships and interactions between the given protein and products of other genes.

Protein fold recognition can serve as a tool in this project. After all, a biologist wants to know protein functions, while structures and sequences are means to this end.

Another project, "Structural Genomics", aims to obtain, using X-rays or NMR, a 3D structure of at least one representative of each protein family (which number about 10 000). When this project is accomplished (at the cost of about $10 000 000 000 for 10 years), it will be possible to recognize the 3D fold of any sequence by homology only, and there will be no practical need for *a priori* protein structure prediction methods – except to satisfy our natural scientific curiosity and the needs of protein design that I am just going to discuss.

In the concluding part of this lecture I want to tell you about *protein engineering*, or more specifically, about the *design* of new proteins.

Oligonucleotide synthesis and recombinant DNA techniques have provided an opportunity to produce genes for proteins that do not exist in nature. X-ray and NMR have made it possible to see 3D protein structures. Powerful computers (and computer graphics) ensure an interactive dialog with these 3D structures: they enable us to modify these structures and to estimate (and see) the consequences. Taken together, these provide "hands", "eyes" and the "brain" of protein engineering, the new field of molecular biology. Its strategic aim is to create proteins with pre-determined structures and functions. Its future role in the creation of new drugs and catalysts, in nanotechnology, etc., can hardly be overestimated.

Directed protein engineering experiments have already answered a number of fundamental questions. Specifically, as concerns protein structure formation, it has been shown that proteins are not "perfect" and that a considerable proportion of mutations ($\sim 20\%$) increase their stability (though the rest $\sim 80\%$ decrease it). It has also been shown that correct folding is not dependent on all the details of side-chain packing within the protein structure: the protein can withstand a lot of point mutations. And the loops have little bearing on the 3D fold choice: if a loop is changed and even deleted, the "wound" at the globule's body is usually healed. The role of directed mutations in studies of protein energetics and folding has been already discussed; and the use of directed mutations in elucidating protein action will be discussed in the next lectures.

It is worth mentioning that protein engineering is also used in the chemical and pharmacological industries for creating, by directed mutations, proteins with increased (or decreased, if necessary) stability and with modified catalytic activity.

Having mastered the modification of natural proteins, engineering turned to the design of new protein molecules.

The problem of protein design is the reverse of the protein structure prediction problem. In prediction, we have to find the best 3D structure for a given sequence. In design, we have to find a sequence capable of folding into the given structure.

The design of artificial proteins can be easier than the prediction of the structures of "natural" proteins (just as the stability of an "artificial" tower can be calculated more easily than that of a "natural" tree: the architectural design is based on standard elements and implies simplicity of calculation of their connections). And protein design, based on the theory of protein structures, can use, as building blocks, only the α-helices, β-structures and loops that are internally most stable and capable of effective sticking.

The idea of protein design emerged at the end of the 1970s after a technique for the creation of new genes had been developed. Close to the end of the 1980s the first new protein molecules were designed and synthesized mainly by trial and error. Their architectures mimicked those of natural proteins, but the sequences designed to stabilize these architectures had no homology with the sequences of natural proteins.

The first such protein, a four-helix bundle, was obtained by DeGrado's group (Fig. 23.6). The design was made in a permanent dialog with experiment. The protein turned out to be helical and globular, as designed, and its structure seemed to be much more heat-resistant than the structure of any natural protein. Later it turned out,

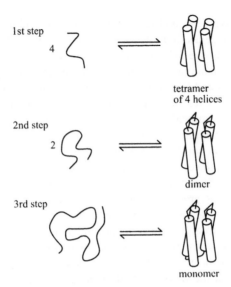

Figure 23.6. The main steps of design of a four-helical bundle performed by DeGrado's group. 1. Design of short helical peptides; selection of those that can form a tetramer. 2. Design of loops for helical hairpins; selection of those hairpins that can form a dimer. 3. Design of the last loop; selection of a tetra-helical monomer. The figure by W. DeGrado is reproduced, with his kind permission, from [4].

though, that the protein does not melt upon heating because it is a molten globule from the very beginning . . . Then its 3D structure was reinforced by the introduction of ion-binding histidines. And this ion-binding protein was as solid as natural proteins.

For a long time all artificial proteins (apart from those reinforced by ion binding) formed a collection of excellent molten globules. They were very compact, they had very good secondary structures – but they were not solid.

Why was a molten globule obtained instead of a solid protein?

It seems that the reason is as follows. Everybody knows how to make stable secondary structures (Leu and Ala in the middle, Glu at the N-end, Lys at the C-end – and you have an α-helix; a lot of Val, Ile, Thr – and you have a β-strand). Everybody also knows how to force these α- and β-structures to stick together (their hydrophobic groups have to form continuous surfaces); and how to rule out aggregation (the opposite surfaces of the α- and β-structures must be composed of polar groups). But nobody has a recipe for forming close side-chain packing in the core of the protein – and therefore they fail to obtain such a packing. And a compact globule with secondary structures but without close packing is just a molten globule . . .

To cope with the insufficient precision of design methods, "rational" design is often supported by the introduction of multiple random mutations and subsequent selection of variants having a "protein-like" activity (for example, those that specifically bind to something). This procedure has been used by Wrighton's group to make

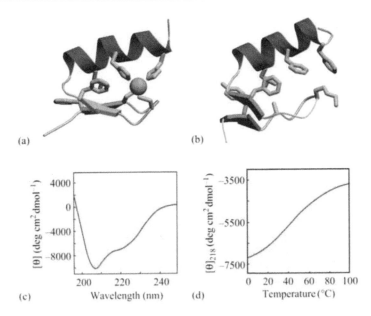

Figure 23.7. (a) Structure of the natural zinc finger (the second module of Zif 268 protein); the Zn ion is shown as a ball. (b) Structure of the artificial FSD-1 protein designed by Dahiyat and Mayo. (c) CD spectrum for FSD-1 at 1°C shows a rich and correct secondary structure. (d) Temperature change of the FSD-1 CD spectrum ellipticity at 218 nm. Adapted from Dahiyat B.I., Mayo S.L., *Science* (1997) **278**: 82–86.

a mini-protein (a dimer of two β-hairpins) that has an erythropoietin hormone activity. It is noteworthy that the natural protein consists of 166 residues, and its artificial analog has 20 residues only.

However, it seems that the rational design of close packing has now become a soluble problem – at least for a small protein.

Dahiyat and Mayo developed an algorithm for sorting out the astronomical number of variants of side group packings and ruling out hopeless variants; in 1997, they designed a small solid protein without any ion binding (Fig. 23.7). The structure of this protein mimics that of a "zinc finger" (a widespread DNA-binding motif), but without the Zn ion forming the structural center of the zinc finger fold. The designed FSD-1 protein has a very low (20%) homology with the natural zinc finger. Nevertheless, it is solid at low temperatures (as shown by NMR). However, it melts within a wide temperature range (Fig. 23.7d), that is, displays a much lower cooperativity than analogous natural proteins.

Some designed architectures, however, do not mimic natural samples. For example, a "non-natural" structure (Fig. 23.8a) was used to design an artificial protein albebetin. It was designed to consist of two $\alpha-\beta-\beta$ repeats (and was named accordingly); although not yet found in nature, this fold satisfies all known principles of protein structure. The sequence was designed in our group, and synthesized and studied by Fedorov, Dolgikh and Kirpichnikov with his team.

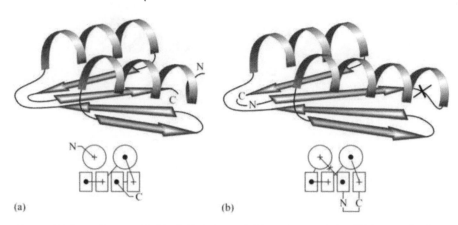

(a)   (b)

Figure 23.8. (a) Designed fold of albebetin and (b) a scheme of the experimentally determined fold of ribosomal protein S6. Topological schemes of these proteins are shown below. An artificial circular permutation of S6, which gives it the albebetin topology, consists in cutting of one loop (×) and making another loop (**N—C**) to connect the N- and C-ends of the natural S6 chain.

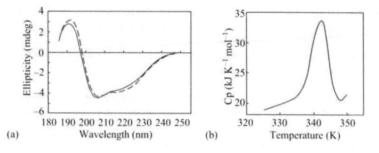

Figure 23.9. (a) CD spectra of albebetin (– –) and albeferon (——); (b) microcalorimetric melting curve of S6 permutant designed to have the albebetin topology. Part (a) is adapted from Dolgikh D.A., Gabrielian A.E., Uversky V.N., Kirpichnikov M.P., *Appl. Biochem. Biotech.* (1996) **61**: 85–96, and part (b) from Abdullaev Z.Kh., Latypov R.F., Badretdinov A.Ya., Dolgikh D.A., Finkelstein A.V., Uversky V.N., Kirpichnikov M.P., *FEBS Lett* (1997) **414**: 243–246.

A structural study of albebetin showed that it has a rich secondary structure (Fig. 23.9a), that it is very compact, that its structure is rather stable against unfolding by urea and that the protein is proteolysis-resistant. However, it does not melt in a cooperative manner, and is a molten rather than a solid globule.

A solid protein with the albebetin topology was obtained in a different way: by circular permutation of the natural protein S6 (its native fold, as well as several other recently solved protein folds, has the albebetin stack of structural segments, but these are differently connected by the chain; see Fig. 23.8b). The modified protein was shown to have a solid structure that melts in a cooperative manner (Fig. 23.9b).

Albebetin was used as a biological activity carrier. The fragment 131–138 of the human $\alpha2$ interferon sequence was attached to it (the resultant protein was called albeferon). This fragment is responsible for the activation of blast-transformation of thymocytes, and the rest of the interferon body serves as a kind of sheath that protects this fragment against proteolysis and does not allow it to work too actively. Experiments show that the albebetin globule works in the same way.

Experiments on functionally active artificial proteins are being carried out by many groups. Fibrous protein models have been designed on the basis of long helical bundles. A "working" model of the membrane protein has been designed on the basis of amphiphilic helices (i.e., those having polar and non-polar surfaces): these helices form membrane pores, and directed mutations of these helices have a drastic effect on the selectivity of these pores. Another "functional protein" has been produced using α-helical polypeptides chemically attached to the heme.

By and large, proteins rapidly transform from the object of respectful and amazed observation into the subject of intense engineering and design (but the amazement still remains . . .).

Part VII

PHYSICAL BACKGROUND OF PROTEIN FUNCTIONS

LECTURE 24

This lecture is devoted to protein functions.

This is a vast subject to discuss, and my lecture will present only a few pictures of functioning proteins, which will emphasize the crucial role of their spatial arrangement in their functioning. I showed some pictures of this kind when talking about membrane proteins. But in this lecture I shall only be talking about globular, water-soluble proteins.

A very rough scheme of protein functioning looks as follows:

$$\text{BIND} \rightarrow \text{TRANSFORM} \rightarrow \text{RELEASE}$$

Remember that some proteins may perform only some of these actions; that the words "BIND" and "RELEASE" may imply binding and releasing a few different molecules; and that the word "TRANSFORM" may mean some chemical transformation, a change in conformation (of both the protein and the substrate), and/or movement of the protein or the substrate in space.

We start with proteins whose main function is BINDing. Among these there are, for example, DNA-binding proteins.

To bind to DNA, an ample portion of the protein surface should be approximately complementary to the double helix surface (Fig. 24.1a). Then protein surface ridges are able to fit deeply into the DNA groove, where protein side groups perform fine recognition of a concrete DNA sequence (Fig. 24.2) and bind to it. All proteins shown in Fig. 24.1 are dimers, and it is in this form that they are complementary to the DNA duplex. Two identical DNA-recognizing α-helices of such a dimer recognize a *palindrome* in the DNA double helix, i.e., such a DNA sequence that preserves the same view after turning by 180° around an axis perpendicular to the duplex, for example,

$$5' \text{ TGTGG} \overset{\bullet}{-\!-\!-\!-\!-} \text{CCACA } 3'$$
$$3' \text{ ACACC} \overset{\bullet}{-\!-\!-\!-\!-} \text{GGTGT } 5'$$

Here, "–" denotes a random DNA sequence between two halves of the palindrome, and the rotation axis is indicated as "•".

In such a protein dimer, DNA-binding helices are mutually *anti*parallel, and the distance between them is close to a period of the DNA double helix, so that the dimer

313

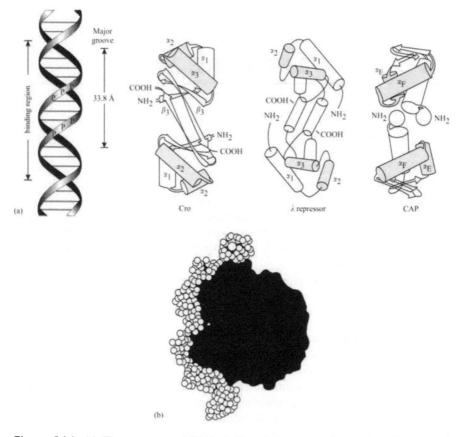

Figure 24.1. (a) The structure of DNA (left) and a number of proteins with a typical DNA-binding motif "helix-turn-helix" (shown in gray). For catabolite activating protein (CAP), only its C-terminal domain is shown. All these proteins are dimers and all of them recognize the major groove of DNA using their helices α_3 (α_F of CAP), the distance between which, in dimers, is close to a period of the DNA double helix (33.8 Å). The drawings by B.W. Matthews are reproduced from [4] with permission. (b) DNA (a lighter helix on the left) bent by CAP dimer (shown in black on the right). CAP association with DNA requires the presence of cyclic AMP (cAMP).

fits onto one side of the DNA double helix. However, different proteins have different tilts of these α-helices with respect to the axis going through their centers, which results in different binding-induced bends in the DNA. Some of these binding-*induced* bends are fairly sharp (Fig. 24.1b).

In some cases protein-to-DNA binding has to be assisted by *co-factors*, which change, or rather, slightly deform, the structure of the protein, thereby making it change from the inactive to the active state.

This is exemplified by trp-repressor (in *E. coli* it represses the operon in charge of the synthesis of RNA that codes proteins necessary for tryptophan synthesis), in which the role of such *co*-factor, or rather *co*-repressor, is played by tryptophan itself (Fig. 24.3). As long as tryptophan is unbound to the protein, the distance between DNA-binding helices in the dimeric trp-repressor is too small (about 28 Å) instead of the required 34 Å, which prevents DNA binding. The protein-bound tryptophan

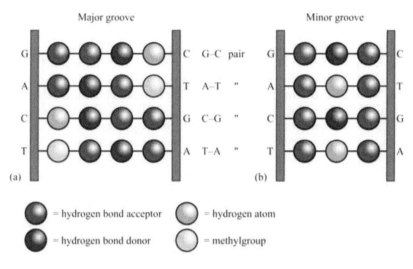

Figure 24.2. Typical patterns created by various functional groups of A–T and C–G pairs in the major and minor grooves of DNA. Reproduced from [1a] with permission.

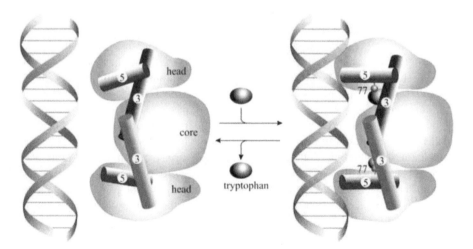

Figure 24.3. The scheme of action of tryptophan (trp) repressor. Against the background giving the general outline of the dimer with the common fused core and two identical heads, only those two helices (3 and 5) are shown between which the co-repressor (amino acid Trp) fits. It moves the 77th residue away, and that, in turn, shifts α-helix 5. Only then can both α-helices 5 bind to DNA. Reproduced from [1a] with permission.

moves the helices apart, so that they become complementary to the groove in the double helix and bind to it. Thus, when there are many tryptophans in the cell, they bind to the repressor, thus blocking further synthesis of Trp-synthesizing proteins, and hence its (tryptophan's) own further synthesis. This mode of regulation is called *negative feedback*.

In this case tryptophan acts both as a *stimulator* of trp-repressor's DNA-binding activity and as an *inhibitor* of the synthesis of proteins required for tryptophan synthesis. Here, stimulation of trp-repressor is *"allosteric"*, since Trp binding to protein occurs *"at another site"*, i.e., at a site different from the site that binds to DNA. The motif "helix-turn-helix" shown in Figs 24.1 and 24.3 is typical, but far from being the only DNA-binding structural motif. To illustrate this statement, I present three other typical motifs in Fig. 24.4. I would like to stress that DNA-binding proteins can pertain to different structural classes (in the drawings you can see α and $\alpha + \beta$ proteins), and that even DNA binding itself can be performed by the α- as well as by the β-structure.

So far we have spoken about the coarse features of the protein structure (their characteristic size is about 10–30 Å) that allow it to fit into the DNA groove. Finer features of the protein surface (their characteristic size is that of an atom, ~ 3 Å) are responsible for recognition of a certain DNA sequence which is to be bound by the protein.

Regrettably, the "general code" used by proteins for selective recognition of DNA fragments is still unknown (if there is a distinct "code" at all), although consideration of the details of each solved DNA–protein contact allows us to understand exactly which H-bonds between protein side groups and nucleotides, and what other of their close contacts, contribute to the occurrence of the DNA–protein contact at this particular site.

The highly selective recognition of other molecules by proteins is clearly exemplified by *immunoglobulins* or *antibodies*, i.e., proteins whose task (in vertebrates) is fine recognition of small-scale *antigenic determinants* (with the characteristic size of an atom or a few atoms) of various molecules. The immunoglobulin-like *receptors of T-cells* similarly recognize small antigenic determinants of specific cells, e.g., of virus-infected cells.

Immunoglobulins are built up from many β-structural domains and relatively small flexible hinges between them (Fig. 24.5a). The diversity of combinations of variable (antigen-binding) domains ensures a great variety of immunoglobulins, and hence, a broad spectrum of their activities, while rigidity of these domains ensures the high selectivity of action of each immunoglobulin. I am not going to recite here the basics of clonal selection theory, which explains the origin of a vast variety of immunoglobulins.

You must remember from other courses that germ cells contain *not* whole genes of light and heavy chains of immunoglobulins but only their fragments. In the genome these fragments are arranged in cassettes: separately for many types of each of three fragments of the heavy-chain variable domain, separately for the light chains; separately for constant domains of each chain, and separately for hinges. During the formation of somatic immune cells these fragments are

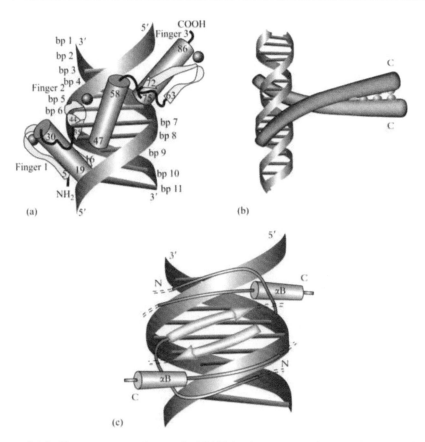

Figure 24.4. Three more motifs typical of DNA-binding proteins. In two of them the key role is played by α-helices: (a), "zinc fingers" (Zn ions are shown as balls). Adapted from [4]; (b), "leucine zipper". Adapted from [1]. In the third motif (met-repressor) (c) the key role is played by the β-hairpin, which specifically binds to the major groove of DNA, while the α-helices αB non-specifically bind to the sugar–phosphate backbone of DNA. Adapted from [1a]. The zinc finger (a domain that can be cut off and isolated from the remaining part of the protein) is the smallest among the known globular proteins, while the leucine zipper is structurally the simplest. When the latter is DNA-unbound, it is simply a dimer composed of parallel α-helices whose narrow hydrophobic surfaces are stuck together along their entire length. The surface-forming side groups are shown as projections. However, each helix has one non-hydrophobic group: it is polar Asn shown as a yellow spot against the helix hydrophobic surface. This Asn ensures formation of a dimer: its replacement by a more hydrophobic residue results in assemblies of not two but three and more helices. N- and/or C-terminus of each chain is denoted.

shuffled (and also in some mysterious way their hyper-variable portions undergo mutations) to build up whole genes of immunoglobulin chains. It is still unknown what exactly introduces these "somatic mutations" and directs them into specific parts of genes. It is also a puzzle why

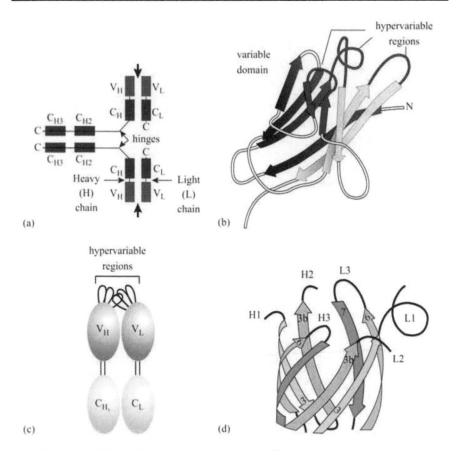

Figure 24.5. (a) The overall structure of an immunoglobulin (IgG). The variable (V) and constant (C) domains of two light (L: having two domains each) and two heavy (H: having four domains each) chains are shown, as well as the C-termini of all these chains. The domains C_{H2} are glycosylated. Bold arrows point to antigen-binding "pockets" between domains V_L and V_H. (b) The structural arrangement of domain V_H; the domains V_L, C_L, C_H and others have almost the same structure. In the V_H domain shown the hyper-variable loops are indicated; hyper-variable loops are also present in domain V_L. Taken together, these two domains form the antigen-binding pocket (c). (d) The antiparallel β-cylinder formed by the β-sheet of domain V_H and the β-sheet of domain V_L. The antigen-binding pocket is formed by projecting from these sheets hyper-variable loops of the heavy and light chains (H1, H2, H3 and L1, L2, L3, respectively). Parts (b) and (c) are reproduced from [1a] with permission. Part (d) is adapted from [1a].

these mutations (which may amount to tens in one domain – too many to be introduced all together with subsequent selection of "survived" proteins) do not explode the protein structure.

What we find important at the moment is that the antigen is recognized jointly by variable domains (V_H and V_L) of the light and heavy chains, or rather by hyper-variable loops fringing the antigen-binding pocket at the interface between these domains (Fig. 24.5b–d). The primary structure of these loops varies from immunoglobulin to immunoglobulin (which underlies the vast variety of immunoglobulin variants). However, each variant has not only a particular amino acid sequence but also strictly fixed conformations of all loops, and the antigen-binding pocket is positioned on the rigid β-cylinder formed by merged antiparallel β-sheets of the variable domains. Therefore, each immunoglobulin molecule can tightly bind only a particular antigen determinant and does not bind to others.

Figure 24.6 shows that the selectivity of binding of antigen determinants is dictated *not* by the *overall* protein arrangement (which serves only as a basis) but rather by the complementary shape of the molecule to be bound to the shape of the relatively small (as compared with the whole domain) antigen-binding pocket. Besides, hydrophobic parts of the molecule to be bound are in contact with hydrophobic parts of the pocket, its charges are complementary to those embedded in the pocket, and also the H-bond donors and acceptors of the antigen are complementary to the pocket's acceptors and donors. All this contributes to the tight binding of only a specific antigen.

A similar location of the active site (in a funnel at the butt-end of a β-cylinder) is observed in many other proteins having nothing in common with immunoglobulins. For example, this is where the active site is usually found in α/β cylinders (which, unlike immunoglobulins, have parallel β-structures).

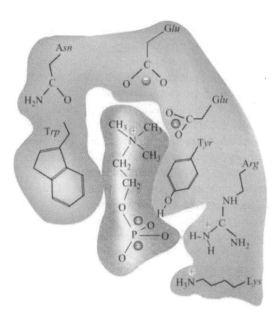

Figure 24.6. Schematic representation of the specific interaction between an antigen (orange) and the antigen-binding "pocket" of an antibody. Mutually approaching charges and formed H-bonds are shown. Reproduced from [1a] with permission.

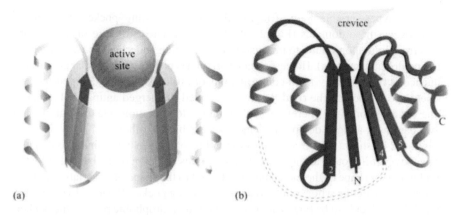

(a) (b)

Figure 24.7. Standard dents in protein globule architectures often dictate the location (but not function) of the active site. (a) The active site lies in the funnel at the top of a β/α-barrel with the parallel β-cylinder; for similar location of the active site in the funnel at the top of the *anti*parallel β-cylinder, see Fig. 24.5d. (b) The active site in the Rossmann fold crevice is formed by moving apart $\beta–\alpha–\beta$ superhelices (in the superhelix $\beta_1–\alpha–\beta_2$ the chain goes away from, while in the superhelix $\beta_4–\alpha–\beta_5$ it goes towards the viewer). Reproduced from [1a] with permission.

In general, when studying proteins, it is easy to see that active sites often occupy "standard defects", i.e., standard dents (determined by the chain fold and not by side groups) in the architectures of protein globules (Fig. 24.7): such a dent automatically provides contacts with many protein side chains at once.

The interface between domains is another frequent place of location of the active site. Fig. 24.8 shows the active site of trypsin-like serine proteases, which is located at the interface between two β-structural domains.

Now it is high time we started discussing enzymes, which are proteins whose main function is to chemically TRANSFORM the molecules bound to them. Enzymes do not create new reactions or change their direction, they "only" accelerate spontaneous processes. But sometimes the acceleration is zillion-fold, many million times stronger than that caused by the most powerful chemical catalysts.

I take the liberty of omitting enzyme classification (it can be found in any biochemical textbook) and focus on how enzymes succeed in providing such a great acceleration of chemical reactions.

Serine proteases are a classical example considered in lectures on simple enzymatic reactions, and I am not going to depart from the tradition.

Serine proteases cut polypeptide chains, i.e., carry out the reaction

$$\text{NH}_2\text{-peptide}' - \overset{\text{H}}{\underset{\text{O}}{\text{C} - \text{N}}}\text{-peptide}'' - \text{COOH} + \text{H}_2\text{O} \ \rightleftharpoons \ \text{NH}_2\text{-peptide}' - \text{COOH} + \text{NH}_2\text{-peptide}'' - \text{COOH}$$

The polypeptide chain hydrolysis reaction proceeds spontaneously when there is enough water in the medium, but it proceeds very *slowly* and may take many years;

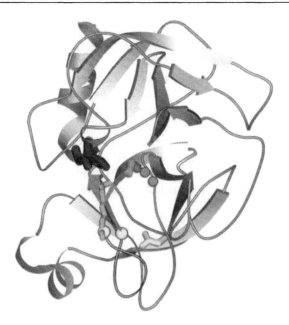

Figure 24.8. Location of the active site in trypsin-like serine proteases. The figure shows parts of the active site: the *catalytic site* with side groups of the "charge-transfer relay triad" [Ser195 (orange), His57 (blue), and Asp102 (crimson)], and the *substrate-binding site* where NH-groups forming the oxyanion hole are shown in green, the non-specific substrate-binding area is shown in light blue and groups lining the specific substrate-binding pocket are shown in yellow. The domain interface is between the pair (His57, Asp102) and other active site parts.

in other words, with free water molecules available, hydrolysis is thermodynamically favorable, although a very high activation barrier has to be overcome. If there are no free water molecules in the medium, the reaction proceeds towards the other side, i.e., towards polypeptide synthesis and water release, and again it is very slow.

However, in the presence of an enzyme, the peptide hydrolysis reaction (or, in the absence of free water molecules, the reverse reaction of peptide synthesis from smaller fragments) takes a fraction of a second, that is, the enzyme decreases the activation barrier dramatically. Let us see how this is done.

First of all, let us consider the main components of the enzyme *active site*. This consists of the *catalytic site* responsible for chemical transformation and of the *substrate-binding site* whose task is to accurately place the substrate under the catalytic cutter (or rather, under the welding/cutting machine, since an enzyme equally accelerates the reaction and its reverse).

In serine proteases, catalysis is carried out by the side chain of one definite serine (whose name "Ser195" has been derived, for historical reasons, from its position in the chymotrypsinogen chain and used for all proteins of the trypsin family; it is this protein, trypsin, that is shown in Fig. 24.8). Let me remind you the chemical formula for a Ser side group is $-CH_2-OH$. However, serine cannot catalyze hydrolysis without certain preparatory steps. It is inactive as long as oxygen is a member of the –OH group and becomes active after H$^+$ removal and oxygen conversion into the –O$^-$ state.

The task of H-atom removal from Ser195 is carried out by the two other members of the "charge transfer triad", His57 (which accepts the H$^+$ atom) and "auxiliary" Asp102. Mutations of these two residues, not to mention mutation or chemical modification of catalytic serine, nearly stop the catalytic activity of serine proteases.

The activated oxygen (–O$^-$) of the Ser195 side group plays the crucial role in catalysis. It attacks the C atom of the treated peptide group (Fig. 24.9) and involves it in a temporary covalent bond (that is longer and not so tight as normal C–O bonds), thereby turning the C atom into the tetrahedral state (when it has covalent bonds not to three but to four atoms: the additional atom is O$^-$ from the Ser195 side group). And

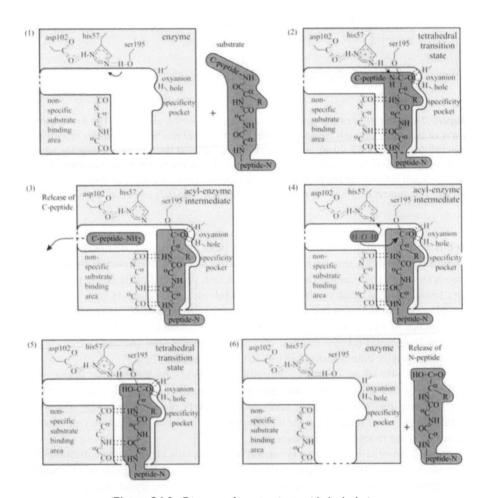

Figure 24.9. Diagram of enzymatic peptide hydrolysis.

subsequent disintegration of this complex results in breaking the C–N bond leading to a breakdown in the treated chain.

The substrate-binding site consists of (Figs 24.8 and 24.9) the *oxyanion hole* (which binds the oxygen of the treated peptide group), of the *non-specific* peptide-binding *area* (which, together with the oxyanion hole, is responsible for the correct positioning of the treated peptide group relative to the activated O atom of Ser195), and of the *specific* substrate-binding *pocket* responsible for recognition of the amino acid preceding the cleavage point.

The general course of the reaction is illustrated by Fig. 24.9. This scheme is a result of long-term studies of many research groups. A lot of different data have been used to build it up: data on catalysis of different substrates and on chemical modifications of the enzyme, results of protein engineering studies (which determined the catalysis-participating points of the enzyme) and studies of enzyme complexes with non-cleavable substrate analogs, as well as X-ray structural work, etc.

What does the enzyme do to accelerate a chemical reaction? To answer this question, let us compare the enzymatic reaction (Fig. 24.9) with a similar spontaneous reaction that occurs in water without any help from an enzyme (Fig. 24.10).

We see that the enzymatic reaction occurs in two stages: first, the C-terminal peptide is cleaved off, and the N-terminal peptide forms a complex with the O atom of serine, and second, this N-peptide changes its bond to the O atom of serine for a bond to the O atom of water. The enzyme-unassisted reaction has only one stage during which the C-terminal peptide is cleaved off, and the N-terminal peptide forms a complex with the O atom of water. Also, we see that all these three reactions proceed via the tetrahedral activated complex, or rather, via the activated complex where the tetrahedral C-atom (which has formed a covalent bond to the O-atom) is near some proton donor (His^+ in the enzyme or OH_3^+ in water).

The fact that the enzymatic reaction proceeds via two activated complexes (and not one as the enzyme-unassisted reaction does) is not an accelerating factor by itself. The

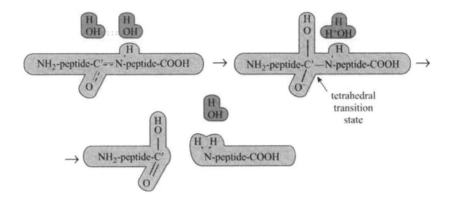

Figure 24.10. Diagram of spontaneous, non-enzymatic peptide hydrolysis in water. The chemical mechanism is similar to that shown in Fig. 24.9.

acceleration happens because the enzyme-bound activated complex is *less* unstable than the activated complex in water (and it is precisely the instability of the activated complex that limits the reaction rate).

In an enzyme, the activated complex (i.e., the transition state) is stabilized as follows.

First, the negative charge of the peptide's $C'-O^-$ group arising upon tetrahedrization of the C'-atom is drawn into the enzyme's oxyanion hole having two protons inside. The hydrogen and electrostatic bond between these positively charged protons and the negatively charged O-atom decreases the energy of the latter and hence the energy of the C'-atom's tetrahedral state (unlike in the non-enzymatic reaction where there is no such pre-arranged "hole"). In other words, the energy (enthalpy) of the transition state decreases, and so-called *enthalpic catalysis* occurs.

Second, near the tetrahedral C'-atom there is the proton-holding His^+, which is approximately as stable as proton-free His^0. In water, the role of the proton donor is played by OH_3^+, the concentration of which, owing to its instability, is extremely low and amounts only to about 10^{-7} mol l^{-1}. The necessity to "fish it out" decreases entropy and hence increases the free energy of the non-enzymatic activated complex. Therefore, entropy hampers the association of all the components of the activated complex in water but does not hamper its association on the enzyme, where all the components are already assembled, the entropy of "fishing out" of the substrate being covered by its sticking to the enzyme's substrate-binding site. This facilitates the association of all the components of the activated complex, decreases its free energy on the enzyme (as compared to that of the non-enzymatic reaction), and results in so-called *entropic catalysis*.

In total, the entropic and enthalpic components of catalysis accelerate the enzymatic reaction by approximately 10^{10} times compared with the non-enzymatic reaction.

If the substrate sticks to the enzyme strongly and the substrate concentration is not too low, entropic catalysis decreases the observed order of the reaction with respect to the substrate concentration. For example, the rate of the non-enzymatic reaction $A + A \rightarrow A_2$ is proportional to the squared concentration of A, while the rate of the same enzymatic reaction where substrate binds to enzyme at the first rapid step depends only on the enzyme concentration and does not depend on the substrate concentration in solution, unless the latter is extremely low. This effect is particularly prominent in the case of reaction involving several molecules and substrates, for example, when a peptide is synthesized from two fragments in the absence of water.

The basic feature of both chemical and enzymatic catalysis is a decrease in the free energy of the transition state, i.e., a decrease of the maximum free energy to be surmounted in the course of the reaction (Fig. 24.11). The free energy may be decreased at the expense of the entropy that is due to association of all the necessary reaction components on the enzyme. Also, the free energy decrease may be achieved (or strengthened) at the expense of the enthalpy of preferential binding of the molecule's transition state and not its initial ("substrate") and/or final ("product") states. By the way, too tight binding of the substrate or product (tighter than that of

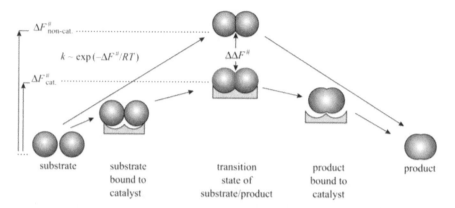

$-\Delta F^{\#}_{non\text{-}cat.}$

$k \sim \exp(-\Delta F^{\#}/RT)$

$\Delta\Delta F^{\#}$

$-\Delta F^{\#}_{cat.}$

| substrate | substrate bound to catalyst | transition state of substrate/product | product bound to catalyst | product |

Figure 24.11. This diagram illustrates the crucial role of catalyst-produced stabilization of the transition (most unstable) state of the substrate/product and stresses the importance of the catalyst's rigidity in decreasing the free energy barrier ($\Delta F^{\#}$) and increasing the reaction rate (k). For the sake of clarity, the figure indicates that the contours of the transition state are complementary to those of the enzyme active site, although actually the key role is usually played by temporary covalent bonds between the substrate/product transition state and the enzyme's catalytic site. The free energies of the initial and final states are not affected by the catalyst. The figure stresses that acceleration of the reaction depends on the $\Delta\Delta F^{\#}$ value, i.e., on how strongly the enzyme binds the transition state.

the transition state) will simply cause *inhibition* of the enzyme, which will lose its "RELEASE" function and quit the game.

In the preferential binding of the transition state the key role may be played either by the binding of its deformed electron system (recall the activated Ser and the "oxyanion hole" of serine proteases) or by preferential binding of the deformed (in the transition state) conformation of the whole molecule.

The latter is vividly illustrated by the catalytic function of artificial "abzymes" (**antibody enzyme**), which are antibodies selected to bind the substrate's transition state and thus to catalyze its chemical conversion (Fig. 24.12). Some abzymes are able to spend almost the entire energy obtained from substrate binding on decreasing the activation barrier, i.e., on deformation of the treated chemical bond. And although the available abzymes are not powerful enzymes (since electron donors and acceptors are not built into them yet, they are only able to accelerate a spontaneous reaction by five orders of magnitude at the most, while this value for natural enzymes is 10–15 orders), the possibility of their creation confirms the importance of preferential binding of the transition state.

The transition state differs from the initial and final states of a chemical reaction by small details, of an Ångström unit in size (Fig. 24.11). To ensure preferential binding of the transition state (rather than the initial and the final ones), the protein must be as rigid as possible. Soft protein, or, rather, a soft active site, is as efficient in catalysis as a rubber razor in shaving . . .

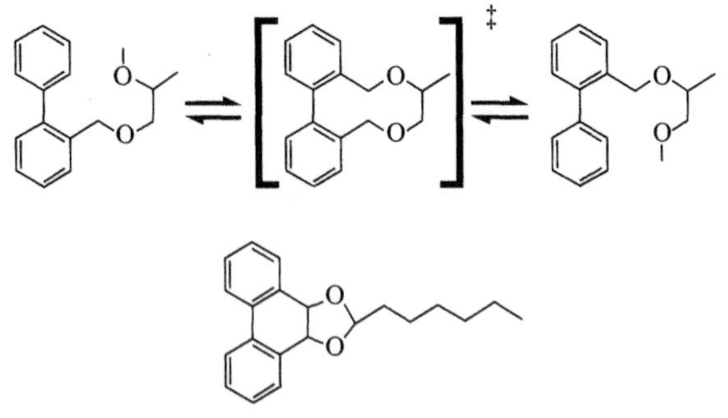

Figure 24.12. A diagram of chemical conversion catalyzed by abzyme, i.e., the antibody selected against the transition state of a slow spontaneous reaction. The transition (‡) state of the given reaction of isomerization has (unlike its initial and final states) a flat three-ring system. To produce antibodies to the transition state of such a shape, the stable molecule (below) with similar rings is injected into the animal's blood.

Here, consideration of the following simple model may be helpful.

Suppose, two molecules (having to form a chemical bond) are drawn into the crevice ("active site") of the protein at the expense of the absorption energy. In the scheme below, these molecules are shown as small balls having the diameter d, and the crevice has the initial width a. If $a < 2d$, the balls are squeezed (which favors bond formation) and the crevice is widened. Both the protein and the molecules are assumed to be elastic (the balls only up to a certain extent, after which they form a chemical bond). Let the protein's elasticity coefficient be k_{pr}, and the ball elasticity coefficient be k_{mol}. The question is: what energy is spent to deform the balls (only this energy can be used to overcome the energy barrier that hinders spontaneous merging of the molecules).

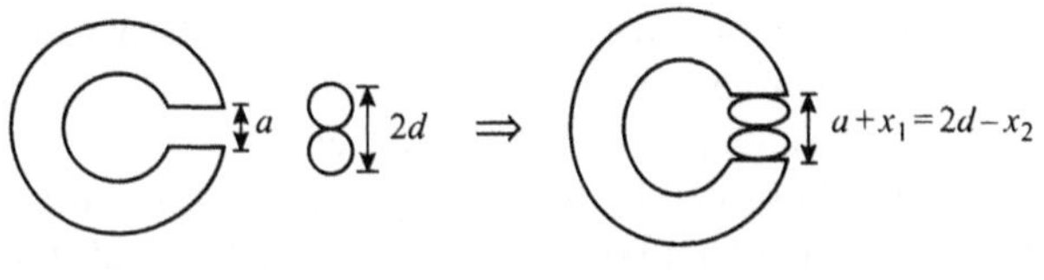

Let us answer this question.

The crevice width grows form a to $a + x_1$ under the pressure of the balls. The height of the two-ball column shrinks from $2d$ to $2d - x_2 = a + x_1$ under the pressure of the crevice. According to Newton's third law, these pressures are equal; that is, $k_{pr}x_1 = k_{mol}x_2$. According to Hooke's law, the energy spent on deformation of protein is $E_{pr} = (1/2)k_{pr}x_1^2$, and the energy spent on deformation of the balls is $E_{mol} = (1/2)k_{mol}x_2^2$. Thus,

$$\frac{E_{mol}}{E_{pr}} = \frac{(1/2)k_{mol}x_2^2}{(1/2)k_{pr}x_1^2} = \frac{(1/k_{mol})(k_{mol}x_2)^2}{(1/k_{pr})(k_{pr}x_1)^2} = k_{pr}/k_{mol} \tag{24.1}$$

and (after some math)

$$E_{mol} = (1/2)k_{mol}(2d - a)^2 \left(\frac{k_{pr}}{k_{pr} + k_{mol}}\right)^2 \tag{24.2}$$

Thus, if the protein is "soft" ($k_{pr} \ll k_{mol}$) the main energy is spent to deform it and it is useless in catalysis. Only a "rigid" protein (with $k_{pr} \gg k_{mol}$) can concentrate most of the absorption energy on deformation of molecules.

Thus, an efficient enzyme must be rigid.

Inner voice: Thus, does "rigid" refer to the active site rather than to the entire protein? And when you talk of "rigidity", do you have mechanical rigidity in mind, or something else?

Lecturer: It is the active site that must be rigid. Rigidity of the entire protein is important for catalysis only so far as it maintains the rigidity (and specificity, see below) of the active site. And I have in mind the energetic rather than the mechanical aspect of rigidity: I mean that a small change in the substrate causes a great change in the energy of its binding to the enzyme. The mechanical picture, see Figs 24.11 and 24.12, is easier to grasp. However, the same effect can be caused by interactions between the enzyme and the substrate valence electrons: the energy of interaction of the electron clouds changes sharply with a small-scale change of their shapes.

Actually, the deformation of rigid chemical bonds of a substrate (which is the essence of any chemical reaction) is more efficient when transient substrate-to-enzyme binding is performed by rigid covalent bonds rather than by soft van der Waals interactions. This explains why abzymes, which use van der Waals interactions only, are weaker catalysts than natural proteins with their chemically active catalytic sites.

Inner voice: Thus, you mean that flexibility of a protein structure plays no role in catalysis?

Lecturer: Here, first of all, one has to distinguish the catalytic act itself, which is not facilitated by flexibility, from the stage of substrate penetration into the active site, where protein flexibility can be required (and when the product leaves this site, too). We will come to this in the next lecture. Second, one has to have in mind the following. If the enzyme's active site does not fit the substrate's transition state perfectly (nothing is perfect. . .), some flexibility would allow this non-perfect site to be suitable for at least a fraction of the time. But even in this case, the active site must fit the transition state better than both initial and final states of the substrate/product.

Inner voice: All your speculations, up to now, dealt with a static or an almost static enzyme. What if the energy necessary for catalysis is stored as kinetic energy of moving parts of the enzyme, for example, as the energy of its vibration?

Lecturer: This is not possible. One can show that kinetic energy dissipates within $\sim 10^{-11}$ s even in a large protein and much faster in a small one, while one act of enzymatic catalysis takes at least $\sim 10^{-6}$ s. This means that the energy released during substrate binding dissipates and converts into heat long before it can be used in catalysis.

Indeed, one part of a protein moves about another in some viscous environment such as water or membrane. This movement is described by the usual equation

$$m\left(\frac{d^2x}{dt^2}\right) = F_{frict} + F_{elast} \tag{24.3}$$

Here m is the particle's mass, d^2x/dt^2 is acceleration, F_{frict} is the force of friction, and F_{elast} is the elasticity force. The mass can be estimated as $m = \rho V$, where ρ is the particle's density and V is its volume. The friction force $F_{frict} = 3\pi D\eta (dx/dt)$, or $\approx 10D\eta (dx/dt)$ according to Stokes' law, where η is the viscosity of the medium, dx/dt is the particle's velocity, and D is the particle's diameter. The F_{elast} value can be estimated (from Hooke's law) as $F_{elast} = -(ES/L)x$, where x is displacement from the equilibrium point, S is the cross-section of an elastic joint between two parts of the protein, L is the length of this joint, and E is the elasticity module. It is reasonable to assume, for the sake of simplicity, that all the linear dimensions are close (see the scheme below), so that $L \approx D$, $S \approx D^2$, $V \approx D^3$.

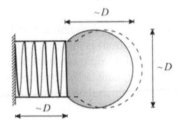

Equation (24.3) determines two characteristic times. One, the most important, we have already discussed. This is t_{kinet}, the time of friction-caused damping of the movement. It can be estimated from the "kinetic" part of eq. (24.3): $m(d^2x/dt^2) \approx 3\pi D\eta (dx/dt)$. The estimate is: $t_{kinet} \approx 0.1\rho D^2/\eta$, since $m \approx \rho D^3$, and $3\pi \approx 10$. As $\rho \approx 1\,g\,cm^{-3}$ for all the molecules we deal with, and $\eta \approx 0.01\,g\,cm^{-1}\,s^{-1}$ for water, $t_{kinet} \approx 10^{-13}\,s\,(D/nm)^2$, where (D/nm) is the particle's diameter expressed in nanometers. This means that the kinetic energy of a small part of protein ($D \approx 1$ nm) dissipates in water within $\sim 10^{-13}$ s, and of a large protein ($D \approx 10$ nm) in $\sim 10^{-11}$ s. In a more viscous environment, e.g., in a membrane, the kinetic energy dissipation is proportionally faster.

A characteristic time of protein vibrations, t_{vibr}, follows from the "oscillation" part of eq. (24.3): $m(d^2x/dt^2) \approx EDx$: $t_{vibr} \approx (m/ED)^{1/2} \approx D(\rho/E)^{1/2}$. At a value of the elasticity module typical for proteins $E \sim 10^{10}\,g\,cm^{-1}\,s^{-2}$, $t_{vibr} \approx 10^{-12}\,s\,(D/nm)$.

Comparing $t_{kinet} \approx 0.1\rho D^2/\eta \approx 10^{-13}\,s\,(D/nm)^2$ to $t_{vibr} \approx D(\rho/E)^{1/2} \approx 10^{-12}\,s\,(D/nm)$, we see that oscillations are absolutely impossible for proteins with $D < 10$ nm: they would be damped by friction before the first oscillation occurs even in water, not to mention more viscous environments like membrane.

The above analysis shows that (1) the potential energy should be put into the substrate (*not* into the protein) to bring it, the substrate, closer to the transition state, and therefore, the protein active site should be rigid; and (2) kinetic energy cannot be used in catalysis.

The theory of catalysis at the expense of preferential binding of transition states was put forward by Holden and Pauling as far back as the 1930s and 1940s. More recently, the transition states for some enzymatic reactions have been probed by protein engineering techniques, i.e., by directed replacement of amino acids in the enzyme active sites. It has become clear (and visible, thanks to X-rays) which residues of the enzyme are responsible for the sticking together of components required to build up the activated complex, and which for preferential binding of exactly and exclusively the transition state (previously hypothetical since its arrangement could not be seen in experiment).

For example, Alan Fersht's study of tyrosyl-tRNA synthase has established the residues (Thr40 and His45, see Fig. 24.13) that bind to ATP γ-phosphate *only* in the transition state. This additional strong binding distinguishes the transition state from bound substrates or products of the reaction.

The transition state is most unstable (by definition!) and therefore cannot be observed directly. So X-rays, which can see only the stable position of the substrate on the enzyme, see the substrate unbound to Thr40 and His45 (the distance is too large); meanwhile, replacement of Thr40 and His45 by small alanines decreases the mutant protein activity by more than a million times, while binding of the substrate by the mutant protein remains almost unaffected. This shows that Thr40 and His45

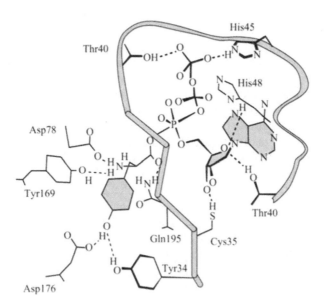

Figure 24.13. The presumed structural arrangement of the transition state for tyrosyl-AMP formation from tyrosine and ATP on the tyrosyl-tRNA synthase enzyme. The tyrosine ring is hatched as well as the adenosine rings. The tyrosine-attacking activated five-coordinate α-phosphorus (P) of the ATP molecule is drawn in the center, while ATP γ-phosphate is at the top. This γ-phosphate forms hydrogen bonds to Thr40 and His45 only in the transition state; a stable, X-ray-observed binding of substrates involves no such bonds. Adapted from [14].

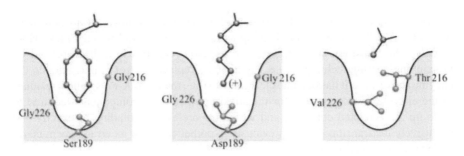

Figure 24.14. The structural arrangement of the specific substrate-binding pocket in various serine proteases. Reproduced from [1a] with permission.

bind to ATP γ-phosphate only in the transition state, at a strained (five-coordinate) state of α-phosphorus attacking the tyrosine's CO-group.

When discussing the enzymatic reaction we cannot but mention its very high specificity.

For example, serine proteases cut a polypeptide only after certain amino acids: chimotrypsin cuts it after aromatic ones, trypsin does so after positively charged residues, elastase only after the smallest. This effect (it is called "*key* (substrate)– *lock* (enzyme)" substrate recognition) is ensured by the neat structure of the *specific* substrate-binding "pocket" (Fig. 24.14). This pocket recognizes the "key" side group of the treated peptide. In trypsin, the key side group is last before the cleavage point. In papain, it is last but one; in thermolysine, it is positioned immediately *after* the treated C–N bond.

The key residues mark the possible cleavage point rather accurately; however, the rate of polypeptide digestion also depends on other residues surrounding this point (this is only natural because they also have some interactions with the protease).

A still more important factor affecting cleavage efficiency and even its possibility, is the polypeptide conformation at the cleavage site. The C–N bond cannot be cut if a fixed substrate conformation does not allow this bond and the key side group to gain their "correct" (for cleavage) positions simultaneously. This is why proteases are not good at digesting proteins with a rigid structure but easily deal with denatured, flexible proteins. Some definite rigid conformation of the polypeptide chain allows it to stick to a definite protease, to block its active site, and to remain there, intact, forever. Many inhibitors use this principle of "unproductive binding".

The proteases that we have discussed in this lecture are efficient but not absolutely accurate enzymes. Presumably, the organism can still tolerate some mistakes in their selection of the cleavage point. But there are proteins that must be absolutely reliable. Such reliability is ensured by the operation of a few active sites in the same protein.

Having dealt with the elementary functions of proteins, we will consider their more complex functions in the next lecture.

LECTURE 25

Having considered elementary protein functions in the previous lecture, we can now proceed to their more complicated functions.

As I have said already, some proteins are supposed to work extremely accurately; this precision in action is achieved by the conjugated work of several active sites of a protein. Specifically, aminoacyl-tRNA synthetases work in this way. They charge tRNAs with the proper amino acids, and precision in their work is crucial for precision of protein biosynthesis. They are allowed to be mistaken only once per many thousands of acts of aminoacyl-tRNA synthesis, while recognition of an amino acid by the substrate-binding cleft is erroneous in about 1% of cases. A "double-sieve" effect used to eliminate these errors (Fig. 25.1) is achieved by the special construction of the enzyme.

Aminoacyl-tRNA hydrolysis is not simply a reversal of the synthesis reaction: both reactions release free phosphates, that is, *both* of them decrease the free energy. Aminoacyl-tRNA synthase has two active sites, one for synthesis and one for hydrolysis, and the substrate-binding cleft of the center of hydrolysis is smaller than the cleft of the aminoacyl-tRNA synthesis site.

The principle of the sieve is to reject particles that are greater than the sieve's mesh.

The first sieve, the sieve at the aminoacyl-tRNA synthesis site, allows tRNA to be charged with the "proper" amino acid and with a small number of those "wrong"

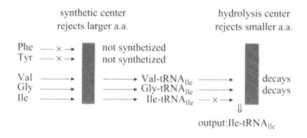

Figure 25.1. A scheme of the "double-sieve" mechanism used by Ile-aminoacyl-tRNA synthase to produce the Ile-charged isoleucine tRNA (tRNA_Ile).

331

amino acids that are no larger than the "proper" one. All amino acids larger in size are rejected by the rigid cleft. Most of the smaller amino acids are rejected too, since their hydrophobicity/hydrophilicity is different from that of the "proper" amino acid. However, selection on a hydrophobicity/hydrophilicity basis is not as strict as selection caused by the necessity to squeeze into a rigid dent of a given size. Thus, the output of the first step is a lot of aminoacyl-tRNAs charged with a "proper" amino acid and some quantity of aminoacyl-tRNAs charged with "wrong" (smaller) amino acids, all larger ones having been rejected by the first sieve.

The second active site has a smaller "mesh" (substrate-binding cleft); it promotes hydrolysis of aminoacyl-tRNAs, but *only* of those charged with amino acids smaller than the "proper" amino acid. The "proper" amino acid is larger, and therefore the second sieve rejects it, and all the others are hydrolyzed and decay. These two sieves provide production of only the correctly charged aminoacyl-tRNAs.

To complete the story about enzyme specificity, it is worthwhile mentioning some important practical aspects of active site studies. (1) Many people are currently studying various substrate-binding pockets and trying to find – with opened eyes, not blindly – the inhibitors that bind to them. Special interest is paid to HIV proteases and other HIV proteins, and to other proteins of most harmful viruses. (2) Directed mutations introduced to the substrate-binding site can modify a "natural" substrate specificity (for example, of serine proteases) and even create a new specificity – to meet the needs of medicine and industry. (3) Great interest is being shown in attempts – they are only attempts so far – to create, on the solid basis of existing protein globules, new active sites with a new, "grafted" activity.

Studies of enzymatic activity have shown (recall the last lecture) that only a small part of a protein globule is involved in the catalytically active site, while the rest, the largest portion of the protein globule, serves only as a solid base that forms and fixes the active site. Therefore, we must not be surprised that proteins with different primary and even tertiary structures can have similar or even identical biochemical functions.

Serine proteases again serve as a classic example. There are two classes of such proteases: the trypsin type and the subtilisin type. They have no similarity either in sequences or in folds (Fig. 25.2). They belong to different structural classes (trypsin is a two-domain β-protein, subtilisin is a single-domain α/β-protein). They have different substrate-binding clefts. They only have similarly positioned key amino acid residues of the catalytic sites (Fig. 25.3), and even this similarity refers mainly to the functional "tips" of the residues.

Moreover, there are proteases that differ from serine proteases even in the catalytic site zone. I have in mind metalloproteases like carboxypeptidase or thermolysin. A key role in their catalytic action is played by a Zn^{2+} ion built into the protein structure by powerful coordinate bonds. Having a small radius because of its lack of outer electrons, a multi-charged metal ion produces a very strong electric field. This allows such an ion, built into the metalloprotease, to activate a water molecule by breaking it into OH^- and H^+, and the activated water disrupts a peptide bond. In doing so, the activated water works approximately in the same way as OH^- in the

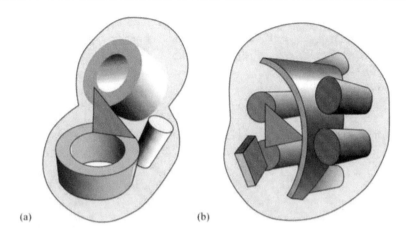

Figure 25.2. Schemes of serine proteases (a) of the trypsin type and (b) of the subtilisin type. α-helices (cylinders) β-sheets and β-barrels are shown against the background of the molecule's contour. The active sites are shown as dark-gray triangles.

non-catalyzed reaction described in the previous lecture. A significant difference is that the enzyme activates exactly the water molecule sitting near the peptide bond to be attacked, while in a non-catalyzed reaction the activated water is to be found among a great many non-activated ones, which takes a lot of time. Simultaneously, the Zn^{2+} charge stabilizes – in metalloproteases – a negative charge at the tetrahedral transition state of the treated peptide; this mimics the oxyanion hole action in serine proteases.

Thus, the same chemical reaction can be catalyzed by completely different proteins, and the same key role of a powerful electric cutter can be played by multi-charged metal ions, or by environment-activated side chains (as in trypsin), or by organic co-factors that accept electrons or protons . . .

On the other hand, as we remember, proteins similarly arranged as α/β-barrels or Rossmann folds can catalyze very different reactions. A classic example of proteins having the same origin and architecture but different functions is the pair lysozyme and α-lactalbumin. These proteins not only have identical, within 1–2 Å, 3D structures, but also display a quite high sequence identity (35%). Nevertheless, one of them (lysozyme) is an enzyme (it cuts oligosaccharides), and the other (α-lactalbumin, which descended from lysozyme only $\sim 100\,000\,000$ years ago, when milk appeared) is not an enzyme at all: it has lost the lyzozyme's active site, and now only modifies the enzymatic action of another protein.

The functional features of proteins are fixed by its enzyme classification, "EC"; the correlation between EC and CATH (protein structure classification) is not striking, that is, in outline, a certain biochemical function is not exclusively performed by proteins of a certain structural class.

However, "large-scale" protein properties (which depend not on a small catalytic site but on a greater area of the protein surface), such as the capability of binding large

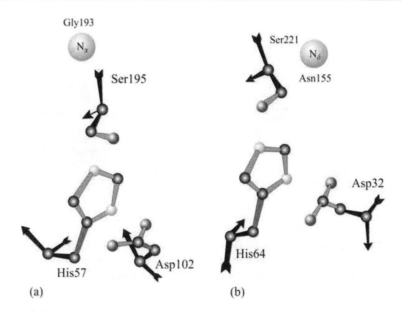

Figure 25.3. Schemes of the active sites of serine proteases of (a) the trypsin type and (b) the subtilisin type. (Similar active sites exist not only in proteases but also in lipases and in some other enzymes.) The main-chain fragments (up to C^β-atoms) are shown in black; the arrows show the main-chain direction. The circle (with N inside) shows the positions of the oxyanion holes. The main-chain N_α–H groups (in trypsin) and the side-chain N_δ–H group (in subtilisin) face these holes. The only invariant feature of the active sites in these two classes of proteins is the position of the tips of the side chains of the catalytic triad (Ser, His, Asp), and with some variation, of the oxyanion hole; all other details are different, including the main chain direction in the catalytic residues His and Asp.

ligands, partially correlate with overall protein architecture. Also, partial correlation is observed between the overall architecture of a protein and its involvement in biological processes.

This is most vividly demonstrated by proteins responsible for transmembrane transport of any kind. Being membrane proteins, they have a most regular transmembrane α- or β-structure. We spoke about them in previous lectures, as well as about fibrous proteins that incorporate huge blocks of secondary structures and usually play a structural role.

The architecture–function correlation, although to a lesser extent, can also be traced in water-soluble globular proteins. For example, half of heme- and DNA-binding proteins are α-proteins, while the proportion of α-proteins among carbohydrate- and nucleotide-binding proteins is only 20%. On the other hand, almost all proteins controlling immunity and cell recognition are β-proteins. Nearly half the carbohydrate-binding proteins are also β-proteins, whereas their proportion among heme-, nucleotide- and DNA-binding proteins only amounts to 10%. Almost all nucleotide-binding proteins belong to the α/β class. They include all 11 proteins of the glycolytic pathway. In general, α/β proteins often (more often than others) appear to be enzymes that are responsible for "cell housekeeping".

All this demonstrates that, although the overall structure of a protein is rather independent of its catalytic activity, the architecture of the protein to a certain extent correlates with its "large-scale" substrate-binding activity, with its environment ("living conditions") and with its involvement in certain biological processes.

So far I have said almost nothing about conformational changes in functioning protein molecules. I have done this on purpose, since so far we have discussed catalysis of one isolated reaction (where substrate/product was the only mobile element). For catalysis of an isolated reaction, conformational changes in a protein can only worsen its catalytic properties. Indeed, efficient catalysis implies preferential binding of the transition state (which must be stronger than the binding of the initial and final states, although they differ from the transition state by as little as one Ångström unit). Catalysis is aimed at either synthesis or disruption of rigid covalent bonds. Any attempt to disrupt the bonds using a "flexible" protein would be similar to an attempt to disrupt a wire using a piece of rubber or a pillow . . .

The following addition would not be out of place here. When the active site does not fit its function ideally (and ideal tuning must be a result of superthorough natural selection), it may require a minor deformation to adopt the active state. Then this deformation is indeed functionally necessary because from time to time it brings the non-ideal protein up to the mark. However, here the necessity (a deformation that compensates for the imperfect geometry of the active site) must be distinguished from virtue (protein perfection).

Of course even the best possible protein undergoes deformation in the course of an elementary catalytic action (just because it is not as hard as a diamond), but, functionally, this may be compared with paper deformation under a pen: the less, the better.

It's quite another matter when a protein must change one action for another; then its deformation providing the transition from one role to another is really of great functional importance. For example, in many proteins penetration of the substrate into the active site requires a slight displacement of site-adjacent side groups. But as soon as the substrate has occupied the site, the catalytic activity of this site does not require any further movement of the side groups. It can be said that mobility is required in preparing the reaction, while rigidity is necessary in its course. This reminds me of the rule by Clausewitz, a military theorist: "Be detached on the march but united in battle".

Sometimes the transition from one protein function to another results simply from substrate movement from one active site to another (recall aminoacyl-tRNAs synthases: there, the entire mobility is executed by the substrate). However, transition from one elementary function to another often occurs through more or less considerable deformation of the protein structure. The ability to deform in such a way is inherent in the construction of the protein molecule. This is what we are going to consider next.

When discussing protein dynamics, we have to distinguish between small and large movements.

Actually, small movements are heat-induced fluctuations. Such fluctuations can be seen against the background of the averaged "static" protein structure using high resolution X-ray crystallography. The fluctuations are seen from diffused X-ray patterns, whose fuzziness grows with temperature. Surface groups of a protein fluctuate with an average amplitude of about 0.5 Å, while the fluctuation amplitude of core groups is a few times less. In loops amino acids fluctuate more strongly than in secondary structures. Side groups fluctuate much more strongly than the main chain. The strongest fluctuations are typical of those long surface side groups, whose ends are not fixed to other groups; fluctuations of these ends are so large that X-rays cannot locate these ends. That is why the protein interior is said to be solid, while its surface resembles a liquid and turns solid only at about $-100\,^{\circ}$C. (I would like to note in parentheses that a semi-liquid surface layer of molecules is generally typical of crystals).

The same conclusion on the combination of "solid" and "liquid" components in protein globule dynamics is supported by γ-resonance ("Mössbauer") spectroscopy. It follows from absorption of monochromatic γ-quanta by Mössbauer nuclei of some heavy metal isotopes (e.g., ^{57}Fe) and allows us to estimate, specifically, the size of a solid part of the protein cemented to these nuclei.

Larger changes in protein structure occur during ligand binding. The relaxation dynamics of such structural deformations are best studied in myoglobin. A laser flash can instantly (within $\sim 10^{-13}$ s) tear the ligand (CO) off the myoglobin heme, and the relaxation towards the initial state can be followed by optical spectra.

The complex kinetics of such relaxation studied for a wide range of temperatures indicate that the native conformation can be achieved through a number of energetic barriers, and that the native conformation comprises a number of subconformations differing from one another in minor details.

Such, and even larger, deformations of proteins that accompany their functioning are studied by crystallizing proteins in various functional states. Also, much information can be obtained using various spectral techniques, chemical modifications, etc.

We know from the previous lecture about one functionally important deformation exemplified by regulation of DNA-binding activity of the trp-repressor. Let us consider functionally important deformations in more detail.

First of all, let us see how protein deformation helps to combine the steps of the cycle BIND $\rightarrow$ TRANSFORM $\rightarrow$ RELEASE. Figure 25.4 illustrates *induced fit* (the term proposed by D. Koshland) of the phosphorylating protein, hexokinase, to its substrates. This protein transfers the phosphate group from ATP to glucose. Judging from chemical principles only, this phosphate group might be alternatively transferred to water, but this never actually happens. In an attempt to understand the reasons, Koshland put forward the following postulates. (1) Prior to binding to the substrate, the enzyme is in the "open" form (in which it can take the substrate from water but cannot phosphorylate it). (2) After substrate binding, the enzyme adopts the "closed", catalytically active form, with all parts of the active site brought together and capable of catalyzing the phosphorylation reaction, but in this form water is removed from the active site and so cannot compete with the substrate for phosphorylation. (3) After the act of catalysis, the enzyme opens again, and the phosphorylated substrate is released.

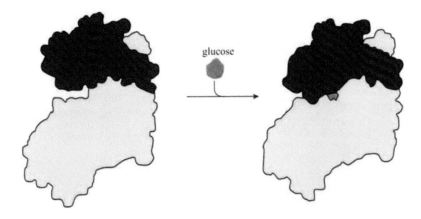

Figure 25.4. Induced fit in hexokinase. In the open form, two domains are separated by a deep crevice, which the glucose can enter. When the glucose enters the crevice, the domains rotate, closing the crevice and expelling water from it, and all the active site components come together. Reproduced from [1a], with permission.

Later this hypothesis received complete experimental verification (Fig. 25.4) but *only* for proteins that need to screen the treated substrate from competing water molecules (although there were numerous attempts to extend the mechanism of induced fit to cover all enzymatic reactions). For example, trypsin has no such need, nor does it have an induced fit for the substrate, because trypsin (as well as chimotrypsin, elastase, subtilisin, etc.) does not undergo deformations and recognizes the substrate using the simplest principle *"key–lock"*.

I should like to draw your attention to the fact that induced fit results from *displacement* of either large blocks (discussed in the previous lecture) or whole protein domains (Fig. 25.4) but *not* from complete rearrangement of the protein fold. In turn, the displacement is mostly a result of small local deformations. (Analogy: muscles contract ("local deformation") and make fingers ("domains") clench into a fist, but the fingers do not turn into teeth or tentacles . . .)

The same happens in all other cases of "conformational rearrangements in proteins", except for rare cases when a whole α-helix or β-strand comes out from its place in the globule and adopts an irregular conformation. One of the greatest rearrangements that I know occurs in calmodulin. This protein is dumbbell-shaped, with α-domains as "heads" kept wide apart by a long α-helix playing the role of the "handle". However, when calmodulin is binding to other proteins, the intact "heads" come together and stick to each other and to the target protein, while the former "handle" (a long α-helix) decays.

Protein domains are mobile not only in space but also in time, in the evolutionary process. As I have already mentioned, domain genes, as a whole, can wander from protein to protein, sometimes combined, sometimes detached. It often happens that in one organism there are several monomeric proteins, while in another they fuse into one multidomain protein.

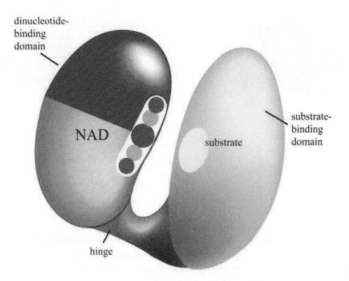

dinucleotide-
binding
domain

NAD

substrate

substrate-
binding
domain

hinge

Figure 25.5. Chains of NAD-dependent dehydrogenases are folded into two independent domains connected by a hinge. One domain is a universal NAD-binding domain; similar domains can be found in many different proteins. The other is a substrate-binding domain; it is individual for each dehydrogenase. Reproduced from [Ia], with permission.

The relative autonomy of domains is well demonstrated by a large family of proteins called dehydrogenases. These proteins catalyze OH-group oxidation and other similar reactions using co-factors NAD^+/NADH that easily accept and release separated protons. However, they deal with different substances: alcohol dehydrogenase treats alcohol (ethanol), lactate dehydrogenase treats lactate, and so on.

Dehydrogenases consist of two domains (Fig. 25.5) with a hinge between them. One domain is the substrate-binding module; it is structurally individual for each dehydrogenase. For example, that of alcohol dehydrogenase contains a β-cylinder, while this domain of glyceraldehyde-3-phosphate dehydrogenase contains a flat β-sheet.

In contrast, the other (NAD-binding) domain is nearly the same for all NAD-dependent dehydrogenases, although in some of them it is located in the N-terminal half of the chain, while in others it is closer to the C-terminus, and despite the fact that the primary structures of NAD-binding domains display no notable homology. The chains of NAD-binding domains form the Rossmann fold (of the α/β class), and the structural similarity of these domains from different dehydrogenases is so close that many of them may be superimposed (including the NAD-binding sites) with an accuracy of 2 Å (presumably, this shows that the kinship is better remembered by the spatial structure than by the primary one).

Thus, the catalytic NAD-binding domain is a universal unit, while structurally different substrate-binding domains provide a variety of actions of dehydrogenases. The active site of dehydrogenase results from the contact of its two halves located in these two domains (Fig. 25.5).

Now I am going to discuss the non-contact (also called *allosteric*) interaction between active sites. Allosteric interactions between various binding and active sites, specifically in oligomeric proteins, play a crucial role in the control and integration of biochemical reactions. A "signal" about the state of one active site is transmitted to another site by means of a deformation of the protein globule that affects the recipient site.

Let us consider (in a strongly simplified form) the best-studied allosteric protein, hemoglobin. This protein is a tetramer, or rather, a complex of two α-chains and two similar to them β-chains (Fig. 25.6a). Its task is to *bind* oxygen in lungs (where it is plentiful), to *transport* it to muscles (where it is scanty) and to *pass* it to muscle myoglobin, which is a monomer protein similar to any of the four subunits of hemoglobin. The active site of myoglobin and of each α and β subunit of hemoglobin is heme, i.e., a ring-shaped co-factor with an iron ion that can bind one O_2 molecule (Fig. 25.6b). The successful action of hemoglobin as an oxygen carrier is determined by allosteric (in this case, concerted) interaction of its four hemes.

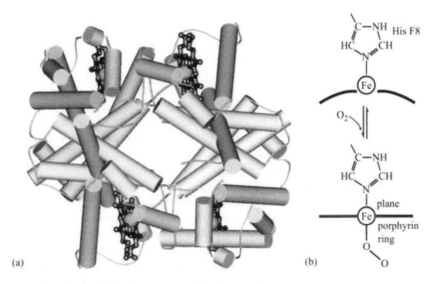

Figure 25.6. (a) The hemoglobin molecule consists of two α- and two β-chains; each chain is colored differently; each chain forms many helices shown as cylinders and includes a heme shown as a violet wire model with an orange iron ion inside. (b) Scheme of oxygen binding to the hemoglobin's heme. Prior to binding O_2, the iron ion (Fe^{2+}) is in a high-spin form with two outer electrons in different orbitals. This makes Fe^{2+} a bit too large to fit into the center of the heme's porphyrin ring, which is therefore somewhat bent. When O_2 is bound, the Fe^{2+} ion acquires the low-spin, more compact form, since the two outer electrons now have opposite spins and are in the same orbital. Now Fe^{2+} can enter the heme ring's center, and the ring acquires a flat form. The shown histidine of the α-helix F coordinates the iron ion. Part (b) is adapted from [14] with minor modifications.

In the heme, an iron ion is fixed by four coordinate bonds $N-Fe^{2+}$. It also participates in one more bond of this kind to the closest histidine of hemo- or myoglobin. This happens to be histidine of the α-helix F. And Fe^{2+} can form yet another coordinate bond to O_2 (or to CO – which would cause carbon monoxide poisoning...).

Analysis of the static structures of both myo- and hemoglobin showed that an oxygen faces a difficulty when approaching a heme, unless there is a simultaneous heat fluctuation of the conformations of a few heme-shielding side groups.

When an oxygen approaches the heme and binds to its Fe^{2+}, the latter changes its spin form and, consequently, the electron envelope. Its diameter becomes a bit smaller, and it enters the center of the heme's porphyrin ring. Its slight displacement (by half an Ångström unit) causes a shift in the position of its coordinating histidine of the α-helix F.

This *electronic-conformational* interaction (implying a rapid change first in the electronic and then in the conformational state) initiates a chain of conformational deformations in hemoglobin. The shift of histidine (through numerous small protein deformations) slightly changes the outline of the O_2-bound subunit, and its contacts with three other subunits weaken. These subunits, still unbound to O_2, begin to relax in their turn (obtaining the same outlines as a protein unaffected by interactions with neighbors and the same shifts of the inner atoms), so that their histidines begin to push the appropriate iron ions into the heme centers. Now oxygen binding to these irons is facilitated. As a result, one oxygen bound to hemoglobin provokes the binding of three more O_2 molecules; and, in the same way, the loss of one oxygen provokes the loss of all others.

In parallel with oxygen binding and releasing, two more important reactions proceed in hemoglobin. These are also connected with allosteric conformational changes. When hemoglobin is in the O_2-free ("deoxy") form, its subunits bind CO_2 (using not the hemes but the N-termini of the chains) and H^+ ions (using histidines of the C-terminal helices that are close in space to the N-termini). When hemoglobin binds O_2 in lungs and changes to its "oxy" form, it loses these four H^+ (which is called the "Bohr effect") and four CO_2 molecules available to be breathed out.

It follows from the above that hemoglobin acts as a protein that binds several O_2 molecules *simultaneously*, and this is reflected by the non-linear, S-shaped plot for the dependence of its (*tetrameric* protein) saturation with oxygen on the oxygen concentration (Fig. 25.7). As to myoglobin (*monomeric* protein), its O_2-saturation plot is free of the S-shaped sag.

Therefore, in the lungs where oxygen is plentiful, hemoglobin becomes oxygen-saturated. In tissues where the (venous) O_2 pressure is relatively low, the tetrameric hemoglobin *releases* O_2, while the monomeric myoglobin *still binds* it and passes it to the muscles (e.g., to the muscles responsible for lung expansion and compression) to take part in oxidative reactions there.

Incidentally, muscles are going to be the concluding topic of this lecture. I should like to describe how cross-striated muscles work, or rather the *mechanochemical* function of proteins underlying their operation.

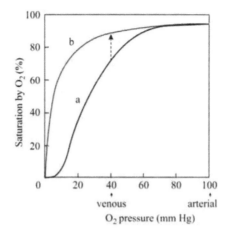

Figure 25.7. Curves of oxygen saturation for tetrameric hemoglobin (a) and for monomeric myoglobin (b). The arterial pressure of oxygen corresponds to its binding in the lungs; the venous pressure corresponds to oxygen's passing from hemoglobin to myoglobin in muscles (which is shown by an arrow). The experimental curve of O_2-to-myoglobin binding corresponds to a first-order reaction. The experimental S-shaped curve of O_2-to-hemoglobin binding corresponds to a reaction of approximately the third order (rather than to a fourth-order reaction, as indicated in the text for simplicity). Adapted from [5] with minor modifications.

In muscle contraction the key role is played by myosin (Fig. 25.8a). It is composed of a fibrous part, "the tail" (a long α-helix), and a globular part, "the head". The helical tails are twisted in couples and these then form a myosin filament. The extended globular heads project from the filament periodically with $\sim 100\,\text{Å}$ pauses between them. They can form bridges to actin filaments (formed by actin, a globular protein, Fig. 25.8b). Moreover, by "burning" ATP these heads can "walk along" the actin filaments. The latter results in muscle contraction.

How do the myosin heads perform this "walking", i.e., how is the chemical potential of ATP converted into mechanical movement?

A "swinging cross-bridge model" satisfactorily describes the muscle contraction. It results from long-term studies by many groups and is based, specifically, on X-ray scattering from the contracting muscle, on electron microscopy of muscles, and on X-ray crystallography of the myosin head in its different states. These studies revealed the best part of the main stages of the muscle contraction process.

As it turned out, the myosin head can adopt either of two macroscopic conformations: the bent, or "45 ° conformation" (R), and the straightened out (at 90° about the myosin fiber), or "attacking" conformation (A).

In each of them, the head can be either actin-bound (a+) or unbound (a−).

Transitions between these $2 \times 2 = 4$ states of the myosin head form a "macroscopic cycle" (Fig. 25.8c): R, a+ $\leftrightarrow$ R, a− $\leftrightarrow$ A, a− $\leftrightarrow$ A, a+ $\leftrightarrow$ $\cdots$ Note that a *directed*

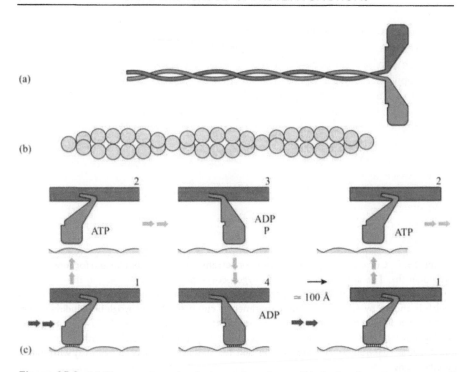

Figure 25.8. (a) The myosin molecule; it consists of two globular heads and a common tail formed by a double α-helical superhelix. The ATP/ADP binding site is schematically shown as a small cleft in the myosin head; actually, the head consists of several globular domains. (b) A fibril formed by actin globules. (c) The mechanochemical cycle. The scheme shows the macroscopic states of myosin and small ligands bound to it at each of the four main stages of this cycle. The myosin head is bound to actin at stages 1 (macroscopic state R,a+) and 4 (A,a+), and not bound at stages 2 (R,a−) and 3 (A,a−). The head is bent at stages 1 (R,a+) and 2 (R,a−), and straightened out at stages 3 (A,a−) and 4 (A,a+). The head changes its bend during the transition from stage 2 to stage 3 and from stage 4 to stage 1; in the latter case, the head is bound to actin, and therefore its bending is accompanied by a "power stroke" that shifts the myosin fiber along the actin fiber by $\sim$ 100 Å. Transition to each next stage consists of two sub-transitions (this is depicted by a double arrow). The "macroscopic" sub-transition is a change of either myosin conformation or its bridging to actin. The "microscopic" sub-transition changes myosin binding to ATP or to the products of its hydrolysis. The order of these two elementary acts is not known in all cases. For example, the loss of phosphate (P) can either precede or follow myosin-to-actin binding during the transition from stage 3 to stage 4.

cycle, and hence, a *directed* motion is possible only when the number of states exceeds two (two states only allow oscillating back and forth).

Also, there are four states of the myosin head relative to "small" (ATP/ADP) ligands: it can be free of these ligands, it can bind ATP, it can bind both products of ATP hydrolysis (ADP and phosphate P), and it can bind only ADP. These four states form a "microscopic cycle" (Fig. 25.8c).

The microscopic cycle is coupled with the macroscopic one. The former provides energy, the latter transforms it into muscle contraction. Such coupling is possible only when each step of the microscopic cycle can take place *only* at a definite stage of the macroscopic cycle. That is, a change of both the head's conformation and the head's binding to actin changes the properties of the site that binds ATP/ADP and catalyzes the ATP $\leftrightarrow$ ADP + P transition. Since the ATP/ADP-binding site (located in the cleft around the turning point in the middle of the myosin head) is 40 Å away from the actin-binding site (in the bottom of the head), the actin binding to myosin changes the state of this site allosterically, through a microscopic change of conformation of the crevice that connects these sites (just as O_2 binding to one heme causes a change in the strength of O_2 binding to another heme of hemoglobin). This coupling of the sites works in both directions, of course: the myosin-ATP and myosin-ADP-P complexes bind weakly to actin, while the actin-binding constant of the myosin-ADP complex is 10 000 times greater.

This large-scale conformational change of the head coupled with alternation of the actin-bound and actin-unbound states of myosin leads to a kind of rowing action (which is a cycle as well: "oar behind in water – oar behind in air – oar in front in air – oar in front in water – oar behind in water – . . .", cf. Fig. 25.8). This "rowing" moves the myosin filament along the actin filament like a boat being moved on water. This analogy should not take us too far, though. Usual rowing assumes continual pushing and pulling efforts. On the contrary, no stage of the myosin's "microscopic rowing" assumes a continual effort. Each of them may occur by a Brownian motion that starts with a spontaneous thermal jump from a state with a higher free energy and ends with coming to the state with a lower free energy (which is the next state along the rowing cycle). Each "rowing" cycle consumes one ATP molecule – but only in the case when the cycle goes in one direction and not in the other (the opposite direction would lead to ATP synthesis). The free energy gained from ATP hydrolysis is spent on muscle contraction.

As a result, myosin walks along actin, muscles contract, lungs pump the air, oxygen binds to hemoglobin and is handed over – via myoglobin – to muscles and other organs, and this is how we breathe and live.

AFTERWORD

To my deep sorrow, I have had to complete this book alone: my teacher and co-author Oleg B. Ptitsyn passed away in 1999 . . .

Therefore, I am completely responsible for any deficiencies in this course of lectures and would be grateful for comments from the readers.

Alexei V. Finkelstein
afinkel@vega.protres.ru

RECOMMENDED READING

[1a] Branden C., Tooze J. *Introduction to protein structure*. New York, London: Garland, 1991.

[1b] ——; 2nd edn., 1999.

[2] Cantor C.R., Schimmel P.R. *Biophysical chemistry*, part 1, chs. 2, 5; part 3, chs. 17, 20, 21. San Francisco, W.H. Freeman, 1980.

[3] Chernavskyii D.S., Chernavskaya N.M. *Protein as a machine. Biological macromolecular constructions* (in Russian). Moscow: Moscow University Publishing, 1999.

[4] Creighton T.E. *Proteins*, 2nd edn. New York: W.H. Freeman, 1991.

[5] Fersht A. *Enzyme structure and mechanism*, 2nd edn., chs. 1, 8–12. New York: W.H. Freeman, 1985.

[6] Fersht A. *Structure and mechanism in protein science: A guide to enzyme catalysis and protein folding*. New York: W.H. Freeman, 1999.

[7] Leninger A.L., Nelson D.L., Cox M.X. *Principles of biochemistry*, 2nd edn., chs. 3–8, 21, 26. New York: Worth, 1993.

[8] Lesk A.M. *Introduction to protein architecture*. Oxford, New York: Oxford University Press, 2001.

[9] Pauling L. *General chemistry*, chs. 1–6, 9–13, 16, 24. New York: W.H. Freeman, 1970.

[10] Perutz M.F. *Protein structure*. New York: W.H. Freeman, 1992.

[11] Rubin A.B. *Biophysics*, v.1, chs. 7–14 (in Russian). Moscow: Publishing House "University", 1999.

[12] Schulz G.E., Schirmer R.H. *Principles of protein structure*. New York, Heidelberg, Berlin: Springer-Verlag, 1979.

[13] Stepanov V.M. *Molecular biology. Protein structures and functions* (in Russian). Moscow: Vysshaya Shkola, 1996.

[14] Stryer L. *Biochemistry*, 4th edn., chs. 1–3, 7–16, 34–36. New York: W.H. Freeman, 1995.

[15] Volkenstein M.V. *Biophysics*, chs. 4, 6 (in Russian). Moscow: Nauka, 1981.

INDEX

Printed and bound by CPI Group (UK) Ltd, Croydon, CR0 4YY

03/10/2024

01040412-0010